AN INTRODUCTION TO BENFORD'S LAW

AN INTRODUCTION TO BENFORD'S LAW

Arno Berger and Theodore P. Hill

PRINCETON UNIVERSITY PRESS
PRINCETON AND OXFORD

Published by Princeton University Press
41 William Street, Princeton, New Jersey 08540

In the United Kingdom: Princeton University Press
6 Oxford Street, Woodstock, Oxfordshire, OX20 1TW

ISBN: 978-0-691-16306-2
Library of Congress Control Number: 2014953765

British Library Cataloging-in-Publication Data is available

This book has been composed in LaTeX

The publisher would like to acknowledge the authors of this volume for providing the
camera-ready copy from which this book was printed.

Printed on acid-free paper. ∞

press.princeton.edu

Printed in the United States of America

10 9 8 7 6 5 4 3 2 1

Contents

Preface

This book is an up-to-date reference on Benford's law, a statistical phenomenon first documented in the nineteenth century. Benford's law, also known as the significant-digit law, is a subject of great beauty, encompassing counterintuitive predictions, deep mathematical theories, and widespread applications ranging from fraud detection to diagnosis and design of mathematical models. Building on over a decade of our joint work, this text is a self-contained comprehensive treatment of the theory of Benford's law that includes formal definitions and proofs, open problems, dozens of basic theorems we discovered in the process of writing that have not before appeared in print, and hundreds of examples. Complementing the theory are overviews of its history, new empirical evidence, and applications.

Inspiration for this project has come first and foremost from the wide variety of lay people and scientists who kept contacting us with basic questions about Benford's law, and repeatedly asked for a good reference text. Not knowing of any, we decided to write one ourselves. Our main goal in doing so has been to assimilate the essential mathematical and statistical aspects of Benford's law, and present them in a way we hope will aid researchers interested in applications, and will also inspire further theoretical advances in this fascinating field.

After a brief overview of the history and empirical evidence of the law, the book makes a smooth progression through the field: basic facts about significant digits, Benford sequences, functions, and random variables; tools from the theory of uniform distribution; scale-, base-, and sum-invariance; one-dimensional dynamical systems and differential equations; powers of matrices, Markov chains, and difference equations; and products, powers, and mixtures of random variables. Two concluding chapters contain summaries of the finitely additive theory of the law, and five general areas of applications. Many of the illustrative examples and verbal descriptions are also intended for the non-theoretician, and are accompanied by figures, graphs, and tables that we hope will be helpful to all readers.

An Introduction to Benford's Law is intended as a reference tool for a broad audience: lay people interested in learning about the history and numerous current applications of this surprising statistical phenomenon; undergraduate students wanting to understand some of the basics; graduate students and professionals in science, engineering, and accounting who are contemplating using or already using Benford's law in their own research; and professional mathematicians and statisticians, both those conducting theoretical or applied research in the field, and those in other areas who want to learn or at least have access to the basic mathematics underlying the subject. Most of the formal statements of theorems are accessible to an advanced undergraduate mathematics student, and although the proofs sometimes require familiarity with more advanced topics such as measure and ergodic theory, they should be accessible to most mathematics and statistics graduate students. As such, we hope the book may also provide a good base for a special topics course or seminar.

We wish to thank the collaborators on our own research on Benford's law, notably Leonid Bunimovich, Gideon Eshun, Steven Evans, Bahar Kaynar, Kent Morrison, Ad Ridder, and Klaus Schürger; the second author also wishes to express his deep gratitude to Lester Dubins, from whom he first learned about Benford's law and who strongly encouraged him to write such a book, and to Amos Tversky for his advice and insights into testing a Benford theory about fabricating data. We gratefully acknowledge Bhisham Bherwani for his excellent copyediting, Kathleen Cioffi and Quinn Fusting at Princeton University Press for the fine administrative support, and especially our editor Vickie Kearn, who has been very enthusiastic about this project from the beginning and has helped us every step of the way. Finally, we both are grateful to Erika Rogers for continued technical and editorial support, assistance in researching the applications, and for designing and maintaining the Benford database [24], which currently contains listings of over 800 research papers, books, newspaper articles, software, and videos.

Comments and suggestions for improvement by readers of this book will be gratefully received.

Arno Berger and Theodore P. Hill, December 2014

Chapter One

Introduction

Benford's law, also known as the *First-digit* or *Significant-digit law,* is the empirical gem of statistical folklore that in many naturally occurring tables of numerical data, the significant digits are not uniformly distributed as might be expected, but instead follow a particular logarithmic distribution. In its most common formulation, the special case of the first significant (i.e., first non-zero) decimal digit, Benford's law asserts that the leading digit is not equally likely to be any one of the nine possible digits $1, 2, \ldots, 9$, but is 1 more than 30% of the time, and is 9 less than 5% of the time, with the probabilities decreasing monotonically in between; see Figure 1.1. More precisely, the exact law for the first significant digit is

$$\mathsf{Prob}(D_1 = d) = \log_{10}\left(1 + \frac{1}{d}\right) \quad \text{for all } d = 1, 2, \ldots, 9; \tag{1.1}$$

here, D_1 denotes the first significant decimal digit, e.g.,

$$D_1(\sqrt{2}) = D_1(1.414) = 1,$$
$$D_1(\pi^{-1}) = D_1(0.3183) = 3,$$
$$D_1(e^\pi) = D_1(23.14) = 2.$$

Hence, the two smallest digits occur as the first significant digit with a combined probability close to 50 percent, whereas the two largest digits together have a probability of less than 10 percent, since

$$\mathsf{Prob}(D_1 = 1) = \log_{10} 2 = 0.3010, \quad \mathsf{Prob}(D_1 = 2) = \log_{10} \frac{3}{2} = 0.1760,$$

and

$$\mathsf{Prob}(D_1 = 8) = \log_{10} \frac{9}{8} = 0.05115, \quad \mathsf{Prob}(D_1 = 9) = \log_{10} \frac{10}{9} = 0.04575.$$

The complete form of Benford's law also specifies the probabilities of occurrence of the second and higher significant digits, and more generally, the *joint* distribution of *all* the significant digits. A general statement of Benford's law that includes the probabilities of all blocks of consecutive initial significant digits is this: For every positive integer m, and for all initial blocks of m significant digits $(d_1, d_2, \ldots, d_m)$, where d_1 is in $\{1, 2, \ldots, 9\}$, and d_j is in $\{0, 1, \ldots, 9\}$ for all $j \geq 2$,

$$\text{Prob}\left(D_1 = d_1, D_2 = d_2, \ldots, D_m = d_m\right) = \log_{10}\left(1 + \left(\sum_{j=1}^{m} 10^{m-j} d_j\right)^{-1}\right),$$
(1.2)

where D_2, D_3, D_4, etc. represent the second, third, fourth, etc. significant decimal digits, e.g.,

$$D_2\left(\sqrt{2}\right) = 4, \quad D_3\left(\pi^{-1}\right) = 8, \quad D_4\left(e^{\pi}\right) = 4.$$

For example, (1.2) yields the probabilities for the individual second significant digits,

$$\text{Prob}(D_2 = d_2) = \sum_{j=1}^{9} \log_{10}\left(1 + \frac{1}{10j + d_2}\right) \quad \text{for all } d_2 = 0, 1, \ldots, 9, \quad (1.3)$$

which also are not uniformly distributed on all the possible second digit values $0, 1, \ldots, 9$, but are strictly decreasing, although they are much closer to uniform than the first digits; see Figure 1.1.

d	0	1	2	3	4	5	6	7	8	9
$\text{Prob}(D_1 = d)$	0	30.10	17.60	12.49	9.69	7.91	6.69	5.79	5.11	4.57
$\text{Prob}(D_2 = d)$	11.96	11.38	10.88	10.43	10.03	9.66	9.33	9.03	8.75	8.49
$\text{Prob}(D_3 = d)$	10.17	10.13	10.09	10.05	10.01	9.97	9.94	9.90	9.86	9.82
$\text{Prob}(D_4 = d)$	10.01	10.01	10.00	10.00	10.00	9.99	9.99	9.99	9.98	9.98

Figure 1.1: Probabilities (in percent) of the first four significant decimal digits, as implied by Benford's law (1.2); note that the first row is simply the first-digit law (1.1).

More generally, (1.2) yields the probabilities for longer blocks of digits as well. For instance, the probability that a number has the same first three significant digits as $\pi = 3.141$ is

$$\text{Prob}\left(D_1 = 3, D_2 = 1, D_3 = 4\right) = \log_{10}\left(1 + \frac{1}{314}\right) = \log_{10}\frac{315}{314} = 0.001380.$$

A perhaps surprising corollary of the general form of Benford's law (1.2) is that the significant digits are dependent, and not independent as one might expect

[74]. To see this, note that (1.3) implies that the (unconditional) probability that the second digit equals 1 is

$$\mathsf{Prob}(D_2 = 1) = \sum_{j=1}^{9} \log_{10}\left(1 + \frac{1}{10j + 1}\right) = \log_{10}\frac{6029312}{4638501} = 0.1138\,,$$

whereas it follows from (1.2) that if the first digit is 1, the (conditional) probability that the second digit also equals 1 is

$$\mathsf{Prob}(D_2 = 1 | D_1 = 1) = \frac{\log_{10} 12 - \log_{10} 11}{\log_{10} 2} = 0.1255\,.$$

Note. Throughout, real numbers such as $\sqrt{2}$ and π are displayed to *four* correct significant decimal digits. Thus an equation like $\sqrt{2} = 1.414$ ought to be read as $1414 \leq 1000 \cdot \sqrt{2} < 1415$, and *not* as $\sqrt{2} = \frac{1414}{1000}$. The only exceptions to this rule are probabilities given in percent (as in Figure 1.1), as well as the numbers Δ and Δ_∞, introduced later; all these quantities only attain values between 0 and 100, and are shown to *two* correct digits after the decimal point. Thus, for instance, $\Delta = 0.00$ means $0 \leq 100 \cdot \Delta < 1$, but not necessarily $\Delta = 0$.

1.1 HISTORY

The first known reference to the logarithmic distribution of leading digits dates back to 1881, when the American astronomer Simon Newcomb noticed "how much faster the first pages [of logarithmic tables] wear out than the last ones," and, after several short heuristics, deduced the logarithmic probabilities shown in the first two rows of Figure 1.1 for the first and second digits [111].

Some fifty-seven years later the physicist Frank Benford rediscovered the law [9], and supported it with over 20,000 entries from 20 different tables including such diverse data as catchment areas of 335 rivers, specific heats of 1,389 chemical compounds, American League baseball statistics, and numbers gleaned from front pages of newspapers and *Reader's Digest* articles; see Figure 1.2 (rows A, E, P, D and M, respectively).

Although P. Diaconis and D. Freedman offer convincing evidence that Benford manipulated round-off errors to obtain a better fit to the logarithmic law [47, p. 363], even the unmanipulated data are remarkably close. Benford's article attracted much attention and, Newcomb's article having been overlooked, the law became known as *Benford's law* and many articles on the subject appeared. As R. Raimi observed nearly half a century ago [127, p. 521],

> This particular logarithmic distribution of the first digits, while not universal, is so common and yet so surprising at first glance that it has given rise to a varied literature, among the authors of which are mathematicians, statisticians, economists, engineers, physicists, and amateurs.

The online database [24] now references more than 800 articles on Benford's law, as well as other resources (books, websites, lectures, etc.).

PERCENTAGE OF TIMES THE NATURAL NUMBERS 1 TO 9 ARE USED AS FIRST
DIGITS IN NUMBERS, AS DETERMINED BY 20,229 OBSERVATIONS

Group	Title	First Digit									Count
		1	2	3	4	5	6	7	8	9	
A	Rivers, Area	31.0	16.4	10.7	11.3	7.2	8.6	5.5	4.2	5.1	335
B	Population	33.9	20.4	14.2	8.1	7.2	6.2	4.1	3.7	2.2	3259
C	Constants	41.3	14.4	4.8	8.6	10.6	5.8	1.0	2.9	10.6	104
D	Newspapers	30.0	18.0	12.0	10.0	8.0	6.0	6.0	5.0	5.0	100
E	Spec. Heat	24.0	18.4	16.2	14.6	10.6	4.1	3.2	4.8	4.1	1389
F	Pressure	29.6	18.3	12.8	9.8	8.3	6.4	5.7	4.4	4.7	703
G	H.P. Lost	30.0	18.4	11.9	10.8	8.1	7.0	5.1	5.1	3.6	690
H	Mol. Wgt.	26.7	25.2	15.4	10.8	6.7	5.1	4.1	2.8	3.2	1800
I	Drainage	27.1	23.9	13.8	12.6	8.2	5.0	5.0	2.5	1.9	159
J	Atomic Wgt.	47.2	18.7	5.5	4.4	6.6	4.4	3.3	4.4	5.5	91
K	$n^{-1}, \sqrt{n}, \cdots$	25.7	20.3	9.7	6.8	6.6	6.8	7.2	8.0	8.9	5000
L	Design	26.8	14.8	14.3	7.5	8.3	8.4	7.0	7.3	5.6	560
M	*Digest*	33.4	18.5	12.4	7.5	7.1	6.5	5.5	4.9	4.2	308
N	Cost Data	32.4	18.8	10.1	10.1	9.8	5.5	4.7	5.5	3.1	741
O	X-Ray Volts	27.9	17.5	14.4	9.0	8.1	7.4	5.1	5.8	4.8	707
P	Am. League	32.7	17.6	12.6	9.8	7.4	6.4	4.9	5.6	3.0	1458
Q	Black Body	31.0	17.3	14.1	8.7	6.6	7.0	5.2	4.7	5.4	1165
R	Addresses	28.9	19.2	12.6	8.8	8.5	6.4	5.6	5.0	5.0	342
S	$n^1, n^2 \cdots n!$	25.3	16.0	12.0	10.0	8.5	8.8	6.8	7.1	5.5	900
T	Death Rate	27.0	18.6	15.7	9.4	6.7	6.5	7.2	4.8	4.1	418
	Average.......	30.6	18.5	12.4	9.4	8.0	6.4	5.1	4.9	4.7	1011
	Probable Error	±0.8	±0.4	±0.4	±0.3	±0.2	±0.2	±0.2	±0.2	±0.3	—

Figure 1.2: Benford's original data from [9]; reprinted courtesy of the American
Philosophical Society.

1.2   EMPIRICAL EVIDENCE

Many tables of numerical data, of course, do *not* follow Benford's law in any
sense. Telephone numbers in a given region typically begin with the same few
digits, and never begin with a 1; lottery numbers in all common lotteries are
distributed uniformly, not logarithmically; and tables of heights of human adults,
whether given in feet or meters, clearly do not begin with a 1 about 30% of the
time. Even "neutral" mathematical data such as square-root tables of integers
do not follow Benford's law, as Benford himself discovered (see row K in Figure
1.2 above), nor do the prime numbers, as will be seen in later chapters.

On the other hand, since Benford's popularization of the law, an abundance
of additional empirical evidence has appeared. In physics, for example, D. Knuth
[90] and J. Burke and E. Kincanon [31] observed that of the most commonly
used physical constants (e.g., the speed of light and the force of gravity listed on
the inside cover of an introductory physics textbook), about 30% have leading
significant digit 1; P. Becker [8] observed that the decimal parts of failure (haz-

ard) rates often have a logarithmic distribution; and R. Buck et al., in studying the values of the 477 radioactive half-lives of unhindered alpha decays that were accumulated throughout the past century, and that vary over many orders of magnitude, found that the frequency of occurrence of the first digits of both measured and calculated values of the half-lives is in "good agreement" with Benford's law [29]. In scientific calculations, A. Feldstein and P. Turner called the assumption of logarithmically distributed mantissas "widely used and well established" [57, p. 241]; R. Hamming labeled the appearance of the logarithmic distribution in floating-point numbers "well-known" [70, p. 1609]; and Knuth observed that "repeated calculations with real numbers will nearly always tend to yield better and better approximations to a logarithmic distribution" [90, p. 262].

Additional empirical evidence of Benford's law continues to appear. M. Nigrini observed that the digital frequencies of certain entries in Internal Revenue Service files are an extremely good fit to Benford's law (see [113] and Figure 1.3); E. Ley found that "the series of one-day returns on the Dow-Jones Industrial Average Index (DJIA) and the Standard and Poor's Index (S&P) reasonably agrees with Benford's law" [98]; and Z. Shengmin and W. Wenchao found that "Benford's law reasonably holds for the two main Chinese stock indices" [148]. In the field of biology, E. Costas et al. observed that in a certain cyanobacterium, "the distribution of the number of cells per colony satisfies Benford's law" [39, p. 341]; S. Docampo et al. reported that "gross data sets of daily pollen counts from three aerobiological stations (located in European cities with different features regarding vegetation and climatology) fit Benford's law" [49, p. 275]; and J. Friar et al. found that "the Benford distribution produces excellent fits" to certain basic genome data [60, p. 1].

Figure 1.3 compares the probabilities of occurrence of first digits predicted by (1.1) to the distributions of first digits in four datasets: the combined data reported by Benford in 1938 (second-to-last row in Figure 1.2); the populations of the 3,143 counties in the United States in the 2010 census [102]; all numbers appearing on the World Wide Web as estimated using a Google search experiment [97]; and over 90,000 entries for Interest Received in U.S. tax returns from the IRS Individual Tax Model Files [113]. To instill in the reader a *quantitative* perception of closeness to, or deviation from, the first-digit law (1.1), for every distribution of the first significant decimal digit shown in this book, the number

$$\Delta = 100 \cdot \max_{d=1}^{9} \left| \mathsf{Prob}(D_1 = d) - \log_{10}\left(1 + \frac{1}{d}\right) \right|$$

will also be displayed. Note that Δ is simply the maximum difference, in percent, between the probabilities of the first significant digits of the given distribution and the Benford probabilities in (1.1). Thus, for example, $\Delta = 0$ indicates exact conformance to (1.1), and $\Delta = 12.08$ indicates that the probability of some digit $d \in \{1, 2, \ldots, 9\}$ differs from $\log_{10}(1 + d^{-1})$ by 12.08%, and the probability of no other digit differs by more than this.

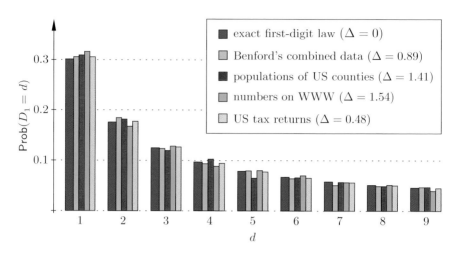

Figure 1.3: Comparisons of four datasets to Benford's law (1.1).

All these statistics aside, the authors also highly recommend that the justifiably skeptical reader perform a simple experiment, such as randomly selecting numerical data from front pages of several local newspapers, or from "a Farmer's Almanack" as Knuth suggests [90], or running a Google search similar to the Dartmouth classroom project described in [97].

1.3 EARLY EXPLANATIONS

Since the empirical significant-digit law (1.1) or (1.2) does not specify a well-defined statistical experiment or sample space, most early attempts to explain the appearance of Benford's law argued that it is "merely the result of our way of writing numbers" [67] or "a built-in characteristic of our number system" [159]. The idea was to first show that the set of real numbers satisfies (1.1) or (1.2), and then suggest that this explains the empirical statistical evidence. A common starting point has been to try to establish (1.1) for the positive integers, beginning with the prototypical set

$$\{D_1 = 1\} = \{1, 10, 11, \ldots, 18, 19, 100, 101, \ldots, 198, 199, 1000, 1001, \ldots\},$$

the set of positive integers with first significant digit 1. The source of difficulty and much of the fascination of the first-digit problem is that the set $\{D_1 = 1\}$ does not have a *natural density* among the integers, that is, the proportion of integers in the set $\{D_1 = 1\}$ up to N, i.e., the ratio

$$\frac{\#\{1 \le n \le N : D_1(n) = 1\}}{N}, \tag{1.4}$$

does not have a limit as N goes to infinity, unlike the sets of even integers or primes, say, which have natural densities $\frac{1}{2}$ and 0, respectively. It is easy to see that the empirical density (1.4) of $\{D_1 = 1\}$ oscillates repeatedly between $\frac{1}{9}$ and $\frac{5}{9}$, and thus it is theoretically possible to assign any number between $\frac{1}{9}$ and $\frac{5}{9}$ as the "probability" of this set. Similarly, the empirical density of $\{D_1 = 9\}$ forever oscillates between $\frac{1}{81}$ and $\frac{1}{9}$; see Figure 1.4.

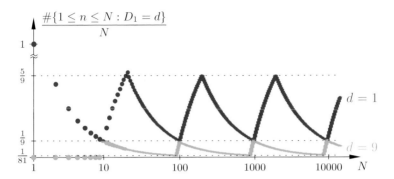

Figure 1.4: The sets $\{D_1 = 1\}$ and $\{D_1 = 9\}$ do not have a natural density (and neither does $\{D_1 = d\}$ for any $d = 2, 3, \ldots, 8$).

Many partial attempts to put Benford's law on a solid logical basis have been made, beginning with Newcomb's own heuristics, and continuing through the decades with various urn model arguments and mathematical proofs; Raimi [127] has an excellent review of these. But as the eminent logician, mathematician, and philosopher C. S. Peirce once observed, "in no other branch of mathematics is it so easy for experts to blunder as in probability theory" [63, p. 273], and the arguments surrounding Benford's law certainly bear that out. Even W. Feller's classic and hugely influential text [58] contains a critical flaw that apparently went unnoticed for half a century. Specifically, the claim by Feller and subsequent authors that "regularity and large spread implies Benford's Law" is fallacious for any reasonable definitions of regularity and spread (measure of dispersion) [21].

1.4 MATHEMATICAL FRAMEWORK

A crucial part of (1.1), of course, is an appropriate interpretation of Prob. In practice, this can take several forms. For sequences of real numbers $(x_1, x_2, \ldots)$, Prob usually refers to the limiting proportion (or relative frequency) of elements in the sequence for which an event such as $\{D_1 = 1\}$ occurs. Equivalently, fix a positive integer N and calculate the probability that the first digit is 1 in an experiment where one of the elements $x_1, x_2, \ldots, x_N$ is selected at random (each with probability $1/N$); if this probability has a limit as N goes to infinity, then the limiting probability is designated $\mathsf{Prob}(D_1 = 1)$. Implicit in this usage of

Prob is the assumption that all limiting proportions of interest actually exist. Similarly, for real-valued functions $f : [0, +\infty) \to \mathbb{R}$, fix a positive real number T, choose a number τ at random uniformly between 0 and T, and calculate the probability that $f(\tau)$ has first significant digit 1. If this probability has a limit, as $T \to +\infty$, then $\mathsf{Prob}(D_1 = 1)$ is that limiting probability.

For a random variable or probability distribution, on the other hand, Prob simply denotes the underlying probability of the given event. Thus, if X is a random variable, then $\mathsf{Prob}(D_1(X) = 1)$ is the probability that the first significant digit of X is 1. Finite datasets of real numbers can also be dealt with this way, with Prob being the empirical distribution of the dataset.

One of the main themes of this book is the robustness of Benford's law. In the context of sequences of numbers, for example, iterations of linear maps typically follow Benford's law exactly; Figure 1.5 illustrates the convergence of first-digit probabilities for the Fibonacci sequence $(1, 1, 2, 3, 5, 8, 13, \dots)$. As will be seen in Chapter 6, not only do iterations of most linear functions follow Benford's law exactly, but iterations of most functions *close* to linear also follow Benford's law *exactly*. Similarly, as will be seen in Chapter 8, powers and products of very general classes of random variables approach Benford's law in the limit; Figure 1.6 illustrates this starting with $U(0, 1)$, the standard random variable uniformly distributed between 0 and 1. Similarly, if random samples from different randomly-selected probability distributions are combined, the resulting meta-sample also typically converges to Benford's law; Figure 1.7 illustrates this by comparing two of Benford's original empirical datasets with the combination of all his data.

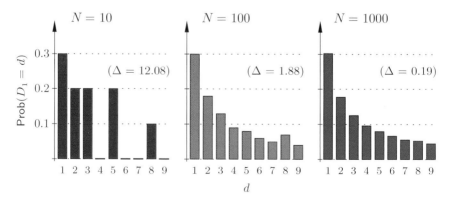

Figure 1.5: Probabilities that a number chosen uniformly from among the first N Fibonacci numbers has first significant digit d.

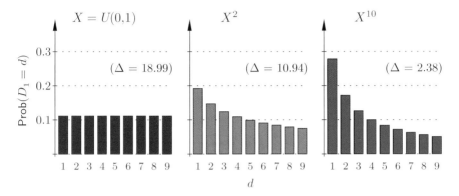

Figure 1.6: First-digit probabilities of powers of a $U(0,1)$ random variable X.

Non-decimal bases

Throughout this book, attention will normally be restricted to decimal (i.e., base-10) significant digits, and when results for more general bases are employed, that will be made explicit. From now on, therefore, $\log x$ will always denote the logarithm base 10 of x, while $\ln x$ is the natural logarithm of x. For convenience, the convention $\log 0 := \ln 0 := 0$ is adopted. Nearly all the results in this book that are stated only with respect to base 10 carry over easily to arbitrary integer bases $b \geq 2$, and the interested reader may find some pertinent details in [15]. In particular, the general form of (1.2) with respect to any such base b is

$$\mathsf{Prob}\left(D_1^{(b)} = d_1, D_2^{(b)} = d_2, \ldots, D_m^{(b)} = d_m\right) = \log_b\left(1 + \left(\sum_{j=1}^m b^{m-j} d_j\right)^{-1}\right),$$
(1.5)

where $\log_b$ denotes the base-b logarithm, and $D_1^{(b)}$, $D_2^{(b)}$, $D_3^{(b)}$, etc. are the first, second, third, etc. significant digits base b, respectively; so in (1.5), d_1 is an integer in $\{1, 2, \ldots, b-1\}$, and for $j \geq 2$, d_j is an integer in $\{0, 1, \ldots, b-1\}$. Note that in the case $m = 1$ and $b = 2$, (1.5) reduces to $\mathsf{Prob}\left(D_1^{(2)} = 1\right) = 1$, which is trivially true because the first significant digit base 2 of every non-zero number is 1.

This book is organized as follows. Chapter 2 contains formal definitions, examples, and graphs of significant digits and the significand (mantissa) function, and also of the probability spaces needed to formulate Benford's law precisely, including the crucial natural domain of "events," the so-called significand σ-algebra. Chapter 3 defines Benford sequences, functions, and random variables, with examples of each. Chapters 4 and 5 contain four of the main mathematical characterizations of Benford's law, with proofs and examples. Chapters 6 and 7 study Benford's law in the context of deterministic processes, including both one- and multi-dimensional discrete-time dynamical systems and algorithms as

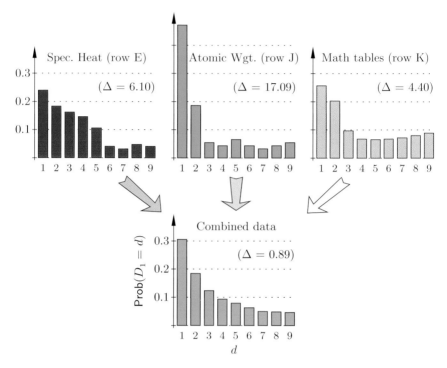

Figure 1.7: Empirical first-digit probabilities in Benford's original data; see Figure 1.2.

well as continuous-time processes generated by differential equations. Chapter 8 addresses Benford's law for random variables and stochastic processes, including products of random variables, mixtures of distributions, and random maps. Chapter 9 offers a glimpse of the complementary theory of Benford's law in the non-traditional context of finitely additive probability theory, and Chapter 10 provides a brief overview of the many applications of Benford's law that continue to appear in a wide range of disciplines.

The mathematical detail in this book is on several levels: The basic explanations, and many of the figures and comments, are intended for a general scientific audience; the formal statements of definitions and theorems are accessible to an undergraduate mathematics student; and the proofs, some of which contain basics of measure and ergodic theory, are accessible to a mathematics graduate student (or a diligent undergraduate).

Chapter Two

Significant Digits and the Significand

Benford's law is a statement about the statistical distribution of significant (decimal) digits or, equivalently, about significands, viz., fraction parts in floating-point arithmetic. Thus, a natural starting point for any study of Benford's law is the formal definition of significant digits and the significand function.

2.1 SIGNIFICANT DIGITS

Informally, the *first significant decimal digit* of a positive real number x is the first non-zero digit appearing in the decimal expansion of x, e.g., the first significant digits of 2015 and 0.2015 are both 2. The second significant digit is the decimal digit following the first significant digit, so the second significant digits of 2015 and 0.2015 are both 0. This informal definition, however, is ambiguous about whether the first significant digit of $19.99\ldots = 20$ is 1 or 2. The convention used throughout this book, as seen in the following formal definition, is that the first significant digit of $19.99\ldots$ is 2. Interchanging the strict and non-strict inequalities in the definition would change that convention, but would not affect the essential conclusions about Benford's law in the following chapters.

DEFINITION 2.1. For every non-zero real number x, the *first significant (decimal) digit* of x, denoted by $D_1(x)$, is the unique integer $j \in \{1, 2, \ldots, 9\}$ satisfying $10^k j \leq |x| < 10^k(j+1)$ for some (necessarily unique) $k \in \mathbb{Z}$.

Similarly, for every $m \geq 2$, $m \in \mathbb{N}$, the mth *significant (decimal) digit* of x, denoted by $D_m(x)$, is defined inductively as the unique integer $j \in \{0, 1, \ldots, 9\}$ such that

$$10^k \left(\sum\nolimits_{i=1}^{m-1} D_i(x)10^{m-i} + j \right) \leq |x| < 10^k \left(\sum\nolimits_{i=1}^{m-1} D_i(x)10^{m-i} + j + 1 \right)$$

for some (necessarily unique) $k \in \mathbb{Z}$; for convenience, $D_m(0) := 0$ for all $m \in \mathbb{N}$.

Note that, by definition, the first significant digit $D_1(x)$ of every non-zero x is not 0, whereas the second, third, etc. significant digits may be any integers in $\{0, 1, \ldots, 9\}$.

EXAMPLE 2.2.

$$D_1(\sqrt{2}) = D_1(-\sqrt{2}) = D_1(10\sqrt{2}) = 1, \quad D_2(\sqrt{2}) = 4, \quad D_3(\sqrt{2}) = 1;$$
$$D_1(\pi^{-1}) = D_1(10\pi^{-1}) = 3, \quad D_2(\pi^{-1}) = 1, \quad D_3(\pi^{-1}) = 8. \qquad \maltese$$

2.2   THE SIGNIFICAND

Informally, the *significand* of a real number is its coefficient when expressed in floating-point ("scientific") notation, so the significand of $2015 = 2.015 \cdot 10^3$ is 2.015. Unlike in the definition of significant digit, there is no ambiguity here — the significand of $19.99\ldots = 20$ is $1.99\ldots = 2$. The formal definition of the significand (function) is this:

DEFINITION 2.3. The *(decimal) significand function* $S : \mathbb{R} \to [1, 10)$ is defined as follows: If $x \neq 0$ then $S(x) = t$, where t is the unique number in $[1, 10)$ with $|x| = 10^k t$ for some (necessarily unique) $k \in \mathbb{Z}$; if $x = 0$ then, for convenience, $S(0) := 0$.

Explicitly, S is given by

$$S(x) = 10^{\log |x| - \lfloor \log |x| \rfloor} \quad \text{for all } x \neq 0 \,;$$

here $\lfloor x \rfloor$ denotes the largest integer less than or equal to the real number x. (The function $\lfloor \cdot \rfloor$ is often referred to as the *floor function*, e.g., $\lfloor \pi \rfloor = \lfloor 3.141 \rfloor = 3$, and $\lfloor -\pi \rfloor = -4$.) Observe that, for all $x \in \mathbb{R}$,

$$S(10^k x) = S(x) = S\big(S(x)\big) \quad \text{for every } k \in \mathbb{Z} \,.$$

Figure 2.1 depicts the graph of the significand function and the graph of its logarithm. Between integers, the latter is a linear function of $\log |x|$. This fact will play an important role in the theory of Benford's law in later chapters.

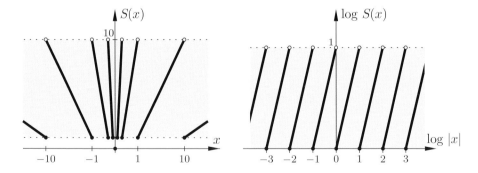

Figure 2.1: Graphs of the significand function S (left) and its logarithm.

Note. The original word used in American English to describe the coefficient of floating-point numbers in computer hardware seems to have been *mantissa*, and this usage remains common among some computer scientists and practitioners. However, this use of the word mantissa is discouraged by the IEEE floating-point standard committee and by many professionals because it conflicts with the preexisting usage of mantissa for the fractional part of a logarithm. In accordance with the IEEE standard, only the term *significand* will be used henceforth in this book. (With the significand $S = S(x)$ as in Definition 2.3, the (traditional) mantissa of x would simply be $\log S(x)$.) The reader should also note that in the literature, the significand is sometimes taken to have values in $[0.1, 1)$ rather than in $[1, 10)$.

EXAMPLE 2.4.

$$S(\sqrt{2}) = S(-\sqrt{2}) = S(10\sqrt{2}) = \sqrt{2} = 1.414\,,$$
$$S(\pi^{-1}) = S(10\pi^{-1}) = 10\pi^{-1} = 3.183\,. \qquad ✠$$

The significand uniquely determines all the significant digits, and vice versa. This relationship is recorded in the following proposition which immediately follows from Definitions 2.1 and 2.3.

PROPOSITION 2.5. *Let x be any real number. Then:*

(i) $S(x) = \sum_{m \in \mathbb{N}} 10^{1-m} D_m(x)$;

(ii) $D_m(x) = \lfloor 10^{m-1} S(x) \rfloor - 10 \lfloor 10^{m-2} S(x) \rfloor$ *for every $m \in \mathbb{N}$.*

Thus, Proposition 2.5(i) expresses the significand of a number as an explicit function of the significant digits of that number, and (ii) expresses the significant digits as a function of the significand.

It is important to note that the definitions of significand and significant digits per se do not involve any decimal *expansion* of x. However, it is clear from Proposition 2.5(i) that the significant digits provide a decimal expansion of $S(x)$, and that a sequence (d_m) in $\{0, 1, \ldots, 9\}$ is the sequence of significant digits of some positive real number if and only if $d_1 \neq 0$, and $d_m \neq 9$ for infinitely many m.

EXAMPLE 2.6. It follows from Proposition 2.5, together with Examples 2.2 and 2.4, that

$$S(\sqrt{2}) = D_1(\sqrt{2}) + 10^{-1} D_2(\sqrt{2}) + 10^{-2} D_3(\sqrt{2}) + \ldots = 1.414 = \sqrt{2}\,,$$

as well as that

$$D_1(\sqrt{2}) = \lfloor \sqrt{2} \rfloor = 1\,,$$
$$D_2(\sqrt{2}) = \lfloor 10\sqrt{2} \rfloor - 10 \lfloor \sqrt{2} \rfloor = 4\,,$$
$$D_3(\sqrt{2}) = \lfloor 100\sqrt{2} \rfloor - 10 \lfloor 10\sqrt{2} \rfloor = 1\,, \quad \text{etc}\,. \qquad ✠$$

Since the significant digits determine the significand, and are in turn determined by it, the informal version (1.2) of Benford's law in Chapter 1 has an immediate and very concise counterpart in terms of the significand function, namely,

$$\mathsf{Prob}(S \leq t) = \log t \quad \text{for all } 1 \leq t < 10\,. \tag{2.1}$$

(Recall that $\log$ denotes the base-10 logarithm throughout.) As noted earlier, the formal versions of (1.2) and (2.1) will be developed in detail below.

2.3   THE SIGNIFICAND σ-ALGEBRA

The informal statements (1.1), (1.2), and (2.1) of Benford's law all involve *probabilities*. Hence, to formulate mathematically precise versions of these statements, it is necessary to reformulate them in the setting of rigorous probability theory.

The fundamental domain of standard modern probability theory is a non-empty set Ω (the "outcome space") together with a σ-algebra of subsets of Ω (the "events"). Recall that a σ-*algebra* $\mathcal{A}$ on Ω is simply a family of subsets of Ω such that $\varnothing \in \mathcal{A}$, and $\mathcal{A}$ is closed under complements and countable unions, that is,

$$A \in \mathcal{A} \implies A^c := \{\omega \in \Omega : \omega \notin A\} \in \mathcal{A}\,,$$

and

$$A_n \in \mathcal{A} \text{ for all } n \in \mathbb{N} \implies \bigcup_{n \in \mathbb{N}} A_n \in \mathcal{A}\,.$$

The largest σ-algebra on a non-empty set Ω is the so-called *power set* of Ω, the set of all subsets of Ω. If Ω is a finite set, this power σ-algebra is often the natural setting for most probabilistic purposes. For larger Ω such as the real line $\mathbb{R}$, however, the power set contains many complicated and non-constructible sets which cause difficulties in mathematical analysis, including standard countably additive probability theory, even with the most common probability distributions such as uniform, exponential, and normal distributions. As detailed below, therefore, instead of the power set, the standard σ-algebra on $\mathbb{R}$ used to define and analyze both continuous and discrete distributions is the so-called *Borel* σ-algebra $\mathcal{B}$, a proper subset (sub-σ-algebra) of the power set of $\mathbb{R}$.

Given any collection $\mathcal{E}$ of subsets of Ω, there is a (unique) smallest σ-algebra on Ω containing $\mathcal{E}$, referred to as the σ-*algebra generated by* $\mathcal{E}$ and denoted by $\sigma(\mathcal{E})$. As indicated above, one of the most important examples of σ-algebras in real analysis and probability theory is the Borel σ-algebra $\mathcal{B}$ on $\mathbb{R}$, which by definition is the σ-algebra generated by all intervals of real numbers, i.e., $\mathcal{B} = \sigma(\{J \subset \mathbb{R} : J \text{ an interval}\})$. If $C \subset \mathbb{R}$ then $\mathcal{B}(C)$ is understood to be the σ-algebra $C \cap \mathcal{B} := \{C \cap B : B \in \mathcal{B}\}$ on C; for brevity, write $\mathcal{B}[a, b)$ instead of $\mathcal{B}([a, b))$, where $[a, b) = \{x \in \mathbb{R} : a \leq x < b\}$, and write $\mathcal{B}^+$ instead of $\mathcal{B}(\mathbb{R}^+)$, where $\mathbb{R}^+ = \{x \in \mathbb{R} : x > 0\}$.

EXAMPLE 2.7. The Borel σ-algebras $\mathcal{B}[0, 1)$ and $\mathcal{B}[1, 10)$, which play important roles in the theory of Benford's law, are simply the smallest σ-algebras on $[0, 1)$ and $[1, 10)$ containing all intervals of the form $[a, b)$, where $0 \leq a \leq b < 1$ and $1 \leq a \leq b < 10$, respectively.   ✠

Just as the power set of $\mathbb{R}$ is simply too large a σ-algebra for much of standard real analysis and probability theory, and is therefore replaced by the smaller Borel σ-algebra $\mathcal{B}$ on $\mathbb{R}$, the σ-algebra $\mathcal{B}$ is itself too large for analyzing the significant digit and significand functions that are essential to formulating and analyzing Benford's law. Instead, a smaller collection of sets, i.e., a proper sub-σ-algebra of $\mathcal{B}$, will be seen to be the natural domain for Benford's law.

In order to describe this new σ-algebra, it is helpful to review the notion of a σ-algebra generated by a function. Given any function $f : \Omega \to \mathbb{R}$, recall that for every subset C of $\mathbb{R}$, the set $f^{-1}(C) \subset \Omega$, called the *pre-image* of C under f, is defined as

$$f^{-1}(C) = \{\omega \in \Omega : f(\omega) \in C\}.$$

The σ-algebra on Ω generated by the collection of sets

$$\mathcal{E} = \{f^{-1}(J) : J \subset \mathbb{R}, \ J \text{ an interval}\}$$

is also referred to as the *σ-algebra generated by f*; it will be denoted by $\sigma(f)$. Thus $\sigma(f)$ is the smallest σ-algebra on Ω that contains all sets of the form $\{\omega \in \Omega : a \leq f(\omega) \leq b\}$, for every $a, b \in \mathbb{R}$. It is easy to check that in fact $\sigma(f) = \{f^{-1}(B) : B \in \mathcal{B}\}$. Similarly, a whole family $\mathcal{F}$ of functions $f : \Omega \to \mathbb{R}$ defines a σ-algebra $\sigma(\mathcal{F})$, via

$$\sigma(\mathcal{F}) = \sigma\left(\bigcup_{f \in \mathcal{F}} \sigma(f)\right) = \sigma\left(\{f^{-1}(J) : J \subset \mathbb{R}, \ J \text{ an interval, } f \in \mathcal{F}\}\right),$$

which is simply the smallest σ-algebra on Ω containing $\{\omega \in \Omega : a \leq f(\omega) \leq b\}$ for all $a, b \in \mathbb{R}$ and all $f \in \mathcal{F}$.

In probability theory, functions $f : \Omega \to \mathbb{R}$ with $\sigma(f) \subset \mathcal{A}$ are called *random variables*. Probability textbooks typically use symbols X, Y, etc., rather than f, g, etc., to denote random variables, and this practice will be followed here also.

EXAMPLE 2.8. **(i)** Let $f : [0, 1] \to \mathbb{R}$ be the so-called *tent map* defined by $f(x) = 1 - |2x - 1|$; e.g., see [45, §1.8]. Since

$$f^{-1}([0, a]) = \left[0, \tfrac{1}{2}a\right] \cup \left[1 - \tfrac{1}{2}a, 1\right] \quad \text{for all } a \in [0, 1],$$

it is clear that $\sigma(f)$ is the proper sub-σ-algebra of $\mathcal{B}[0, 1]$ consisting of all Borel sets in $[0, 1]$ that are symmetric about $x = \frac{1}{2}$, i.e., with $1 - A$ denoting the set $\{1 - a : a \in A\}$,

$$\sigma(f) = \{A \in \mathcal{B}[0, 1] : A = 1 - A\}.$$

(ii) Let X be a Bernoulli random variable, i.e., a random variable on $(\mathbb{R}, \mathcal{B})$ taking only the values 0 and 1. Then $\sigma(X)$ is the sub-σ-algebra of $\mathcal{B}$ given by

$$\sigma(X) = \{\varnothing, \{0\}, \{1\}, \{0, 1\}, \mathbb{R}, \mathbb{R}\backslash\{0\}, \mathbb{R}\backslash\{1\}, \mathbb{R}\backslash\{0, 1\}\};$$

here and throughout, $A\backslash B = A \cap B^c$ is the set of all elements in A that are not in B. ✠

One key step in formulating Benford's law precisely is identifying the correct σ-algebra. As it turns out, in the significant digit framework there is only one natural candidate, which, although strictly smaller than $\mathcal{B}$, is nevertheless both intuitive and easy to describe. Informally, this σ-algebra is simply the collection of all subsets of $\mathbb{R}$ that can be described in terms of the significand function.

DEFINITION 2.9. The *significand σ-algebra* $\mathcal{S}$ is the σ-algebra on $\mathbb{R}^+$ generated by the significand function S, i.e., $\mathcal{S} = \mathbb{R}^+ \cap \sigma(S)$.

The importance of the σ-algebra $\mathcal{S}$ comes from the fact that for every event $A \in \mathcal{S}$ and every $x > 0$, knowing the significand $S(x)$ of x is enough to decide whether $x \in A$ or $x \notin A$. Worded slightly more formally, this observation reads as follows.

LEMMA 2.10. *For every function $f : \mathbb{R}^+ \to \mathbb{R}$ the following statements are equivalent:*

(i) *f is completely determined by S, i.e., there exists a function $\varphi : [1, 10) \to \mathbb{R}$ with $\sigma(\varphi) \subset \mathcal{B}[1, 10)$ such that $f(x) = \varphi\big(S(x)\big)$ for all $x \in \mathbb{R}^+$;*

(ii) *$\sigma(f) \subset \mathcal{S}$.*

PROOF. Assume (i) and let $J \subset \mathbb{R}$ be any interval. Then $B = \varphi^{-1}(J) \in \mathcal{B}$ and $f^{-1}(J) = S^{-1}\big(\varphi^{-1}(J)\big) = S^{-1}(B) \in \mathcal{S}$, showing that $\sigma(f) \subset \mathcal{S}$.

Conversely, if $\sigma(f) \subset \mathcal{S}$ then $f(10x) = f(x)$ for all $x > 0$. To see this, assume by way of contradiction that $f(x_0) < f(10x_0)$ for some $x_0 > 0$, let

$$A = f^{-1}\left(\left[f(x_0) - 1, \tfrac{1}{2}\big(f(x_0) + f(10x_0)\big)\right]\right) \in \sigma(f) \subset \mathcal{S},$$

and note that $x_0 \in A$ whereas $10x_0 \notin A$. Since $A = S^{-1}(B)$ for some $B \in \mathcal{B}$, this leads to the contradiction that $S(x_0) \in B$ and $S(x_0) = S(10x_0) \notin B$. Hence $f(10x) = f(x)$ for all $x > 0$, and by induction $f(10^k x) = f(x)$ for all $k \in \mathbb{Z}$. Given $t \in [1, 10)$, pick any $x > 0$ with $S(x) = t$ and define $\varphi(t) = f(x)$. Since any two choices of x differ by a factor 10^k for some $k \in \mathbb{Z}$, $\varphi : [1, 10) \to \mathbb{R}$ is well-defined, and $\varphi\big(S(x)\big) = f(x)$ for all $x > 0$. Moreover, for any interval $J \subset \mathbb{R}$ and $t \in [1, 10)$, $\varphi(t) \in J$ if and only if $t \in \bigcup_{k \in \mathbb{Z}} 10^k f^{-1}(J)$. By assumption, the latter set belongs to $\mathcal{S}$, which in turn shows that $\sigma(\varphi) \subset \mathcal{B}[1, 10)$.   ■

Thus Lemma 2.10 states that the significand σ-algebra $\mathcal{S}$ is the family of all events $A \subset \mathbb{R}^+$ that can be described completely in terms of their significands, or, equivalently (by Theorem 2.11 below), in terms of their significant digits. For example, the set A_1 of positive numbers whose first significant digit is 1 and whose third significant digit is not 7, i.e.,

$$A_1 = \{x > 0 : D_1(x) = 1, D_3(x) \neq 7\},$$

belongs to $\mathcal{S}$, as do the set A_2 of all $x > 0$ whose significant digits are all 5 or 6, i.e.,

$$A_2 = \{x > 0 : D_m(x) \in \{5, 6\} \text{ for all } m \in \mathbb{N}\},$$

and the set A_3 of numbers whose significand is rational,

$$A_3 = \{x > 0 : S(x) \in \mathbb{Q}\}.$$

On the other hand, the interval $[1, 2]$, for instance, does not belong to $\mathcal{S}$. This follows from the next theorem, which provides a useful characterization of the significand sets, i.e., the members of the family $\mathcal{S}$. For its formulation, for every $x \in \mathbb{R}$ and every set $C \subset \mathbb{R}$, let $xC = \{xc : c \in C\}$.

THEOREM 2.11 ([74]). *For every $A \in \mathcal{S}$,*

$$A = \bigcup\nolimits_{k \in \mathbb{Z}} 10^k S(A), \tag{2.2}$$

where $S(A) = \{S(x) : x \in A\} \subset [1, 10)$. Moreover,

$$\mathcal{S} = \mathbb{R}^+ \cap \sigma(D_1, D_2, D_3, \ldots) = \left\{ \bigcup\nolimits_{k \in \mathbb{Z}} 10^k B : B \in \mathcal{B}[1, 10) \right\}. \tag{2.3}$$

PROOF. By definition,

$$\mathcal{S} = \mathbb{R}^+ \cap \sigma(S) = \mathbb{R}^+ \cap \{S^{-1}(B) : B \in \mathcal{B}\} = \mathbb{R}^+ \cap \{S^{-1}(B) : B \in \mathcal{B}[1, 10)\}.$$

Thus, given any $A \in \mathcal{S}$, there exists a set $B \in \mathcal{B}[1, 10)$ with

$$A = \mathbb{R}^+ \cap S^{-1}(B) = \bigcup\nolimits_{k \in \mathbb{Z}} 10^k B.$$

Since $S(A) = B$, it follows that (2.2) holds for all $A \in \mathcal{S}$.

To prove (2.3), first observe that by Proposition 2.5(i) the significand function S is completely determined by the significant digits $D_1, D_2, D_3, \ldots$, so $\sigma(S) \subset \sigma(D_1, D_2, D_3, \ldots)$ and hence $\mathcal{S} \subset \mathbb{R}^+ \cap \sigma(D_1, D_2, D_3, \ldots)$. Conversely, by Proposition 2.5(ii), every D_m is determined by S; thus $\sigma(D_m) \subset \sigma(S)$ for all $m \in \mathbb{N}$, showing that $\sigma(D_1, D_2, D_3, \ldots) \subset \sigma(S)$ as well. To verify the remaining equality in (2.3), note that for every $A \in \mathcal{S}$, $S(A) \in \mathcal{B}[1, 10)$ and hence $A = \bigcup_{k \in \mathbb{Z}} 10^k B$ for $B = S(A)$, by (2.2). Conversely, every set of the form $\bigcup_{k \in \mathbb{Z}} 10^k B = \mathbb{R}^+ \cap S^{-1}(B)$ with $B \in \mathcal{B}[1, 10)$ clearly belongs to $\mathcal{S}$. ∎

Note that for every $A \in \mathcal{S}$ there is a *unique* set $B \in \mathcal{B}[1, 10)$ such that $A = \bigcup_{k \in \mathbb{Z}} 10^k B$, and (2.2) shows that in fact $B = S(A)$.

EXAMPLE 2.12. The set A_4 of positive numbers with first significant digit 1 and all other significant digits 0,

$$A_4 = \{10^k : k \in \mathbb{Z}\} = \{\ldots, 0.01, 0.1, 1, 10, 100, \ldots\},$$

belongs to $\mathcal{S}$. This can be seen either by observing that A_4 is the set of positive reals with significand exactly equal to 1, i.e., $A_4 = \mathbb{R}^+ \cap S^{-1}(\{1\})$, or by noting that $A_4 = \{x > 0 : D_1(x) = 1, D_m(x) = 0 \text{ for all } m \geq 2\}$, or by using (2.3) and the fact that $A_4 = \bigcup_{k \in \mathbb{Z}} 10^k \{1\}$ and $\{1\} \in \mathcal{B}[1, 10)$. ✠

EXAMPLE 2.13. The singleton set $\{1\}$ and the interval $[1, 2]$ do not belong to $\mathcal{S}$, since the number 1 cannot be distinguished from the number 10 using only significant digits, as both have first significant digit 1 and all other significant digits 0. Nor, for the same reason, can the interval $[1, 2]$ be distinguished from the interval $[10, 20]$. Formally, neither of these sets is of the form $\bigcup_{k \in \mathbb{Z}} 10^k B$ for any $B \in \mathcal{B}[1, 10)$. ✠

Although the significand function and σ-algebra above were defined in the setting of *real* numbers, the same concepts carry over immediately to the most fundamental setting of all, namely, the set of *positive integers* $\mathbb{N}$. In this case, the induced σ-algebra $\mathcal{S}_{\mathbb{N}}$ on $\mathbb{N}$ is interesting in its own right. Note that, as in the case of the significand σ-algebra, $\mathcal{S}_{\mathbb{N}}$ does not include all subsets of positive integers, i.e., it is a smaller σ-algebra than the power set of $\mathbb{N}$.

EXAMPLE 2.14. The restriction $\mathcal{S}_{\mathbb{N}}$ of $\mathcal{S}$ to subsets of $\mathbb{N}$, i.e.,

$$\mathcal{S}_{\mathbb{N}} = \{\mathbb{N} \cap A : A \in \mathcal{S}\},$$

is a σ-algebra on $\mathbb{N}$. A characterization of $\mathcal{S}_{\mathbb{N}}$ analogous to that of $\mathcal{S}$ given in Theorem 2.11 is as follows: Denote by $\mathbb{N}_{10}$ the set of all positive integers not divisible by 10, i.e., $\mathbb{N}_{10} = \mathbb{N}\backslash 10\mathbb{N}$. Then

$$\mathcal{S}_{\mathbb{N}} = \left\{ A \subset \mathbb{N} : A = \bigcup_{k \in \mathbb{N}_0} 10^k B \text{ for some } B \subset \mathbb{N}_{10} \right\}.$$

A typical member of $\mathcal{S}_{\mathbb{N}}$ is

$$\{271, 2710, 3141, 27100, 31410, 271000, 314100, \ldots\}.$$

For instance, note that the set $\{31410, 314100, 3141000, \ldots\}$ does *not* belong to $\mathcal{S}_{\mathbb{N}}$ since 31410 is indistinguishable from 3141 in terms of significant digits, because all significant digits of both numbers are identical, so if the former number were to belong to $A \in \mathcal{S}_{\mathbb{N}}$ then the latter would too. Note also that the corresponding significand function on $\mathbb{N}$ still only takes values in $[1, 10)$, as before, but may never be an irrational number. In fact, the possible values of S on $\mathbb{N}$ are even more restricted: $S(n) = t$ for some $n \in \mathbb{N}$ if and only if $t \in [1, 10)$ and $10^k t \in \mathbb{N}$ for some integer $k \geq 0$. ✠

The next lemma establishes some basic closure properties of the significand σ-algebra that will be essential later in studying characteristic aspects of Benford's law such as scale- and base-invariance. To formulate these properties concisely, for every $C \subset \mathbb{R}^+$ and $n \in \mathbb{N}$, let $C^{1/n} = \{x > 0 : x^n \in C\}$.

LEMMA 2.15. *The following properties hold for the significand σ-algebra $\mathcal{S}$:*

(i) $\mathcal{S}$ *is self-similar with respect to multiplication by integer powers of* 10*, i.e.,* $10^k A = A$ *for every* $A \in \mathcal{S}$ *and* $k \in \mathbb{Z}$;

(ii) $\mathcal{S}$ *is closed under multiplication by scalars, i.e.,* $aA \in \mathcal{S}$ *for every* $A \in \mathcal{S}$ *and* $a > 0$;

(iii) $\mathcal{S}$ *is closed under integral roots, i.e.,* $A^{1/n} \in \mathcal{S}$ *for every* $A \in \mathcal{S}$ *and* $n \in \mathbb{N}$.

Informally, property (i) says that every significand set remains unchanged when multiplied by an integer power of 10 — reflecting the simple fact that shifting the decimal point keeps all the significant digits, and hence the set itself, unchanged; (ii) asserts that if every element of a set expressible solely in terms of significant digits is multiplied by a positive constant, then the new set is also expressible by significant digits; similarly, (iii) states that the collection of square (cube, fourth, etc.) roots of the elements of every significand set is also expressible in terms of significant digits alone.

PROOF. (i) This is clear from (2.2) since $S(10^k A) = S(A)$ for every $k \in \mathbb{Z}$.
(ii) Given $A \in \mathcal{S}$, by (2.3) there exists $B \in \mathcal{B}[1, 10)$ such that $A = \bigcup_{k \in \mathbb{Z}} 10^k B$. In view of (i), assume without loss of generality that $1 < a < 10$. Then

$$aA = \bigcup_{k \in \mathbb{Z}} 10^k aB = \bigcup_{k \in \mathbb{Z}} 10^k \Big(\big(aB \cap [a, 10)\big) \cup \big(\tfrac{1}{10} aB \cap [1, a)\big) \Big)$$
$$= \bigcup_{k \in \mathbb{Z}} 10^k C,$$

with $C = \big(aB \cap [a, 10)\big) \cup \big(\tfrac{1}{10} aB \cap [1, a)\big) \in \mathcal{B}[1, 10)$, showing that $aA \in \mathcal{S}$.
(iii) Since intervals of the form $[1, 10^s]$ with $0 < s < 1$ generate $\mathcal{B}[1, 10)$, i.e., since $\mathcal{B}[1, 10) = \sigma\big(\{[1, 10^s] : 0 < s < 1\}\big)$, it is enough to verify the claim for the special case $A = \bigcup_{k \in \mathbb{Z}} 10^k [1, 10^s]$ for every $0 < s < 1$. In this case,

$$A^{1/n} = \bigcup_{k \in \mathbb{Z}} 10^{k/n} \big[1, 10^{s/n}\big] = \bigcup_{k \in \mathbb{Z}} 10^k \bigcup_{j=0}^{n-1} \big[10^{j/n}, 10^{(j+s)/n}\big]$$
$$= \bigcup_{k \in \mathbb{Z}} 10^k C,$$

with $C = \bigcup_{j=0}^{n-1} \big[10^{j/n}, 10^{(j+s)/n}\big] \in \mathcal{B}[1, 10)$. Hence $A^{1/n} \in \mathcal{S}$. ∎

By Theorem 2.11, the significand σ-algebra $\mathcal{S}$ is the same as the significant digit σ-algebra $\sigma(D_1, D_2, D_3, \ldots)$, so the closure properties established in Lemma 2.15 carry over to sets determined by significant digits. Note also that $\mathcal{S}$ is *not* closed under taking integer powers: If $A \in \mathcal{S}$ and $n \in \mathbb{N}$, then it is easy to check that $A^n \in \mathcal{S}$ if and only if

$$S(A)^n = B \cup 10B \cup \ldots \cup 10^{n-1}B \quad \text{for some } B \in \mathcal{B}[1, 10).$$

The next example illustrates closure under multiplication by a scalar and under integral roots, as well as lack of closure under powers.

EXAMPLE 2.16. Let A_5 be the set of positive real numbers with first significant digit less than 3, i.e.,

$$A_5 = \{x > 0 : D_1(x) < 3\} = \{x > 0 : 1 \le S(x) < 3\} = \bigcup_{k \in \mathbb{Z}} 10^k [1, 3).$$

Then

$$2A_5 = \big\{x > 0 : D_1(x) \in \{2, 3, 4, 5\}\big\} = \{x > 0 : 2 \le S(x) < 6\}$$
$$= \bigcup_{k \in \mathbb{Z}} 10^k [2, 6) \in \mathcal{S},$$

and also

$$A_5^{1/2} = \left\{ x > 0 : S(x) \in [1, \sqrt{3}) \cup [\sqrt{10}, \sqrt{30}) \right\}$$

$$= \bigcup_{k \in \mathbb{Z}} 10^k \left([1, \sqrt{3}) \cup [\sqrt{10}, \sqrt{30}) \right) \in \mathcal{S},$$

whereas, on the other hand, clearly

$$A_5^2 = \bigcup_{k \in \mathbb{Z}} 10^{2k}[1, 9) \notin \mathcal{S},$$

since $[1, 9) \subset A_5^2$ but $[10, 90) \not\subset A_5^2$; see Figure 2.2.   ✠

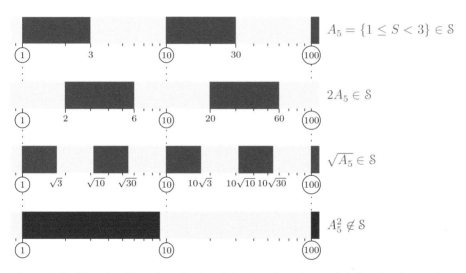

Figure 2.2: The significand σ-algebra $\mathcal{S}$ is closed under multiplication by scalars and under integral roots but not under integral powers (bottom), as seen here for the set A_5 of Example 2.16.

EXAMPLE 2.17. Recall the significand σ-algebra $\mathcal{S}_{\mathbb{N}}$ on the positive integers defined in Example 2.14. Unlike its continuous counterpart $\mathcal{S}$, the family $\mathcal{S}_{\mathbb{N}}$ is not even closed under multiplication by a positive integer, since, for example,

$$A_6 = \mathbb{N} \cap \{ x > 0 : S(x) = 2 \} = \{ 2, 20, 200, \ldots \} \in \mathcal{S}_{\mathbb{N}},$$

but

$$5A_6 = \{ 10, 100, 1000, \ldots \} \notin \mathcal{S}_{\mathbb{N}}.$$

Of course, this does not rule out the fact that some events determined by significant digits, i.e., some members of $\mathcal{S}_{\mathbb{N}}$, still belong to $\mathcal{S}_{\mathbb{N}}$ after multiplication by an integer. For example, if

$$A_7 = \{ n \in \mathbb{N} : D_1(n) = 1 \} = \{ 1, 10, 11, \ldots, 18, 19, 100, 101, \ldots \} \in \mathcal{S}_{\mathbb{N}}$$

then

$$3A_7 = \{3, 30, 33, \ldots, 54, 57, 300, 303, \ldots\} \in \mathcal{S}_\mathbb{N}\,.$$

It is easy to see that, more generally, $\mathcal{S}_\mathbb{N}$ is closed under multiplication by $m \in \mathbb{N}$ precisely if the greatest common divisor $\gcd(m, 10)$ equals 1, that is, whenever m and 10 have no nontrivial common factor. Moreover, like $\mathcal{S}$, the σ-algebra $\mathcal{S}_\mathbb{N}$ is closed under integral roots: If $A = \bigcup_{k \in \mathbb{N}_0} 10^k B$ with $B \subset \mathbb{N}_{10}$ then $A^{1/n} = \bigcup_{k \in \mathbb{N}_0} 10^k B^{1/n} \in \mathcal{S}_\mathbb{N}$. With A_7 from above, for instance,

$$A_7^{1/2} = \left\{ n \in \mathbb{N} : S(n) \in \left[1, \sqrt{2}\right) \cup \left[\sqrt{10}, \sqrt{20}\right) \right\}$$
$$= \{1, 4, 10, 11, 12, 13, 14, 32, 33, \ldots, 43, 44, 100, 101, \ldots\} \in \mathcal{S}_\mathbb{N}\,.$$

Thus many of the conclusions drawn later for positive real numbers carry over to positive integers in a straightforward way.

Chapter Three

The Benford Property

In order to translate the informal versions (1.1), (1.2), and (2.1) of Benford's law into more precise formal statements, it is necessary to specify exactly what the Benford property means in various mathematical contexts. For the purpose of this book, the objects of interest fall mainly into three categories: sequences of real numbers, real-valued functions defined on $[0, +\infty)$, and probability distributions and random variables.

Since (1.1), (1.2), and (2.1) are statements about probabilities, it is useful to first review the formal definition of probability. Given a non-empty set Ω and a σ-algebra $\mathcal{A}$ of subsets of Ω, recall that a *probability measure* on $(\Omega, \mathcal{A})$ is a function $\mathbb{P} : \mathcal{A} \to [0, 1]$ such that $\mathbb{P}(\varnothing) = 0$, $\mathbb{P}(\Omega) = 1$, and

$$\mathbb{P}\left(\bigcup_{n \in \mathbb{N}} A_n \right) = \sum_{n \in \mathbb{N}} \mathbb{P}(A_n)$$

whenever the sets $A_n \in \mathcal{A}$ are disjoint. The natural probabilistic interpretation of $\mathbb{P}$ is that, for every $A \in \mathcal{A}$, the number $\mathbb{P}(A) \in [0, 1]$ is the probability that the event $\{\omega \in \Omega : \omega \in A\}$ occurs.

Two of the most important examples of probability measures are the *discrete uniform distribution* on a non-empty finite set $A \subset \Omega$, where the probability of any set $B \subset \Omega$ is simply

$$\frac{\#(B \cap A)}{\#A},$$

and its continuous counterpart the *uniform distribution* $\lambda_{a,b}$ on a finite interval $[a, b)$ with $a < b$, technically referred to as the (*normalized*) *Lebesgue measure* on $[a, b)$, or more precisely on $\big([a, b), \mathcal{B}[a, b)\big)$, given by

$$\lambda_{a,b}([c, d]) = \frac{d - c}{b - a} \quad \text{for every } [c, d] \subset [a, b). \tag{3.1}$$

In advanced analysis courses, it is shown that (3.1) does indeed entail a unique, consistent definition of $\lambda_{a,b}(B)$ for *every* $B \in \mathcal{B}[a, b)$, and $\lambda_{a,b}\big([a, b)\big) = 1$.

Another example of a probability measure, on any $(\Omega, \mathcal{A})$, is the *Dirac measure* (or *point mass*) concentrated at some $\omega \in \Omega$, symbolized by δ_ω. In this case, $\delta_\omega(A) = 1$ if $\omega \in A$, and $\delta_\omega(A) = 0$ otherwise. Throughout, probability measures on $(\Omega, \mathcal{A})$ with $\Omega \subset \mathbb{R}$ and $\mathcal{A} \subset \mathcal{B}$ will typically be denoted by capital Roman letters P, Q, etc.

3.1   BENFORD SEQUENCES

A sequence $(x_n) = (x_1, x_2, \ldots)$ of real numbers is a (base-10) *Benford sequence* if, as $N \to \infty$, the limiting proportion of indices $n \leq N$ for which x_n has first significant digit d exists and equals $\log(1 + d^{-1})$ for all $d \in \{1, 2, \ldots, 9\}$, and similarly for the limiting proportions of the occurrences of all other finite blocks of initial significant digits. The formal definition is as follows.

DEFINITION 3.1. A sequence (x_n) of real numbers is a *Benford sequence*, or *Benford* for short, if

$$\lim_{N \to \infty} \frac{\#\{1 \leq n \leq N : S(x_n) \leq t\}}{N} = \log t \quad \text{for all } t \in [1, 10),$$

or, equivalently, if for all $m \in \mathbb{N}$, all $d_1 \in \{1, 2, \ldots, 9\}$, and all $d_j \in \{0, 1, \ldots, 9\}$, $j \geq 2$,

$$\lim_{N \to \infty} \frac{\#\{1 \leq n \leq N : D_j(x_n) = d_j \text{ for } j = 1, 2, \ldots, m\}}{N}$$
$$= \log \left(1 + \left(\sum_{j=1}^{m} 10^{m-j} d_j \right)^{-1} \right).$$

As will be shown below, the sequence of powers of 2, i.e., $(2^n) = (2, 4, 8, \ldots)$ is Benford. Note, however, that (2^n) is not base-2 Benford since the second significant digit base 2 of 2^n is 0 for every n, i.e., $D_2^{(2)}(n) \equiv 0$, whereas the generalized version (1.5) of Benford's law requires that

$$\mathsf{Prob}\left(D_2^{(2)} = 0\right) = 1 - \mathsf{Prob}\left(D_2^{(2)} = 1\right) = \log_2 3 - 1 > \tfrac{1}{2}.$$

Similarly, as will be seen below, the powers of 3, i.e., $(3^n) = (3, 9, 27, \ldots)$, also form a Benford sequence, and so do the sequence of factorials $(n!)$ and the sequence of Fibonacci numbers (F_n). Common sequences that are not Benford include the positive integers (n), the powers of 10, i.e., (10^n), and the sequence of logarithms $(\log n)$.

The notion of Benford sequence given in Definition 3.1 offers a natural interpretation of Prob in the informal expressions (1.1)–(1.3): A sequence of real numbers $(x_n) = (x_1, x_2, \ldots)$ is Benford if, when one of the first N entries $x_1, x_2, \ldots, x_N$ is chosen (uniformly) at random, the probability that its first significant digit is d approaches the Benford probability $\log(1 + d^{-1})$ as $N \to \infty$ for every $d \in \{1, 2, \ldots, 9\}$, and similarly for all other blocks of significant digits.

EXAMPLE 3.2. Two specific sequences of positive integers will be used repeatedly to illustrate key concepts concerning Benford's law: the Fibonacci numbers and the prime numbers. Both sequences play prominent roles in many areas of mathematics.

(i) As will be seen in Example 4.18 below, the sequence of Fibonacci numbers $(F_n) = (1, 1, 2, 3, 5, 8, 13, \ldots)$, where every entry is the sum of its two predecessors, and $F_1 = F_2 = 1$, is Benford. The first $N = 10^2$ elements of the Fibonacci sequence already conform very well to the first-digit version (1.1) of Benford's law, with Prob being interpreted as relative frequency; see Figures 1.5 and 3.1. The conformance becomes even better if the first $N = 10^4$ elements are considered; see Figure 3.3.

(ii) Example 4.17(v) below will establish that the sequence of prime numbers $(p_n) = (2, 3, 5, 7, 11, 13, 17, \ldots)$ is *not* Benford. Figure 3.2 illustrates how poorly the first hundred prime numbers conform to (1.1). The conformance is even worse if the first ten thousand primes are considered; see Figure 3.3.   ✠

$$F_1, F_2, \ldots, F_{100}$$

1	121393	20365011074	3416454622906707
1	196418	32951280099	5527939700884757
2	317811	53316291173	8944394323791464
3	514229	86267571272	14472334024676221
5	832040	139583862445	23416728348467685
8	1346269	225851433717	37889062373143906
13	2178309	365435296162	61305790721611591
21	3524578	591286729879	99194853094755497
34	5702887	956722026041	160500643816367088
55	9227465	1548008755920	259695496911122585
89	14930352	2504730781961	420196140727489673
144	24157817	4052739537881	679891637638612258
233	39088169	6557470319842	1100087778366101931
377	63245986	10610209857723	1779979416004714189
610	102334155	17167680177565	2880067194370816120
987	165580141	27777890035288	4660046610375530309
1597	267914296	44945570212853	7540113804746346429
2584	433494437	72723460248141	12200160415121876738
4181	701408733	117669030460994	19740274219868223167
6765	1134903170	190392490709135	31940434634990099905
10946	1836311903	308061521170129	51680708854858323072
17711	2971215073	498454011879264	83621143489848422977
28657	4807526976	806515533049393	135301852344706746049
46368	7778742049	1304969544928657	218922995834555169026
75025	12586269025	2111485077978050	354224848179261915075

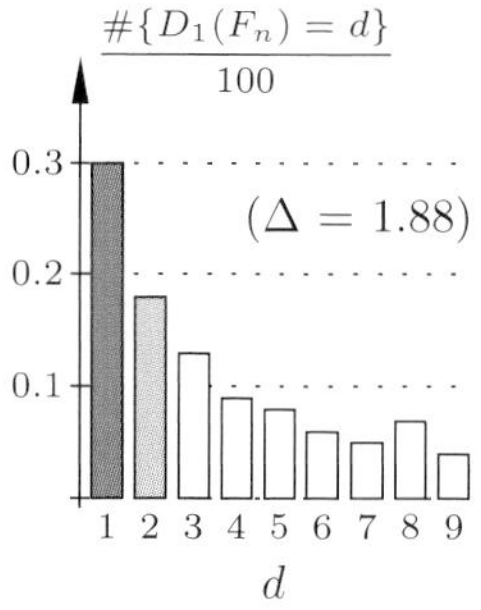

$$\frac{\#\{D_1(F_n) = d\}}{100}$$

$(\Delta = 1.88)$

Figure 3.1: The first one hundred Fibonacci numbers conform to the first digit law (1.1) quite well.

Alternative notions of Benford sequences

The notion of Benford sequence specified in Definition 3.1 is based on the classical concept of *natural density* of subsets of the positive integers. Recall that a subset C of $\mathbb{N}$ is said to have (*natural*) *density* ρ if

$$\lim_{N \to \infty} \frac{\#\{1 \leq n \leq N : n \in C\}}{N} = \rho. \qquad (3.2)$$

For example, the set of even positive integers has density $\frac{1}{2}$; the set of integral multiples of 3, i.e., the set $\{3n : n \in \mathbb{N}\}$, has density $\frac{1}{3}$; and the set of prime

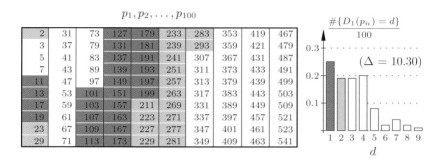

Figure 3.2: The first one hundred prime numbers conform poorly to the first digit law (1.1).

numbers and all finite sets have density 0. In these terms, Definition 3.1 says that a sequence of real numbers (x_n) is Benford if and only if the density of the set $\{n \in \mathbb{N} : S(x_n) \le t\}$ exists and is $\log t$ for all $t \in [1, 10)$. For instance, since the Fibonacci sequence (F_n) is Benford (see Example 4.18 below), this means that the limiting proportion of Fibonacci numbers that start with a 1 exists and is $\log 2$. In other words, the density of $\{n \in \mathbb{N} : D_1(F_n) = 1\}$ is exactly $\log 2$. Not all subsets of $\mathbb{N}$ have natural densities, however, as was seen in Chapter 1 with the set $C = \{n \in \mathbb{N} : D_1(n) = 1\}$ of positive integers with leading digit 1, for which

$$\liminf_{N \to \infty} \frac{\#\{1 \le n \le N : n \in C\}}{N} = \frac{1}{9}$$
$$< \frac{5}{9} = \limsup_{N \to \infty} \frac{\#\{1 \le n \le N : n \in C\}}{N},$$

so the limit in (3.2) does not exist. Thus, in particular, the sequence of positive integers (n) is not Benford in the natural density sense of Definition 3.1.

Apart from natural density there are many other important notions of density in use, especially in analysis and number theory; see [68, 104] for recent surveys. To understand how these notions are related, note first that (3.2) can equivalently be written as

$$\lim_{N \to \infty} \frac{1}{N} \sum_{n=1}^{N} \mathbb{1}_C(n) = \rho, \tag{3.3}$$

where $\mathbb{1}_C$ is the *indicator* (or *characteristic*) *function* of $C \subset \mathbb{N}$, i.e., $\mathbb{1}_C(n) = 1$ if $n \in C$, and $\mathbb{1}_C(n) = 0$ otherwise. Thus the concept of natural density is based on the arithmetic mean (Cesàro average), assigning the uniform weight $1/N$ to each of the first N terms of the sequence $(\mathbb{1}_C(n))$. While this is arguably the simplest notion of density, often alternative definitions are considered that are based on non-uniform weights. Two such alternative densities in particular, the *logarithmic density* and the H_∞-*density*, have been studied extensively in

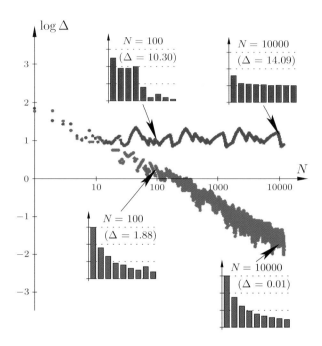

Figure 3.3: The Fibonacci sequence (blue data) is Benford, hence the deviation Δ from (1.1) goes to 0 as $N \to \infty$; this is clearly not the case for the sequence of prime numbers (red data).

the context of Benford sequences [104, 126, 127, 141]. Both of these densities assign lighter weights to later terms, and unlike the uniformly weighted natural density underlying the notion of Benford sequence in Definition 3.1, both of these alternative densities assign the value $\log 2$ to the set of positive integers with leading digit 1.

The logarithmic (or harmonic) density, a powerful tool in analytic number theory, replaces the uniform weights $1/N$ in (3.3) with the strictly decreasing weights $1/n \cdot 1/(1 + \frac{1}{2} + \ldots + \frac{1}{N})$. Recall that $\lim_{N \to \infty} (1 + \frac{1}{2} + \ldots + \frac{1}{N})/\ln N = 1$. Accordingly, a set $C \subset \mathbb{N}$ is said to have *logarithmic density* ρ if

$$\lim_{N \to \infty} \frac{1}{\ln N} \sum_{n=1}^{N} \frac{1}{n} \mathbb{1}_C(n) = \rho \,. \tag{3.4}$$

It is well known (e.g., [104]) that the natural density is strictly stronger than the logarithmic density. This means that every set with natural density ρ has logarithmic density ρ as well, but there are subsets C of $\mathbb{N}$ which have a logarithmic density yet do not have a natural density, i.e., the limit in (3.4) does exist whereas the limit in (3.3) does not. A prominent example is the set of positive integers with leading digit 1, i.e., $C = \{n \in \mathbb{N} : D_1(n) = 1\}$, whose logarithmic density is $\log 2$ whereas, as seen above, its natural density does not exist. In terms of logarithmic density, a sequence of real numbers (x_n) is sometimes called *logarithmic Benford* (e.g., [104]) if, for all $t \in [1, 10)$, the logarithmic density of $\{n \in \mathbb{N} : S(x_n) \leq t\}$ is $\log t$. For example, in contrast to the fact that neither the sequence of positive integers (n) nor the sequence of prime numbers (p_n) is Benford (see Figures 3.2, 3.3, and Example 4.17(v) below), both sequences are logarithmic Benford (e.g., [104, 127]). On the other hand, since the natural density is stronger than the logarithmic density, every sequence that is Benford, such as (2^n) and (F_n), is automatically logarithmic Benford as well.

A second alternative density that, like the logarithmic density, also assigns the "Benford value" $\log t$ to the set of positive integers with significands not exceeding t, is the so-called H_∞-density (e.g., [59, 104, 141]). The H_∞-density, an iterated averaging method, is defined inductively as follows: For any set $C \subset \mathbb{N}$, let

$$H_{0,n} = \mathbb{1}_C(n) \ \text{ for } n \in \mathbb{N} \quad \text{and} \quad H_{m,n} = \frac{1}{n} \sum\nolimits_{j=1}^{n} H_{m-1,j} \ \text{ for } m, n \in \mathbb{N}.$$

By definition, the set C has H_∞-*density* ρ if

$$\lim\nolimits_{m \to \infty} \liminf\nolimits_{n \to \infty} H_{m,n} = \rho = \lim\nolimits_{m \to \infty} \limsup\nolimits_{n \to \infty} H_{m,n}.$$

It can be shown that the natural density is strictly stronger than the H_∞-density, which in turn is strictly stronger than the logarithmic density [104]. As with the logarithmic density, the sequence of positive integers (n) is also H_∞-*Benford* in the sense that every set $\{n \in \mathbb{N} : S(n) \leq t\}$ has H_∞-density $\log t$. Conversely, as with logarithmic density, every Benford sequence is also H_∞-Benford. Some sequences such as (10^n), of course, are not Benford in any of these senses.

As will become apparent in subsequent chapters, one of the simplest but also most fundamental examples in this book is the sequence (a^n) with $a > 0$. For every such sequence, the notions of logarithmic Benford and H_∞-Benford are both equivalent to Definition 3.1. Another basic example is the sequence (a^{b^n}) with $b > 1$. As will be explained in Section 6.4, this sequence is Benford for (Lebesgue) almost every, but not every, $a > 0$. Again this situation would remain completely unchanged if, instead of Definition 3.1, the notion of logarithmic Benford or H_∞-Benford were used throughout.

Many other notions of density exist, and can also be used to define alternative concepts of Benford sequences. Arithmetic averages, however, play a central role in probability and statistics, as evidenced by the strong law of large numbers, the classical ergodic theorems, and the empirical distribution functions appearing in the fundamental Glivenko–Cantelli Theorem. Thus it is the arithmetic average, with its uniform weights, that is used exclusively in this book for the definition of Benford sequence. Extensions to settings of logarithmic density, H_∞-density, and other notions of density are beyond the scope of this book, and a possible topic for future research.

3.2 BENFORD FUNCTIONS

Benford's law also appears frequently in real-valued functions such as those arising as solutions of initial value problems for differential equations; see Chapters 6 and 7. Thus, the starting point here is to define what it means for a function to follow Benford's law.

In direct analogy with the terminology for sequences, a function f is a (base-10) Benford function, or simply Benford, if the limiting proportion of time $\tau < T$ that the first digit of $f(\tau)$ equals d_1 is exactly $\log(1 + d_1^{-1})$, and more generally, if for all t in $[1, 10)$, the proportion of time $\tau < T$ that the significand of f is less than or equal to t approaches $\log t$ as $T \to +\infty$.

To formalize this notion, first recall that a function $f : \mathbb{R} \to \mathbb{R}$ is (*Borel*) *measurable* if $f^{-1}(J)$ is a Borel set, i.e., $f^{-1}(J) \in \mathcal{B}$ for every interval $J \subset \mathbb{R}$. With the terminology introduced in Section 2.3, this is equivalent to saying that every set in the σ-algebra generated by f is a Borel set, i.e., $\sigma(f) \subset \mathcal{B}$. Slightly more generally, for any set Ω and any σ-algebra $\mathcal{A}$ on Ω, a function $f : \Omega \to \mathbb{R}$ is (*Borel*) *measurable* if $\sigma(f) \subset \mathcal{A}$.

The collection of Borel measurable functions $f : \mathbb{R} \to \mathbb{R}$ contains all functions of practical interest. For example, every piecewise continuous function (meaning that f has at most countably many discontinuities) is measurable. Thus every polynomial, trigonometric, and exponential function is Borel measurable, and so is every probability density function of any relevance. In fact, it is a difficult exercise to produce a function that is *not* measurable, or a set $C \subset \mathbb{R}$ that is *not* a member of $\mathcal{B}$, and this can be done only in a non-constructive way. For all practical purposes, therefore, the reader may simply read "set" for "Borel set," and "function" for "Borel measurable function."

Recall that given a set Ω and a σ-algebra $\mathcal{A}$ on Ω, a *measure* μ on $(\Omega, \mathcal{A})$ is a function $\mu : \mathcal{A} \to [0, +\infty]$ that has all the properties of a probability measure, except that $\mu(A)$ may also be smaller or larger than 1, and even infinite. A very important example of a measure is the so-called *Lebesgue measure* on $(\mathbb{R}, \mathcal{B})$, denoted by λ here and throughout. The basic defining property of λ is that $\lambda([a, b]) = b - a$ for every interval $[a, b] \subset \mathbb{R}$, that is, λ is an extension of the informal notion of length. This measure λ is related to the probability measures $\lambda_{a,b}$ considered in (3.1), since, e.g.,

$$\lambda(B) = \lim_{N \to \infty} 2N \lambda_{-N,N}(B \cap [-N, N]) \quad \text{for every } B \in \mathcal{B}.$$

It is customary to also use the symbol λ, often without a subscript, to denote the restriction of Lebesgue measure to $(C, \mathcal{B}(C))$, with the Borel set C being clear from the context. The next definition formalizes the notion of a function having the Benford property.

DEFINITION 3.3. A (Borel measurable) function $f : [0, +\infty) \to \mathbb{R}$ is *Benford* if

$$\lim_{T \to +\infty} \frac{\lambda(\{\tau \in [0, T) : S(f(\tau)) \le t\})}{T} = \log t \quad \text{for all } t \in [1, 10),$$

or, equivalently, if for all $m \in \mathbb{N}$, all $d_1 \in \{1, 2, \ldots, 9\}$, and all $d_j \in \{0, 1, \ldots, 9\}$, $j \geq 2$,

$$\lim_{T \to +\infty} \frac{\lambda(\{\tau \in [0, T) : D_j(f(\tau)) = d_j \text{ for } j = 1, 2, \ldots, m\})}{T}$$

$$= \log\left(1 + \left(\sum_{j=1}^{m} 10^{m-j} d_j\right)^{-1}\right).$$

Directly analogous to the probabilistic interpretation of a Benford sequence, the definition of a Benford function given in Definition 3.3 also offers a natural probabilistic interpretation: A function $f : [0, +\infty) \to \mathbb{R}$ is Benford if, when a time τ is chosen (uniformly) at random in $[0, T)$, the probability that the first digit of $f(\tau)$ equals d approaches $\log(1 + d^{-1})$ as $T \to +\infty$, for every $d \in \{1, 2, \ldots, 9\}$, and similarly for all other blocks of significant digits.

As will be seen in Example 4.9 below, the function $f(t) = e^{at}$ is Benford whenever $a \neq 0$, but the functions $f(t) = t$ and $f(t) = (\sin t)^2$ are not Benford.

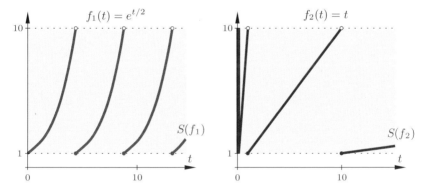

Figure 3.4: The function f_1 is Benford, but the function f_2 is not.

3.3   BENFORD DISTRIBUTIONS AND RANDOM VARIABLES

Benford's law appears in a wide variety of statistics and probability settings, including: mixtures of random samples; stochastic models such as geometric Brownian motion that are of great importance for modeling real-world processes; and products of independent, identically distributed random variables. (The term *independent, identically distributed* will henceforth be abbreviated *i.i.d.*, in accordance with standard practice.) This section lays the foundations for analyzing the Benford property for probability distributions and random variables.

Recall from Section 2.3 that a probability space is a triple $(\Omega, \mathcal{A}, \mathbb{P})$ where Ω is a non-empty set (the *set of outcomes*), $\mathcal{A}$ is a σ-algebra on Ω (the *family of events*), and $\mathbb{P}$ is a probability measure on $(\Omega, \mathcal{A})$. A (real-valued) *random variable* X on $(\Omega, \mathcal{A}, \mathbb{P})$ is simply a Borel measurable function $X : \Omega \to \mathbb{R}$, and its *distribution* P_X is the probability measure on $(\mathbb{R}, \mathcal{B})$ defined by

$$P_X\big((-\infty, x]\big) = \mathbb{P}(X \leq x) \quad \text{for all } x \in \mathbb{R}.$$

A probability measure on $(\mathbb{R}, \mathcal{B})$ will be referred to as a *Borel probability measure* on $\mathbb{R}$. Again, since all subsets of $\mathbb{R}$ of any practical interest are Borel sets, the modifier "Borel" will be suppressed unless there is a potential for confusion, i.e., the reader may read "probability measure on $\mathbb{R}$" for "Borel probability measure on $\mathbb{R}$." Every probability measure P on $\mathbb{R}$ is uniquely determined by its *distribution function F_P*, defined as

$$F_P(x) = P\big((-\infty, x]\big) \quad \text{for all } x \in \mathbb{R}.$$

As is well known and easy to check, the function F_P is right-continuous and non-decreasing, with $\lim_{x \to -\infty} F_P(x) = 0$ and $\lim_{x \to +\infty} F_P(x) = 1$. For the sake of notational simplicity, write F_X instead of F_{P_X} for every random variable X. The probability measure P, or any random variable X with $P_X = P$, is *continuous* (or *atomless*) if $P(\{x\}) = 0$ for every $x \in \mathbb{R}$, or, equivalently, if the function F_P is continuous. The probability measure P is *absolutely continuous* (*a.c.*) if, for any $B \in \mathcal{B}$, $P(B) = 0$ whenever $\lambda(B) = 0$. By the Radon–Nikodym Theorem, this is equivalent to P having a *density*, i.e., there exists a measurable function $f_P : \mathbb{R} \to [0, +\infty)$ such that

$$P([a, b]) = \int_a^b f_P(x)\, dx \quad \text{for all } [a, b] \subset \mathbb{R}. \tag{3.5}$$

Again, for simplicity, write f_X instead of f_{P_X} for every a.c. random variable X. Note that (3.5) implies $\int_{-\infty}^{+\infty} f_P(x)\, dx = 1$. Every a.c. probability measure on $(\mathbb{R}, \mathcal{B})$ is continuous, but not vice versa; e.g., see [36].

DEFINITION 3.4. A Borel probability measure P on $\mathbb{R}$ is *Benford* if

$$P\big(\{x \in \mathbb{R} : S(x) \leq t\}\big) = P\big(S^{-1}(\{0\} \cup [1, t])\big) = \log t \quad \text{for all } t \in [1, 10).$$

A random variable X on a probability space $(\Omega, \mathcal{A}, \mathbb{P})$ is *Benford* if P_X is Benford, i.e., if

$$F_{S(X)}(t) = \mathbb{P}(S(X) \leq t) = P_X\big(\{x \in \mathbb{R} : S(x) \leq t\}\big) = \log t \quad \text{for all } t \in [1, 10),$$

or, equivalently, if $S(X)$ is an absolutely continuous random variable with density $f_{S(X)}(t) = t^{-1} \log e$ for $t \in [1, 10)$. In terms of its significant digits, X is Benford if and only if

$$\mathbb{P}\big(D_j(X) = d_j \text{ for } j = 1, 2, \ldots, m\big) = \log\left(1 + \left(\sum_{j=1}^m 10^{m-j} d_j\right)^{-1}\right),$$

for all $m \in \mathbb{N}$, all $d_1 \in \{1, 2, \ldots, 9\}$, and all $d_j \in \{0, 1, \ldots, 9\}$, $j \geq 2$.

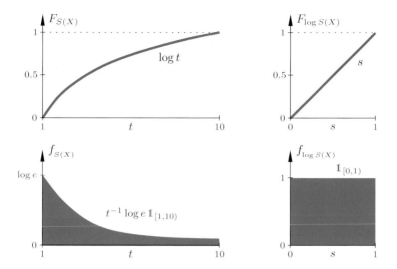

Figure 3.5: The distribution functions (top) and densities of $S(X)$ and $\log S(X)$, respectively, for every Benford random variable X.

EXAMPLE 3.5. If X is a Benford random variable on some probability space $(\Omega, \mathcal{A}, \mathbb{P})$, then

$$\mathbb{P}(D_1(X) = 1) = \mathbb{P}(1 \le S(X) < 2) = \log 2 = 0.3010\,,$$
$$\mathbb{P}(D_1(X) = 9) = \log \tfrac{10}{9} = 0.04575\,,$$
$$\mathbb{P}(D_1(X) = 3, D_2(X) = 1, D_3(X) = 4) = \log \tfrac{315}{314} = 0.001380\,. \qquad \maltese$$

It follows easily from Definition 3.4 that for every Benford random variable X, the significant digits $D_1(X), D_2(X), \ldots$ of X are asymptotically independent and uniformly distributed, i.e.,

$$\lim_{n \to \infty} \mathbb{P}\big(D_{n+1}(X) = d_1, D_{n+2}(X) = d_2, \ldots, D_{n+m}(X) = d_m\big) = 10^{-m}$$
for all $m \in \mathbb{N}$ and all $d_1, d_2, \ldots d_m \in \{0, 1, \ldots, 9\}\,.$ \qquad (3.6)

There is nothing special about the Benford distribution in this sense, as (3.6) holds for every random variable with a density [76, Cor. 4.4].

As the following example shows, there are many Benford probability measures on the positive real numbers, and correspondingly many positive random variables that are Benford.

EXAMPLE 3.6. (i) If U is a random variable uniformly distributed on $[0, 1)$, then the random variable $X = 10^{U}$ and its probability distribution P_X are both Benford, i.e., $X = S(X)$ is absolutely continuous with density $t^{-1} \log e$ for

$t \in [1, 10)$. This follows since

$$\mathbb{P}(S(X) \leq t) = \mathbb{P}(X \leq t) = \mathbb{P}(10^U \leq t) = \mathbb{P}(U \leq \log t)$$
$$= \log t \quad \text{for all } t \in [1, 10)\,.$$

On the other hand, $Y = 2^U$ is clearly not Benford, since

$$\mathbb{P}(S(Y) > 2) = 0 \neq 1 - \log 2\,.$$

(ii) More generally, for every integer k, the probability measure P_k with density $f_k(x) = x^{-1} \log e$ on $[10^k, 10^{k+1})$ is Benford, and so is $\frac{1}{2}(P_k + P_{k+1})$. In fact, every convex combination of the $(P_k)_{k \in \mathbb{Z}}$, i.e., every probability measure $\sum_{k \in \mathbb{Z}} q_k P_k$ with $0 \leq q_k \leq 1$ for all k and $\sum_{k \in \mathbb{Z}} q_k = 1$, is Benford.

(iii) The absolutely continuous random variable X with density

$$f_X(x) = \begin{cases} x^{-2}(x - 1) \log e & \text{if } x \in [1, 10)\,, \\ 10 x^{-2} \log e & \text{if } x \in [10, 100)\,, \\ 0 & \text{otherwise}\,, \end{cases}$$

is Benford, since the random variable $S(X)$ is also absolutely continuous, with density $t^{-1} \log e$ for $t \in [1, 10)$, even though X is not of the type of distribution considered in (ii). ✠

DEFINITION 3.7. The *Benford distribution* $\mathbb{B}$ is the unique probability measure on $(\mathbb{R}^+, \mathcal{S})$ with

$$\mathbb{B}(S \leq t) = \mathbb{B}\left(\bigcup\nolimits_{k \in \mathbb{Z}} 10^k[1, t] \right) = \log t \quad \text{for all } t \in [1, 10)\,,$$

or, equivalently, for all $m \in \mathbb{N}$, all $d_1 \in \{1, 2, \ldots, 9\}$, and all $d_j \in \{0, 1, \ldots, 9\}$, $j \geq 2$,

$$\mathbb{B}\big(D_j = d_j \text{ for } j = 1, 2, \ldots, m\big) = \log \left(1 + \left(\sum\nolimits_{j=1}^m 10^{m-j} d_j \right)^{-1} \right).$$

Given any probability P on $(\mathbb{R}, \mathcal{B})$, let $|P|$ denote the probability given by

$$|P|(B) = P\big(\{x \in \mathbb{R} : |x| \in B\}\big) \quad \text{for all } B \in \mathcal{B}\,.$$

Clearly, $|P|$ is concentrated on $[0, +\infty)$, i.e., $|P|\big([0, +\infty)\big) = 1$, and

$$F_{|P|}(x) = \begin{cases} 0 & \text{if } x < 0\,, \\ F_P(x) - F_P(-x) + P(\{-x\}) & \text{if } x \geq 0\,; \end{cases}$$

in particular, therefore, if P is continuous or a.c., then so is $|P|$, its density in the latter case being $f_P(x) + f_P(-x)$, for all $x \geq 0$, where f_P is the density of P.

YBP Library Services

ERGER, ARNO, 1968-

NTRODUCTION TO BENFORD'S LAW.

                         Cloth   248 P.
RINCETON: PRINCETON UNIV PRESS, 2015

UTH: UNIVERSITY OF ALBERTA.

CCN 2014953765
  ISBN 0691163065   **Library PO#** GENERAL APPROVAL

		List	75.00	USD
5461 UNIV OF TEXAS/SAN ANTONIO		Disc	17.0%	
App. Date 6/24/15 MGS.APR	6108-11	Net	62.25	USD

UBJ: 1. DISTRIBUTION (PROBABILITY THEORY)
. PROBABILITY MEASURES.

LASS QA273.6   DEWEY# 519.24   LEVEL ADV-AC

YBP Library Services

ERGER, ARNO, 1968-

NTRODUCTION TO BENFORD'S LAW.

                         Cloth   248 P.
RINCETON: PRINCETON UNIV PRESS, 2015

UTH: UNIVERSITY OF ALBERTA.

LCCN 2014953765
  ISBN 0691163065   **Library PO#** GENERAL APPROVAL

		List	75.00	USD
5461 UNIV OF TEXAS/SAN ANTONIO		Disc	17.0%	
App. Date 6/24/15 MGS.APR	6108-11	Net	62.25	USD

UBJ: 1. DISTRIBUTION (PROBABILITY THEORY)
. PROBABILITY MEASURES.

LASS QA273.6   DEWEY# 519.24   LEVEL ADV-AC

The next theorem characterizes Benford random variables in terms of the Benford distribution and the moments (expected values of positive integral powers) of $S(X)$. To this end, recall that the *expectation*, or *expected* (or *mean*) *value* of X is

$$\mathbb{E}[X] = \int_{\Omega} X \, \mathrm{d}\mathbb{P} = \int_{\mathbb{R}} x \, \mathrm{d}P_X(x) \,,$$

provided that this integral exists. More generally, for every measurable function $g : \mathbb{R} \to \mathbb{R}$, the expectation of the random variable $g(X)$ is

$$\mathbb{E}[g(X)] = \int_{\Omega} g(X) \, \mathrm{d}\mathbb{P} = \int_{\mathbb{R}} g(x) \, \mathrm{d}P_X(x) \,.$$

In particular, if $\mathbb{E}[X]$ exists, then $\operatorname{var} X := \mathbb{E}\big[(X - \mathbb{E}X)^2\big]$ is the *variance* of X.

THEOREM 3.8. *Let X be a random variable. The following are equivalent:*

(i) X *is Benford;*

(ii) $\mathbb{P}(|X| \in A) = \mathbb{B}(A)$ *for all $A \in \mathcal{S}$;*

(iii) $\mathbb{E}[S(X)^n] = (10^n - 1)n^{-1} \log e$ *for all $n \in \mathbb{N}$.*

PROOF. The equivalence of (i) and (ii) follows immediately from Definitions 3.4 and 3.7. Statement (iii) follows from (i) by direct calculation, since Definition 3.4 implies that $S(X)$ has density $t^{-1} \log e$ on $[1, 10)$. Statement (iii) implies (i) since the distribution of every bounded random variable such as $S(X)$ is uniquely determined by its moments; e.g., see [37, Exc. 6.6.5]. ∎

Note that the Benford distribution $\mathbb{B}$ is a probability distribution on the significant digits, or the significand, of the underlying data, and not on the raw data themselves. That is, $\mathbb{B}$ is a probability measure on the family of sets defined by the base-10 significand, i.e., on $(\mathbb{R}^+, \mathcal{S})$, but not on the larger $(\mathbb{R}^+, \mathcal{B}^+)$ or the still larger $(\mathbb{R}, \mathcal{B})$. For example, the probability $\mathbb{B}(\{1\})$ is not defined, simply because the singleton set $\{1\}$ cannot be defined in terms of significant digits or significands alone, and hence does not belong to the domain of $\mathbb{B}$.

In the framework of Examples 2.14 and 2.17, it is tempting to call an $\mathbb{N}$-valued random variable X, i.e., $\mathbb{P}(X \in \mathbb{N}) = 1$, or its distribution P_X, *Benford on $\mathbb{N}$* if

$$\mathbb{P}(S(X) \le t) = P_X\big(\{n \in \mathbb{N} : S(n) \le t\}\big) = \log t \quad \text{for all } t \in [1, 10) \,. \tag{3.7}$$

However, no such random variable exists! To see this, for every $n \in \mathbb{N}_{10}$ let $A_n = \bigcup_{k \in \mathbb{N}_0} 10^k \{n\} \in \mathcal{S}_{\mathbb{N}}$, and note that $\mathbb{N}$ equals the disjoint union of the sets A_n, and that $S(A_n) = \{10^{\langle \log n \rangle}\}$; here $\langle \log n \rangle \in [0, 1)$ denotes the fractional part of $\log n$, that is, $\langle \log n \rangle = \log n - \lfloor \log n \rfloor$. Thus the random variable $S(X)$ is concentrated on the countable set $B := S(\mathbb{N}) = \{10^{\langle \log n \rangle} : n \in \mathbb{N}_{10}\}$, that is, $\mathbb{P}(S(X) \in B) = 1$. On the other hand, (3.7) implies that $S(X)$ is continuous, and so $\mathbb{P}(S(X) \in B) = 0$. This contradiction shows that a random variable X

supported on $\mathbb{N}$, i.e., with $\mathbb{P}(X \in \mathbb{N}) = 1$, cannot be Benford. However, given $\varepsilon > 0$, it is not hard to find a random variable X_ε supported on $\mathbb{N}$ with

$$\left| \mathbb{P}(S(X_\varepsilon) \le t) - \log t \right| < \varepsilon \quad \text{for all } t \in [1, 10) \,. \tag{3.8}$$

The next example shows two such random variables, each supported on a finite number of points. The first is distributed non-uniformly on an interval of positive integers, and the second is distributed uniformly on a non-uniform range of positive integers.

EXAMPLE 3.9. **(i)** Fix $N \in \mathbb{N}$ and let Y_N be a random variable defined by

$$\mathbb{P}(Y_N = j) = \frac{1}{j c_N} \quad \text{for } j = 10^N, 10^N + 1, \ldots, 10^{N+1} - 1,$$

where $c_N = \sum_{j=10^N}^{10^{N+1}-1} j^{-1}$. Note that Y_N has $9 \cdot 10^N$ possible outcomes and may be thought of as a discrete approximation of the Benford random variable with distribution P_N in Example 3.6(ii). From

$$P_{S(Y_N)} = c_N^{-1} \sum_{j=10^N}^{10^{N+1}-1} j^{-1} \delta_{S(j)} = c_N^{-1} \sum_{j=1}^{10^{N+1}-10^N} \frac{1}{10^N + j - 1} \delta_{1+10^{-N}(j-1)} \,,$$

together with the elementary estimate $\ln \dfrac{M+1}{L} < \sum_{j=L}^{M} j^{-1} < \ln \dfrac{M}{L-1}$, valid for all $L, M \in \mathbb{N}$ with $2 \le L < M$, it is straightforward to deduce that, for all $1 \le t < 10$,

$$\left| \mathbb{P}(S(Y_N) \le t) - \log t \right| < - \log(1 - 10^{-N}) < \tfrac{1}{2} \cdot 10^{-N} \,.$$

Thus (3.8) holds for $X_\varepsilon = Y_N$ with $N > |\log \varepsilon|$.

(ii) Fix $N \in \mathbb{N}$, and let Z_N be the random variable uniformly distributed on the N positive integers $\lfloor 10^{N+j/N} \rfloor$, where $j = 0, 1, \ldots, N-1$, that is,

$$\mathbb{P}\left(Z_N = \left\lfloor 10^{N+j/N} \right\rfloor \right) = \frac{1}{N} \quad \text{for all } j = 0, 1, \ldots, N-1 \,.$$

Note that $S(\lfloor 10^{N+j/N} \rfloor) \approx S(10^{N+j/N}) = 10^{j/N}$, and hence it is plausible that $\log S(Z_N)$ is approximately uniform on $[0, 1)$ for large N. More precisely,

$$10^{j/N} - 10^{-N} \le S\left(\left\lfloor 10^{N+j/N} \right\rfloor \right) \le 10^{j/N} \quad \text{for all } j = 1, 2, \ldots, N-1 \,,$$

and therefore, for every N,

$$\frac{\#\{1 \le j \le N - 1 : j \le N \log t\}}{N} \le \mathbb{P}(S(Z_N) \le t)$$

$$\le \frac{1}{N} + \frac{\#\{1 \le j \le N - 1 : j \le N \log(t + 10^{-N})\}}{N} \,,$$

which in turn implies that, for all $1 \leq t < 10$,

$$\left| \mathbb{P}(S(Z_N) \leq t) - \log t \right| \leq \log(1 + 10^{-N}) + N^{-1} \leq N^{-1} \log 11 \,.$$

Hence, given $\varepsilon > 0$, (3.8) is satisfied with $X_\varepsilon = Z_N$ and $N > \varepsilon^{-1} \log 11$. This example is reminiscent of the distribution shown in [23, Thm. 3.22] to be the most scale-invariant probability distribution on N points; cf. Section 5.1. ✠

None of the standard continuous probability distributions (e.g., uniform, exponential, normal, etc.) are Benford, and the next example illustrates this with five common distributions. Here, as well as on several later occasions, the quantity $\Delta_\infty := 100 \cdot \sup_{1 \leq t < 10} \left| F_{S(X)}(t) - \log t \right|$ is used to quantify the deviation from, or conformance to, Benford's law for a random variable X. Note that $\Delta_\infty \in [0, 100]$ is a continuous counterpart of Δ, with $\Delta_\infty = 0$ if and only if X is Benford, and $\Delta_\infty = 100$ if and only if $\mathbb{P}(S(X) \leq 1) = 1$. Many other metrics, such as $\int_1^{10} \left| F_{S(X)}(t) - \log t \right| dt$, could also be used to quantify the deviation from Benford's law.

EXAMPLE 3.10. **(i)** If $X = U(0, 1)$, i.e., X is uniformly distributed on $(0, 1)$, so $P_X = \lambda_{0,1}$, then for every $1 \leq t < 10$,

$$F_{S(X)}(t) = \mathbb{P}(S(X) \leq t) = \lambda_{0,1} \left(\bigcup_{k \in \mathbb{Z}} 10^k[1, t] \right) = \sum_{n \in \mathbb{N}} 10^{-n}(t - 1)$$
$$= \tfrac{1}{9}(t - 1) \not\equiv \log t \,,$$

showing that $S(X)$ is uniform on $[1, 10)$, and hence X is not Benford. In particular, $\mathbb{P}(D_1(X) = 1) = \tfrac{1}{9} < \log 2$.

(ii) If $X = \exp(1)$, i.e., X has the exponential distribution with mean 1, then its distribution function is $F_{\exp(1)}(x) = \max\{0, 1 - e^{-x}\}$, so

$$\mathbb{P}(D_1(X) = 1) = \mathbb{P}\left(X \in \bigcup_{k \in \mathbb{Z}} 10^k[1, 2) \right) = \sum_{k \in \mathbb{Z}} \left(e^{-10^k} - e^{-2 \cdot 10^k} \right)$$
$$> \left(e^{-1/10} - e^{-2/10} \right) + \left(e^{-1} - e^{-2} \right) + \left(e^{-10} - e^{-20} \right)$$
$$= 0.3186 > \log 2 \,,$$

and hence $\exp(1)$ is not Benford either. An explicit calculation shows that, for every $1 \leq t < 10$,

$$F_{S(X)}(t) = \sum_{k \in \mathbb{Z}} \left(F_X(10^k t) - F_X(10^k) \right) = \sum_{k \in \mathbb{Z}} \left(e^{-10^k} - e^{-10^k t} \right) \,.$$

Since $F_{S(X)}(t) \not\equiv \log t$, the random variable X is not Benford. Numerically, one finds $\Delta_\infty = 3.05$; see also Figure 3.6. Thus, even though X is not exactly Benford, it is close to being Benford in the sense that $\left| \mathbb{P}(S(X) \leq t) - \log t \right|$ is small for all $t \in [1, 10)$; see [56, 96, 108] for a detailed analysis of the exponential distribution's relation to Benford's law.

(iii) If X has the Pareto distribution with parameter 1, i.e., $\mathbb{P}(X > x) = x^{-1}$ for all $x \geq 1$, then for all $1 \leq t < 10$,

$$F_{S(X)}(t) = \mathbb{P}\left(X \in \bigcup_{k=0}^{\infty}[10^k, 10^k t]\right) = \sum_{k=0}^{\infty}\left(10^{-k} - (10t)^{-k}\right) = \tfrac{10}{9}(t-1)/t \,,$$

so X is not Benford.

(iv) If X has the arcsin distribution, i.e., $\mathbb{P}(X \leq x) = \dfrac{2}{\pi}\arcsin\sqrt{x}$ for all $0 \leq x < 1$, then for every $1 \leq t < 10$,

$$
\begin{aligned}
F_{S(X)}(t) = \mathbb{P}(S(X) \leq t) &= \mathbb{P}\left(X \in \bigcup_{n \in \mathbb{N}} 10^{-n}[1, t]\right) \\
&= \frac{2}{\pi}\sum_{n=1}^{\infty}\left(\arcsin\left(10^{-n/2}\sqrt{t}\right) - \arcsin\left(10^{-n/2}\right)\right) \\
&= \frac{2}{\pi}\sum_{k=0}^{\infty}\frac{(2k)!}{2^{2k}(k!)^2(2k+1)} \cdot \frac{t^{k+1/2}-1}{10^{k+1/2}-1}\,,
\end{aligned}
$$

and hence in particular

$$
\begin{aligned}
F_{S(X)}\left(\sqrt{10}\right) &= \frac{2}{\pi}\sum_{k=0}^{\infty}\frac{(2k)!}{2^{2k}(k!)^2(2k+1)} \cdot \frac{1}{10^{k/2+1/4}+1} \\
&< \frac{2}{\pi}\sum_{k=0}^{\infty}\frac{(2k)!}{2^{2k}(k!)^2(2k+1)}10^{-(k/2+1/4)} \\
&= \frac{2}{\pi}\arcsin\left(10^{-1/4}\right) = 0.3801 \, < \, \log\sqrt{10}\,,
\end{aligned}
$$

so X is not Benford.

(v) If X is standard normal then, for every $t \in [1, 10)$,

$$\mathbb{P}(S(X) \leq t) = 2\sum_{k \in \mathbb{Z}}\left(\Phi(10^k t) - \Phi(10^k)\right),$$

where Φ is the distribution function of X, that is,

$$\Phi(x) = F_X(x) = \mathbb{P}(X \leq x) = \frac{1}{\sqrt{2\pi}}\int_{-\infty}^{x} e^{-\frac{1}{2}\tau^2}\,\mathrm{d}\tau\,, \quad x \in \mathbb{R}\,.$$

A numerical calculation indicates that $\Delta_{\infty} = 6.05$. Though larger than in the exponential case, the deviation of X from Benford's law is still rather small; see Figure 3.6. ✠

As is suggested by the special cases in Example 3.10, none of the familiar classical probability distributions or random variables, such as uniform, exponential, Pareto, normal, beta, binomial, or gamma distributions, are exactly Benford. On the other hand, some standard parametrized families of distributions are arbitrarily close to being Benford for *some* values of the parameters, whereas other families of distributions are bounded strictly away from being Benford for *all* values of the parameters. The next two theorems illustrate this with the families of Pareto and positive uniform distributions, respectively.

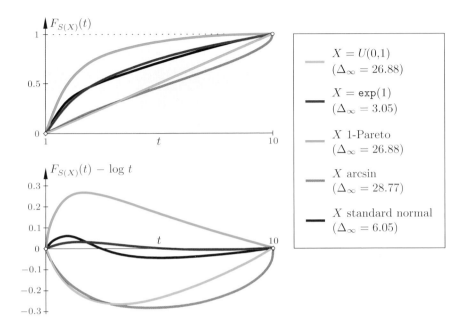

Figure 3.6: The distribution functions of $S(X)$ and their deviation from $\log t$ (bottom), where X has a uniform, exponential, Pareto, arcsin, and standard normal distribution, respectively; see Example 3.10.

THEOREM 3.11. *Let X_α be a Pareto random variable with parameter $\alpha > 0$, i.e., $\mathbb{P}(X_\alpha > x) = x^{-\alpha}$ for all $x \geq 1$. Then $\max_{1 \leq t < 10} \left| F_{S(X_\alpha)}(t) - \log t \right| \to 0$ monotonically as $\alpha \searrow 0$.*

PROOF. Note first that for every $1 < t < 10$, the function $h_t : \mathbb{R} \to \mathbb{R}$ defined as

$$
h_t(x) = \begin{cases} \dfrac{t^x - 1}{10^x - 1} & \text{if } x \neq 0\,, \\ \log t & \text{if } x = 0\,, \end{cases}
$$

is continuous and strictly decreasing, and

$$
\lim_{x \to -\infty} h_t(x) = 1 \quad \text{and} \quad \lim_{x \to +\infty} h_t(x) = 0\,.
$$

As in Example 3.10(iii), observe that, for all $1 < t < 10$ and $\alpha > 0$,

$$
F_{S(X_\alpha)}(t) = \mathbb{P}\left(X_\alpha \in \bigcup_{k=0}^\infty [10^k, 10^k t] \right) = \sum_{k=0}^\infty \left(10^{-\alpha k} - 10^{-\alpha k} t^{-\alpha} \right)
$$

$$
= \frac{t^{-\alpha} - 1}{10^{-\alpha} - 1} = h_t(-\alpha)\,.
$$

For every $1 < t < 10$, therefore, $F_{S(X_\alpha)}(t) - \log t > 0$ is strictly decreasing to 0 as $\alpha \searrow 0$, and so is $\max_{1 \leq t < 10}(F_{S(X_\alpha)}(t) - \log t)$. ∎

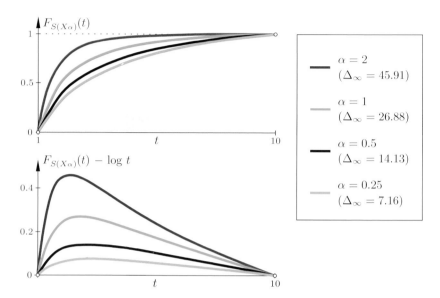

Figure 3.7: For a Pareto random variable X_α, the deviation $F_{S(X_\alpha)}(t) - \log t$ goes to zero monotonically as $\alpha \searrow 0$; see Theorem 3.11.

As the following corollary of Theorem 3.11 shows, the distribution functions of the significands of powers X_α^n for any Pareto random variable X_α approach the Benford distribution function $\log t$ as $n \to \infty$, and do so monotonically. Later, in Corollary 8.8, it will be seen that in fact the distribution functions of $S(X^n)$ approach $\log t$ for *every* random variable X that has a density. In general, however, this convergence is not monotone.

COROLLARY 3.12. *Let X be a Pareto random variable. Then, for every $1 \le t < 10$, $F_{S(X^n)}(t) - \log t$ decreases to 0 monotonically as $n \to \infty$.*

PROOF. Let X be Pareto with parameter $\alpha > 0$. For every $x \ge 1$, therefore, $\mathbb{P}(X^n > x) = \mathbb{P}(X > x^{1/n}) = x^{-\alpha/n}$, showing that X^n is Pareto with parameter α/n. Hence the conclusion follows from Theorem 3.11. ∎

In contrast to Theorem 3.11, no uniform distribution is even close to Benford's law, no matter how large its range or where it is centered. This statement can be quantified explicitly for positive uniform distributions as follows.

THEOREM 3.13 ([12]). *For every uniformly distributed positive random variable X,*

$$\max_{1 \le t < 10} \left| F_{S(X)}(t) - \log t \right| \ge \tfrac{1}{18} + \tfrac{1}{2}(\log 9 - \log e + \log \log e) = 0.1344,$$

and this bound is sharp.

PROOF. Let X be a uniformly distributed positive random variable. First note that $F_{S(X)}(t) = F_{\langle \log X \rangle}(\log t)$ for all $1 \leq t < 10$, and hence

$$\max_{1 \leq t < 10} \left| F_{S(X)}(t) - \log t \right| = \max_{0 \leq s \leq 1} \left| F_{\langle \log X \rangle}(s) - s \right|.$$

A complete argument will be given here only for the case where X is uniformly distributed on $(0, T)$, with $T > 0$. In this case,

$$F_{\log X}(x) = \begin{cases} \dfrac{10^x}{T} & \text{if } x < \log T, \\ 1 & \text{if } x \geq \log T, \end{cases}$$

and with $\delta := \langle \log T \rangle$, it follows that, for all $0 \leq s \leq 1$,

$$F_{\langle \log X \rangle}(s) = \sum_{k \in \mathbb{Z}} \left(F_{\log X}(s + k) - F_{\log X}(k) \right)$$

$$= \begin{cases} \frac{10}{9} \left(10^{s-\delta} - 10^{-\delta} \right) & \text{if } 0 \leq s < \delta, \\ 1 - \frac{10}{9} \left(10^{-\delta} - 10^{s-\delta-1} \right) & \text{if } \delta \leq s \leq 1. \end{cases}$$

For convenience, let $G_\delta(s) = F_{\langle \log X \rangle}(s) - s$, and note that G_δ is a continuous function, differentiable everywhere except at $s = \delta$, with $G_\delta(0) = G_\delta(1) = 0$. Moreover, G_δ is strictly convex on the two intervals $[0, \delta]$ and $[\delta, 1]$, and

$$G'_\delta(0+) = \frac{10}{9}(10^\delta \log e)^{-1} - 1 = G'_\delta(1-), \quad 0 < \delta < 1.$$

This shows that G_δ attains its maximal value at $s = \delta$. Again for convenience, let

$$\psi(\delta) = G_\delta(\delta) = \frac{10}{9}\left(1 - 10^{-\delta}\right) - \delta,$$

and observe that ψ is continuous, with $\psi(0) = \psi(1) = 0$. Since $\psi''(\delta) < 0$ for all $0 \leq \delta < 1$, the function ψ is strictly concave and attains its maximal value at a unique $0 < \delta^* < 1$. Concretely, $\delta^* = 1 - \log 9 - \log \log e = 0.4079$. Assume now that $0 \leq \delta \leq \delta^*$. In this case, the function G_δ has a non-positive minimum at $s = 1 + \delta - \delta^* > \delta$, with

$$G_\delta(1 + \delta - \delta^*) = 1 - \frac{10}{9}\left(10^{-\delta} - 10^{-\delta^*}\right) - 1 - \delta + \delta^* = \psi(\delta) - \psi(\delta^*);$$

hence $\max_{0 \leq s \leq 1} G_\delta(s) = \psi(\delta)$ as well as $-\min_{0 \leq s \leq 1} G_\delta(s) = \psi(\delta^*) - \psi(\delta)$. Similarly, if $\delta^* < \delta < 1$ then G_δ has a negative minimum at $s = \delta - \delta^* < \delta$, with

$$G_\delta(\delta - \delta^*) = \frac{10}{9}\left(10^{-\delta^*} - 10^{-\delta}\right) - \delta + \delta^* = \psi(\delta) - \psi(\delta^*).$$

In summary, therefore, for every $0 \leq \delta < 1$,

$$\max_{0 \leq s \leq 1} G_\delta(s) = \psi(\delta), \quad -\min_{0 \leq s \leq 1} G_\delta(s) = \psi(\delta^*) - \psi(\delta),$$

and consequently

$$\max_{0 \leq s < 1} \left| F_{\langle \log X \rangle}(s) - s \right| = \max_{0 \leq s \leq 1} |G_\delta(s)| = \max\left\{ \psi(\delta), \psi(\delta^*) - \psi(\delta) \right\}$$

$$\geq \tfrac{1}{2}\psi(\delta^*) = \tfrac{1}{18} + \tfrac{1}{2}(\log 9 - \log e + \log \log e),$$

with equality holding exactly for the two values of δ with $\psi(\delta) = \frac{1}{2}\psi(\delta^*)$. This establishes the claim whenever X is uniformly distributed on $(0, T)$ for some $T > 0$. Similar arguments show that $\max_{0 \leq s \leq 1} |F_{\langle \log X \rangle}(s) - s| \geq \frac{1}{2}\psi(\delta^*)$ holds more generally for *every* uniformly distributed positive random variable X; the interested reader is referred to [12] for details.   ∎

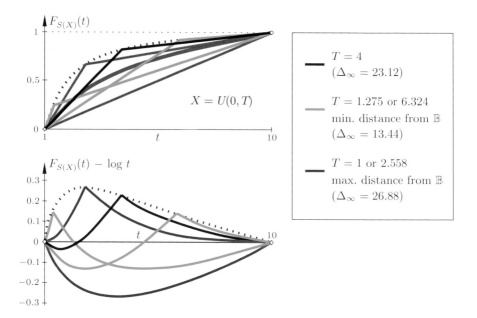

Figure 3.8: Distribution functions of $S(X)$ for the random variable $X = U(0, T)$ and different values of $1 \leq T < 10$. These distribution functions consist of two linear parts, with the "break-point" located on the dotted curve (top left). Note that the maximum vertical distance between any such distribution function and the (blue) Benford curve $\log t$ is bounded away from zero. In fact, the diagram also depicts those situations for which this maximum vertical distance is minimal (green) and maximal (red), respectively.

Remark. It is easy to see that $\max_{1 \leq t < 10} |F_{S(X)}(t) - \log t| \geq \delta$ holds with the appropriate $\delta > 0$ whenever the random variable X is uniformly distributed (on an open interval that may contain 0), normal, or exponential. Unlike in Theorem 3.13, however, the best possible (that is, largest) value of δ is not known to the authors in any of these cases.

The next lemma provides a convenient framework for studying probabilities on the significand σ-algebra by translating them into probability measures on the classical space of Borel subsets of $[0, 1)$, that is, on $([0, 1), \mathcal{B}[0, 1))$. For a proper formulation, observe that for every function $f : \Omega \to \mathbb{R}$ with $\sigma(f) \subset \mathcal{A}$

and every probability measure $\mathbb{P}$ on $(\Omega, \mathcal{A})$, f and $\mathbb{P}$ together induce a probability measure $\mathbb{P}_f$ on $(\mathbb{R}, \mathcal{B})$ in a natural way, namely, by setting

$$\mathbb{P}_f(B) = \mathbb{P}\big(f^{-1}(B)\big) \quad \text{for all } B \in \mathcal{B}. \tag{3.9}$$

For instance, if X is a (real-valued) random variable on the probability space $(\Omega, \mathcal{A}, \mathbb{P})$ then $\mathbb{P}_X$ is simply the distribution of X, i.e., $\mathbb{P}_X = P_X$. (Note that some textbooks denote $\mathbb{P}_f$ by $\mathbb{P} \circ f^{-1}$, and this notation will be used later in this book also.) The special case of interest for significands is $(\Omega, \mathcal{A}) = (\mathbb{R}^+, \mathcal{S})$ and $f = \log S$.

LEMMA 3.14. *The function* $\ell : \mathbb{R}^+ \to [0, 1)$ *defined by* $\ell(x) = \log S(x)$ *establishes a one-to-one and onto correspondence (measure isomorphism) between probability measures on* $(\mathbb{R}^+, \mathcal{S})$ *and probability measures on* $\big([0, 1), \mathcal{B}[0, 1)\big)$.

PROOF. From $\ell^{-1}([a, b]) = S^{-1}\big([10^a, 10^b]\big)$ for all $0 \leq a < b < 1$, it follows that $\sigma(\ell) = \mathbb{R}^+ \cap \sigma(S) = \mathcal{S}$, and hence $\mathbb{P}_\ell$ as defined by (3.9), with any probability $\mathbb{P}$ on $(\mathbb{R}^+, \mathcal{S})$, is a well-defined probability measure on $\big([0, 1), \mathcal{B}[0, 1)\big)$.

Conversely, given any probability measure P on $\big([0, 1), \mathcal{B}[0, 1)\big)$ and any A in $\mathcal{S}$, let $B \in \mathcal{B}[0, 1)$ be the unique set for which $A = \bigcup_{k \in \mathbb{Z}} 10^k 10^B$, where $10^B = \{10^s : s \in B\}$, and define

$$P^\ell(A) = P(B).$$

It is readily confirmed that $\ell(A) = B$, $\ell^{-1}(B) = A$, and P^ℓ is a well-defined probability measure on $(\mathbb{R}^+, \mathcal{S})$. Moreover,

$$(P^\ell)_\ell(B) = P^\ell\big(\ell^{-1}(B)\big) = P^\ell(A) = P(B) \quad \text{for all } B \in \mathcal{B}[0, 1),$$

showing that $(P^\ell)_\ell = P$, and hence *every* probability measure on $\big([0, 1), \mathcal{B}[0, 1)\big)$ is of the form $\mathbb{P}_\ell$ with the appropriate $\mathbb{P}$. On the other hand,

$$(\mathbb{P}_\ell)^\ell(A) = \mathbb{P}_\ell(B) = \mathbb{P}\big(\ell^{-1}(B)\big) = \mathbb{P}(A) \quad \text{for all } A \in \mathcal{S};$$

thus $(\mathbb{P}_\ell)^\ell = \mathbb{P}$, and the correspondence $\mathbb{P} \mapsto \mathbb{P}_\ell$ is one-to-one as well. In summary, $\mathbb{P} \leftrightarrow \mathbb{P}_\ell$ is bijective. ∎

From the proof of Lemma 3.14 it is clear that a bijective correspondence between probability measures on $(\mathbb{R}^+, \mathcal{S})$ and on $\big([0, 1), \mathcal{B}[0, 1)\big)$, respectively, could have been established in many other ways as well, e.g., by using the function $\widetilde{\ell}(x) = \frac{1}{9}(S(x) - 1)$ instead of ℓ; with this choice of $\widetilde{\ell}$, for instance, $(\delta_{10^k})_{\widetilde{\ell}} = \delta_0 = (\delta_{10^k})_\ell$ for all $k \in \mathbb{Z}$. The special role of ℓ defined in Lemma 3.14 only becomes apparent through its relation to Benford's law. To see this, recall that $\mathbb{B}$ is the (unique) probability measure on $(\mathbb{R}^+, \mathcal{S})$ with

$$\mathbb{B}(\{x > 0 : S(x) \leq t\}) = \mathbb{B}\left(\bigcup_{k \in \mathbb{Z}} 10^k[1, t]\right) = \log t \quad \text{for all } 1 \leq t < 10.$$

In view of (2.1), the probability measure $\mathbb{B}$ on $(\mathbb{R}^+, \mathcal{S})$ is the most natural formalization of Benford's law. On the other hand, it will become clear in subsequent chapters that on $\big([0,1), \mathcal{B}[0,1)\big)$, the uniform distribution $\lambda_{0,1}$ has many special properties that make it play a very important role. The relevance of the specific choice for ℓ in Lemma 3.14, therefore, is that $\mathbb{B}_\ell = \lambda_{0,1}$. (Notice, for example, that with the function $\widetilde{\ell} = \frac{1}{9}(S-1)$ from above, the distribution function of $\mathbb{B}_{\widetilde{\ell}}$ is $\log(9s+1)$, and so clearly $\mathbb{B}_{\widetilde{\ell}} \neq \lambda_{0,1}$.) The reader will learn shortly why, for a deeper understanding of Benford's law, the relation $\mathbb{B}_\ell = \lambda_{0,1}$ is very beneficial indeed.

Chapter Four

The Uniform Distribution and Benford's Law

The uniform distribution characterization of Benford's law is undoubtedly the most basic and powerful of all characterizations, largely because the mathematical theory of uniform distribution modulo one is very well developed; see [50, 93] for authoritative surveys. This chapter records and develops tools from that theory which will be used throughout this book to establish Benford behavior of sequences, functions, and random variables.

4.1 UNIFORM DISTRIBUTION CHARACTERIZATION OF BENFORD'S LAW

Here and throughout, denote by $\langle x \rangle$ the *fractional part* of any real number x, that is, $\langle x \rangle = x - \lfloor x \rfloor$. For example, $\langle \pi \rangle = \langle 3.1415 \rangle = 0.1415 = \pi - 3$. Recall that λ denotes Lebesgue measure on $(\mathbb{R}, \mathcal{B})$.

DEFINITION 4.1. A sequence $(x_n) = (x_1, x_2, \ldots)$ of real numbers is *uniformly distributed modulo one*, abbreviated henceforth as *u.d.* mod 1, if

$$\lim_{N \to \infty} \frac{\#\{1 \le n \le N : \langle x_n \rangle \le s\}}{N} = s \quad \text{for all } s \in [0, 1) \,;$$

a (Borel measurable) function $f : [0, +\infty) \to \mathbb{R}$ is *u.d.* mod 1 if

$$\lim_{T \to +\infty} \frac{\lambda(\{t \in [0, T) : \langle f(t) \rangle \le s\})}{T} = s \quad \text{for all } s \in [0, 1) \,;$$

a random variable X on a probability space $(\Omega, \mathcal{A}, \mathbb{P})$ is *u.d.* mod 1 if

$$\mathbb{P}(\langle X \rangle \le s) = s \quad \text{for all } s \in [0, 1) \,;$$

and a probability measure P on $(\mathbb{R}, \mathcal{B})$ is *u.d.* mod 1 if

$$P(\{x : \langle x \rangle \le s\}) = P\left(\bigcup_{k \in \mathbb{Z}} [k, k+s] \right) = s \quad \text{for all } s \in [0, 1) \,.$$

Remark. A function that is u.d. mod 1 in the sense of Definition 4.1 is often called *continuously uniformly distributed modulo one* (c.u.d. mod 1) in the literature.

The next elementary result, implicit in [9] and [111] and explicitly stated in [46], is one of the main tools in the theory of Benford's law because it allows application of the powerful theory of uniform distribution modulo one. (Recall the conventions $\log 0 = 0$ and $S(0) = 0$.)

THEOREM 4.2 (Uniform distribution characterization). *A sequence of real numbers (or a Borel measurable function, a random variable, a Borel probability measure) is Benford if and only if the decimal logarithm of its absolute value is uniformly distributed modulo one.*

PROOF. Let X be a random variable. Then, for all $s \in [0, 1)$,

$$\mathbb{P}(\langle \log |X| \rangle \leq s) = \mathbb{P}\left(\log |X| \in \bigcup_{k \in \mathbb{Z}} [k, k + s] \right)$$

$$= \mathbb{P}\left(|X| \in \bigcup_{k \in \mathbb{Z}} [10^k, 10^{k+s}] \right) + \mathbb{P}(X = 0) = \mathbb{P}(S(X) \leq 10^s).$$

Hence, by Definitions 3.4 and 4.1, the random variable X is Benford if and only if $\mathbb{P}(S(X) \leq 10^s) = \log 10^s = s$ for all $s \in [0, 1)$, i.e., if and only if $\log |X|$ is u.d. mod 1. The proofs for sequences, functions, and probability measures are completely analogous. ∎

The following proposition from the basic theory of uniform distribution mod 1, together with Theorem 4.2, will be instrumental in establishing the Benford property for many sequences, functions, and random variables.

PROPOSITION 4.3. (i) *The sequence (x_n) is u.d. mod 1 if and only if $(kx_n + b)$ is u.d. mod 1 for every integer $k \neq 0$ and every $b \in \mathbb{R}$. Also, the sequence (x_n) is u.d. mod 1 if and only if (y_n) is u.d. mod 1 whenever $\lim_{n \to \infty} |y_n - x_n| = 0$.*

(ii) *The function f is u.d. mod 1 if and only if $kf(t) + b$ is u.d. mod 1 for every integer $k \neq 0$ and every $b \in \mathbb{R}$.*

(iii) *The random variable X is u.d. mod 1 if and only if $kX + b$ is u.d. mod 1 for every integer $k \neq 0$ and every $b \in \mathbb{R}$.*

PROOF. (i) The "if" part is obvious with $k = 1$, $b = 0$. For the "only if" part, assume that (x_n) is u.d. mod 1, and recall that $\lambda_{0,1}$ denotes Lebesgue measure on $([0, 1), \mathcal{B}[0, 1))$. Note first that

$$\lim_{N \to \infty} \frac{\#\{n \leq N : \langle x_n \rangle \in C\}}{N} = \lambda_{0,1}(C)$$

whenever C is a finite union of intervals. Let k be a non-zero integer and observe that, for any $0 < s < 1$,

$$\{x : \langle kx \rangle \leq s\} = \begin{cases} \left\{ x : \langle x \rangle \in \bigcup_{j=0}^{k-1} \left[\frac{j}{k}, \frac{j+s}{k} \right] \right\} & \text{if } k > 0 \,, \\[2mm] \left\{ x : \langle x \rangle \in \bigcup_{j=0}^{|k|-1} \left[\frac{j+1-s}{|k|}, \frac{j+1}{|k|} \right] \right\} & \text{if } k < 0 \,. \end{cases}$$

Consequently,

$$\lim_{N\to\infty} \frac{\#\{n \le N : \langle kx_n\rangle \le s\}}{N} = \begin{cases} \lambda_{0,1}\left(\bigcup_{j=0}^{k-1}\left[\frac{j}{k}, \frac{j+s}{k}\right]\right) & \text{if } k > 0, \\ \lambda_{0,1}\left(\bigcup_{j=0}^{|k|-1}\left[\frac{j+1-s}{|k|}, \frac{j+1}{|k|}\right]\right) & \text{if } k < 0, \end{cases}$$

$$= \begin{cases} k \cdot \frac{s}{k} & \text{if } k > 0, \\ |k| \cdot \frac{s}{|k|} & \text{if } k < 0, \end{cases}$$

$$= s,$$

showing that (kx_n) is u.d. mod 1. Similarly, note that, for any $b, s \in (0,1)$,

$$\{x : \langle x + b\rangle \le s\} = \begin{cases} \{x : \langle x\rangle \in [0, s-b] \cup [1-b, 1)\} & \text{if } s \ge b, \\ \{x : \langle x\rangle \in [1-b, 1+s-b]\} & \text{if } s < b. \end{cases}$$

Thus, assuming without loss of generality that $0 \le b < 1$,

$$\lim_{N\to\infty} \frac{\#\{n \le N : \langle x_n + b\rangle \le s\}}{N} = \begin{cases} \lambda_{0,1}\big([0, s-b]\cup[1-b, 1)\big) & \text{if } s \ge b, \\ \lambda_{0,1}\big([1-b, 1+s-b]\big) & \text{if } s < b, \end{cases}$$

$$= s,$$

and hence $(x_n + b)$ is also u.d. mod 1. The second assertion is clear from the definition. The proofs of (ii) and (iii) are completely analogous. ∎

It follows immediately from Theorem 4.2 and Proposition 4.3 that powers and reciprocals of Benford sequences, functions, and random variables are also Benford.

THEOREM 4.4. *Let (x_n), f, and X be a Benford sequence, function, and random variable, respectively. Then, for all $a \in \mathbb{R}$ and $k \in \mathbb{Z}$ with $ak \ne 0$, the sequence (ax_n^k), the function af^k, and the random variable aX^k are also Benford.*

Without the hypothesis that $k \in \mathbb{Z}$, the conclusion of Theorem 4.4 may fail. In general, the Benford property is not preserved under taking roots, in contrast to the situation for the significand σ-algebra, which is closed under roots but not under powers; see Lemma 2.15 and Example 2.16.

EXAMPLE 4.5. Let (x_n) be the sequence $\big(S(2^n)\big) = (2, 4, 8, 1.6, 3.2, \ldots)$, let f be the function $f(t) = S(e^t)$, and let X be the random variable $X = 10^{U(0,1)}$. Then none of $\big(\sqrt{x_n}\big)$, $\sqrt{f}$, and $\sqrt{X}$ is Benford, since all their significands are less than $\sqrt{10}$. On the other hand, since $S\big(S(x)\big) = S(x)$ for all x, (x_n) is Benford by Example 4.7(i) below, f is Benford by Example 4.9(i), and X is Benford by Example 3.6(i). ✠

4.2   UNIFORM DISTRIBUTION OF SEQUENCES AND FUNCTIONS

This section focuses on tools and techniques useful in analyzing the Benford behavior of sequences and functions. The first two propositions record, for ease of reference, several basic results from the theory of uniform distribution of sequences and functions.

PROPOSITION 4.6. *Let $(x_n) = (x_1, x_2, \ldots)$ be a sequence of real numbers.*

(i) *If $\lim_{n\to\infty}(x_{n+1} - x_n) = \theta$ for some irrational θ, then (x_n) is u.d.* mod 1.

(ii) *If (x_n) is periodic, i.e., $x_{n+p} = x_n$ for some $p \in \mathbb{N}$ and all n, then $(n\theta + x_n)$ is u.d.* mod 1 *if and only if θ is irrational.*

(iii) *The sequence (x_n) is u.d.* mod 1 *if and only if $(x_n + a\log n)$ is u.d.* mod 1 *for all $a \in \mathbb{R}$.*

(iv) *If (x_n) is u.d.* mod 1 *and non-decreasing, then $(x_n/\log n)$ is unbounded.*

(v) *If f is differentiable for $t \geq t_0$, if $f'(t)$ tends to zero monotonically as $t \to +\infty$, and if $t|f'(t)| \to +\infty$ as $t \to +\infty$, then the sequence $\big(f(n)\big)$ is u.d.* mod 1.

(vi) *Suppose $\lim_{n\to\infty} n(y_{n+1} - y_n) = 0$ for the sequence (y_n) of real numbers. Then (x_n) is u.d.* mod 1 *if and only if $(x_n + y_n)$ is u.d.* mod 1.

PROOF. Conclusion (i) is a classical result of van der Corput [93, Thm. I.3.3]; (ii) follows directly from Weyl's criterion [93, Thm. I.2.1]; (iii) is [11, Lem. 2.8]; (iv) is [15, Lem. 2.4.(i)]; (v) is Fejér's Theorem [93, Cor. I.2.1]; and (vi) follows from [130, Bsp. 3.I] using a suitable C^1-function f that satisfies $f(n) = y_n$ for all n, e.g., $f(t) = y_{\lfloor t\rfloor} + (y_{\lfloor t\rfloor+1} - y_{\lfloor t\rfloor})\big(3\langle t\rangle^2 - 2\langle t\rangle^3\big)$.   ∎

The converse of (i) is not true in general: (x_n) may be u.d. mod 1 even if $(x_{n+1} - x_n)$ has a rational limit. Also, in (ii) the sequence $(n\theta)$ cannot be replaced by an arbitrary uniformly distributed sequence (θ_n), i.e., $(\theta_n + x_n)$ may not be u.d. mod 1 even though (θ_n) is u.d. mod 1 and (x_n) is periodic.

EXAMPLE 4.7. **(i)** The sequences $(n\pi) = (\pi, 2\pi, \ldots)$, $(n\log \pi)$, and $(n\log 2)$ are all u.d. mod 1, by Proposition 4.6(i). Thus, by Theorem 4.2, the sequences $(10^{n\pi})$, (π^n), and (2^n) are all Benford sequences.

Similarly, $(x_n) = \big(n\sqrt{2}\big)$ is u.d. mod 1, whereas $\big(x_n\sqrt{2}\big) = (2n) = (2, 4, \ldots)$ clearly is not, since $\langle 2n\rangle = 0$ for all n. Thus the requirement in Proposition 4.3(i) that k be an integer cannot be dropped.

For an analogous example using random variables, let X be uniform on $\big(0, \frac{2}{\log 2}\big)$. Since a random variable that is uniform on $(0, a)$ for some $a > 0$ is u.d. mod 1 if and only if $a \in \mathbb{N}$, X is not u.d. mod 1, but $X\log 2$ is. By Theorem 4.2, therefore, the random variable 2^X is Benford, but 10^X is not.

(ii) The sequence $(\log n)$ is not u.d. mod 1. A short calculation shows that, for every $s \in [0, 1)$, the sequence $\left(N^{-1} \# \{1 \leq n \leq N : \langle \log n \rangle \leq s\}\right)_{N \in \mathbb{N}}$ has

$$\tfrac{1}{9}(10^s - 1) \quad \text{and} \quad \tfrac{10}{9}(1 - 10^{-s})$$

as its limit inferior and limit superior, respectively. Thus, by Theorem 4.2, the sequence of positive integers (n) is not Benford, and neither is (an^b) for any $a, b \in \mathbb{R}$, by Theorem 4.4.

(iii) The sequence $\left(\sqrt{a + bn}\right)$ is u.d. mod 1 for all $a, b > 0$, as follows easily from Proposition 4.6(v) using the function $f(t) = \sqrt{a + bt}$. Hence the sequence $\left(10^{\sqrt{a+bn}}\right)$ is Benford for all $a, b > 0$. ✠

The concepts of uniform distribution for sequences and functions are closely related, and in order to decide whether a Borel measurable function is u.d. mod 1, the following result is often useful; part (i) is [93, Thm. I.9.6(b)], whereas part (ii) follows directly from Proposition 4.6(iv) together with [93, Thm. I.9.7].

PROPOSITION 4.8. *Let* $f : [0, +\infty) \to \mathbb{R}$ *be Borel measurable.*

(i) *If, for some* $\delta_0 > 0$*, the sequence* $\left(f(n\delta)\right)$ *is u.d.* mod 1 *for almost all* $0 < \delta < \delta_0$*, then the function* f *is u.d.* mod 1 *as well.*

(ii) *If* $\limsup_{t \to +\infty} |f(t)| / \log t < +\infty$*, and* f *is continuously differentiable and monotone, then* f *is not u.d.* mod 1.

EXAMPLE 4.9. **(i)** The function $f(t) = at + b$ with real a, b is u.d. mod 1 if and only if $a \neq 0$. Clearly, if $a = 0$ then f is constant and hence not u.d. mod 1. On the other hand, if $a \neq 0$ then $\left(f(n\delta)\right) = (a\delta n + b)$ is u.d. mod 1 if (and only if) $a\delta$ is irrational, by Proposition 4.6(ii), and hence for all but countably many $\delta > 0$. Thus f is u.d. mod 1 by Proposition 4.8(i). This can also be confirmed directly by an elementary calculation: If $a > 0$ then $\langle at + b \rangle \leq s$ if and only if $t \in \left[\frac{k-b}{a}, \frac{k-b+s}{a}\right]$ for some $k \in \mathbb{Z}$. Note that each of the intervals $\left[\frac{k-b}{a}, \frac{k-b+s}{a}\right]$ has the same length $\frac{s}{a}$. Thus, given $T > 0$ and $s \in [0, 1)$,

$$\frac{s}{a}(\lfloor aT \rfloor - 2) \leq \lambda\left(\{t \in [0, T) : \langle at + b \rangle \leq s\}\right) \leq \frac{s}{a}(\lfloor aT \rfloor + 2),$$

and since $\lim_{T \to +\infty} \frac{s}{aT}(\lfloor aT \rfloor \pm 2) = s$, the function f is u.d. mod 1. The argument for the case $a < 0$ is similar.

As a consequence, although the function $g(t) = at$ is not Benford for any a, the function $g(t) = e^{at}$ is Benford whenever $a \neq 0$, via Theorem 4.2, since $\log g(t) = at \log e$ is u.d. mod 1; see Figure 4.1.

(ii) The function $f(t) = \log |at + b|$ is not u.d. mod 1 for any $a, b \in \mathbb{R}$. Indeed, if $a = 0$ then f is constant and hence not u.d. mod 1. On the other hand, if $a \neq 0$ then $\lim_{t \to +\infty} f(t) / \log t = 1$, and f is continuously differentiable and monotone for $t > |b|/|a|$. Hence Proposition 4.8(ii) shows that f is not u.d. mod 1. Again, this can also be seen by means of an elementary calculation: For

$a \neq 0$ essentially the same calculation as the one in Example 4.7(ii) above shows that, for every $s \in [0, 1)$,

$$\liminf_{T \to +\infty} \frac{\lambda\big(\{t \in [0, T] : \langle \log |at + b| \rangle \leq s\}\big)}{T} = \tfrac{1}{9}(10^s - 1)\,,$$

and

$$\limsup_{T \to +\infty} \frac{\lambda\big(\{t \in [0, T] : \langle \log |at + b| \rangle \leq s\}\big)}{T} = \tfrac{10}{9}(1 - 10^{-s})\,.$$

Again, by Theorem 4.2, this implies that $g(t) = at + b$ is not Benford for any $a, b \in \mathbb{R}$.

Similarly, $f(t) = -\log(1 + t^2)$ is not u.d. mod 1, so $g(t) = (1 + t^2)^{-1}$ is not Benford; see Figure 4.1. In fact, it is clear that if g is continuously differentiable and monotone with $\lim_{t \to +\infty} |g(t)| = +\infty$, but $\limsup_{t \to +\infty} |g(t)|/t^a < +\infty$ for some $a > 0$, then g is not Benford. Thus, the function $g(t) = at^b$, for instance, is not Benford for any $a, b \in \mathbb{R}$; recall the analogous conclusion for sequences in Example 4.7(ii).

(iii) The function $f(t) = e^t$ is u.d. mod 1. To see this, let $T > 0$ and $N = \lfloor e^T \rfloor$, and recall that $t - \tfrac{1}{2}t^2 \leq \ln(1 + t) \leq t$ for all $t \geq 0$. Given $0 \leq s < 1$, it follows from

$$\lambda\left(\{t \in [0, T] : \langle e^t \rangle \leq s\}\right) = \sum_{n=1}^{N-1} \ln\left(1 + \frac{s}{n}\right) + \min\left\{T - \ln N, \ln\left(1 + \frac{s}{N}\right)\right\}$$

that

$$\frac{s \sum_{n=1}^{N-1} n^{-1} - \tfrac{1}{2}s^2 \sum_{n=1}^{N-1} n^{-2}}{\ln(N + 1)} \leq \frac{\lambda\left(\{t \in [0, T] : \langle e^t \rangle \leq s\}\right)}{T}$$

$$\leq \frac{s \sum_{n=1}^{N-1} n^{-1} + \ln(1 + N^{-1})}{\ln N}\,,$$

and hence indeed $\lim_{T \to +\infty} T^{-1}\lambda\big(\{t \in [0, T] : \langle e^t \rangle \leq s\}\big) = s$. Alternatively, note that $\big(f(n\delta)\big) = (e^{n\delta})$, and Proposition 4.14 below shows that this sequence is u.d. mod 1 for almost all $\delta > 0$. Hence Proposition 4.8(i) again implies that f is u.d. mod 1. In fact, af is u.d. mod 1 for every real $a \neq 0$, and so by Theorem 4.2, the function $g(t) = e^{e^t} = 10^{e^t \log e}$, for instance, is Benford.

(iv) For the function $f(t) = (\sin t)^2$, it is straightforward to check that, given any $0 \leq s < 1$,

$$\lim_{T \to +\infty} \frac{\lambda\big(\{t \in [0, T] : \langle (\sin t)^2 \rangle \leq s\}\big)}{T} = \frac{2}{\pi} \arcsin \sqrt{s}\,.$$

Thus, asymptotically $\langle f \rangle$ is not uniform on $(0, 1)$ but rather arcsin-distributed; see Example 3.10(iv). Theorem 4.2 implies that $g(t) = 10^{(\sin t)^2}$ is not Benford. This also follows easily from Theorem 4.10 below.

(v) For the function $f(t) = \log\big((\sin t)^2\big)$, it follows from (iv) that the asymptotic distribution of $\langle f \rangle$ has density

$$\frac{\mathrm{d}}{\mathrm{d}s}\left(\frac{2}{\pi}\sum\nolimits_{n=1}^{\infty}\left(\arcsin 10^{(s-n)/2} - \arcsin 10^{-n/2}\right)\right) = \frac{\ln 10}{\pi}\sum\nolimits_{n=1}^{\infty}\frac{1}{\sqrt{10^{n-s}-1}}$$
$$> \frac{\ln 10}{\pi}\cdot\frac{10^{s/2}}{10^{1/2}-1},$$

for $0 \le s < 1$. Thus clearly f is not u.d. mod 1, showing that $g(t) = (\sin t)^2$ is not Benford; see Figure 4.1 and Theorem 4.10 below. ✠

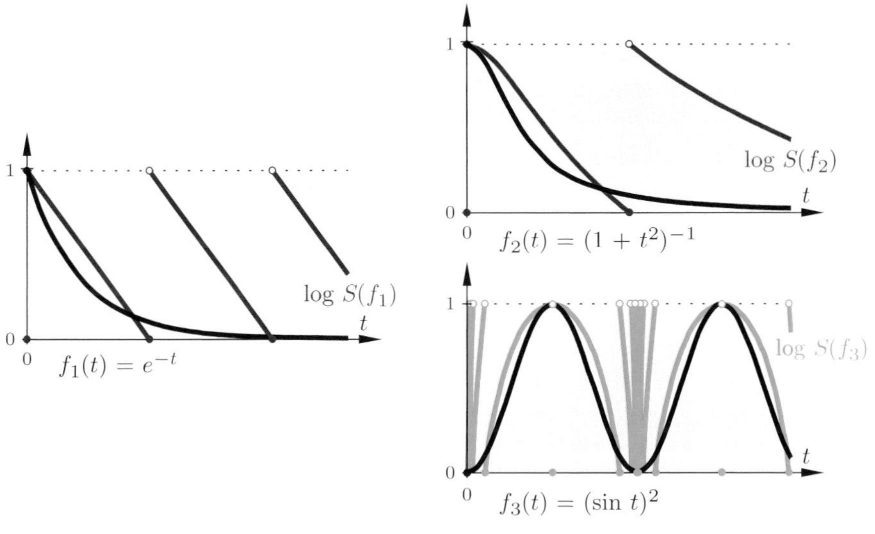

Figure 4.1: While the function f_1 is Benford, the functions f_2, f_3 are not; see Example 4.9.

The only Benford functions encountered thus far are functions like e^{-t} and e^{e^t}, which converge to either 0 or $\pm\infty$ as $t \to +\infty$. Much later, it will be seen that $e^t\cos(\pi t)$, for example, is also Benford whereas $10^t\cos(\pi t)$ is not; see Example 7.57 below. On the other hand, a function f may be Benford even if $\liminf_{t\to+\infty} f(t)$ and $\limsup_{t\to+\infty} f(t)$ are different but are both finite, as is the case if f is bounded and periodic but not constant. As the next theorem suggests, however, such Benford functions are quite rare. Recall that $f : [0, +\infty) \to \mathbb{R}$ is *periodic*, with *period* $p > 0$, if $f(t + p) = f(t)$ for all $t \ge 0$. For instance, all functions in Example 4.9(iv, v) are periodic with period $p = \pi$.

THEOREM 4.10. *Let $f : [0, +\infty) \to \mathbb{R}$ be a (Borel measurable) periodic function with period $p > 0$. Then the following are equivalent:*

(i) *f is Benford;*

(ii) $\int_0^p |f(t)|^{2\pi i n \log e}\, \mathrm{d}t = 0$ *for all* $n \in \mathbb{N}$;

(iii) $f(U)$ *is a Benford random variable, where* U *is uniform on* $(0, p)$.

Moreover, if the function f *is differentiable then it is not Benford.*

PROOF. By [93, Exc. I.9.9], a periodic function $g : [0, +\infty) \to \mathbb{R}$ with period $p > 0$ is u.d. mod 1 if and only if $\int_0^p e^{2\pi i n g(t)}\mathrm{d}t = 0$ for all $n \in \mathbb{N}$. With $g = \log|f|$, the equivalence of (i) and (ii) then follows from Theorem 4.2.

To see the equivalence of (i) and (iii), note that for all $1 \le t < 10$,

$$\lim_{T \to +\infty} \frac{\lambda\big(\{\tau \in [0, T) : S\big(f(\tau)\big) \le t\}\big)}{T} = \frac{\lambda\big(\{\tau \in [0, p) : S\big(f(\tau)\big) \le t\}\big)}{p}$$
$$= \mathbb{P}\big(S\big(f(U)\big) \le t\big),$$

where the first equality follows from the fact that $S(f) : [0, +\infty) \to [1, 10)$ is periodic with period p, and the second equality follows from the random variable U being uniform on $(0, p)$.

To prove the last assertion, let f be differentiable and periodic with period $p > 0$. Without loss of generality, assume that f is not constant. It follows that $m := \max_{t \in [0, p]} |f(t)| > 0$, and, since $|f|$ is continuous, $|f(t_0)| = m$, where $t_0 = \min\{t \in [0, p] : |f(t)| = m\}$. Since f is not constant, it can be assumed that $t_0 > 0$. (Otherwise, replace f by $f(\cdot + c)$ with the appropriate $c > 0$.) Furthermore, since $S(10^k f) = S(|f|)$ for all $k \in \mathbb{Z}$, it can also be assumed that $1 < m \le 10$ and $f(t_0) > 0$. For every $n \in \mathbb{N}$, let

$$t_n = \sup\big\{0 \le t \le t_0 : f(t) \le m - \tfrac{1}{n}\big\},$$

with the convention $\sup \varnothing := 0$. Clearly, the sequence (t_n) is non-decreasing with $t_n \le t_0$, and is hence convergent, with $t_\infty := \lim_{n \to \infty} t_n$ and $t_\infty \le t_0$. If $t_\infty < t_0$ then $S\big(f(t)\big) = S(m)$ for all $t_\infty < t < t_0$. With the random variable U being uniform on $(0, p)$, therefore, $S\big(f(U)\big)$ equals $S(m)$ with probability at least $(t_0 - t_\infty)/p > 0$, and hence $f(U)$ is not Benford. If, on the other hand, $t_\infty = t_0$ then $t_n < t_0$ for all n, by the continuity of f. In this case, for every $n \in \mathbb{N}$ consider the open interval $J_n := (f(t_n), m) = (m - \frac{1}{n}, m)$ and note that $S\big(f(U)\big) \in J_n$ whenever $U \in (t_n, t_0)$. Thus, if $f(U)$ were Benford, then, for all sufficiently large n,

$$\frac{f(t_0) - f(t_n)}{t_0 - t_n} = \frac{m - S\big(f(t_n)\big)}{t_0 - t_n}$$

$$\ge \frac{1}{(t_0 - t_n)\log e} \log\left(\frac{m}{S\big(f(t_n)\big)}\right) = \frac{\mathbb{P}\big(S\big(f(U)\big) \in J_n\big)}{(t_0 - t_n)\log e}$$

$$\ge \frac{\mathbb{P}\big(U \in (t_n, t_0)\big)}{(t_0 - t_n)\log e} = \frac{1}{p \log e} > 0,$$

which is impossible because f is differentiable at t_0 with vanishing derivative; see Example 7.16(iii) for a similar but somewhat less general argument. Thus $f(U)$ is not Benford, so by the equivalence of (i) and (iii), f is not Benford. ∎

EXAMPLE 4.11. The following three functions are all periodic with period $p = 1$, and all satisfy both $\min_{t \geq 0} f_j(t) = 1$ and $\sup_{t \geq 0} f_j(t) = 10$:

$$f_1(t) = 1 + 9\langle t \rangle, \quad f_2(t) = \tfrac{11}{2} + \tfrac{9}{2}\sin(2\pi t), \quad f_3(t) = 10^{|1 - 2\langle t + \frac{1}{4}\rangle|}.$$

Note that f_1 is discontinuous whereas f_2 and f_3 are continuous. With U being uniform on $(0, 1)$, clearly $f_1(U)$ is uniform on $(1, 10)$. As the latter random variable is not Benford by Theorem 3.13, Theorem 4.10 shows that f_1 is not Benford. Also, the function f_2 is differentiable and hence not Benford, again by Theorem 4.10. On the other hand, the random variable $\langle U + \frac{1}{4}\rangle$ is uniform on $(0, 1)$, and so is $|1 - 2\langle U + \frac{1}{4}\rangle|$. Hence f_3 is Benford; see Figure 4.2. Note that f_3 is smooth on $[0, 1]$ except at two points.   ✠

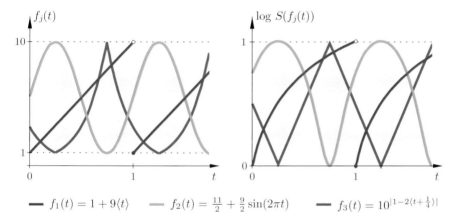

$$f_1(t) = 1 + 9\langle t \rangle \qquad f_2(t) = \tfrac{11}{2} + \tfrac{9}{2}\sin(2\pi t) \qquad f_3(t) = 10^{|1 - 2\langle t + \frac{1}{4}\rangle|}$$

Figure 4.2: Three periodic functions with period $p = 1$. Only f_3 is Benford; see Example 4.11.

As was seen in Examples 4.7 and 4.9, the sequence (2^n) and the function $f(t) = e^t$ are both Benford. The next theorem shows that all sequences and functions sufficiently close to such exponential Benford sequences and functions are themselves exactly Benford. The robustness of the Benford property, of which this is a simple example, will be a recurrent theme throughout this book.

THEOREM 4.12. (i) *Let (a_n) and (b_n) be sequences of real numbers with $|a_n| \to +\infty$ as $n \to \infty$, and such that $\lim_{n \to \infty} |a_n/b_n|$ exists and is positive. Then (a_n) is Benford if and only if (b_n) is Benford.*

(ii) *Let $f, g : [0, +\infty) \to \mathbb{R}$ be (Borel measurable) functions with $|f(t)| \to +\infty$ as $t \to +\infty$, and such that $\lim_{t \to +\infty} |f(t)/g(t)|$ exists and is positive. Then f is Benford if and only if g is Benford.*

PROOF. To prove (i), let (a_n) and (b_n) be sequences of real numbers with $|a_n| \to +\infty$, and, by Theorem 4.4, assume without loss of generality that $\lim_{n\to\infty} |a_n/b_n| = 1$. Thus $|b_n| \to +\infty$ and $\log |a_n| - \log |b_n| = \log |a_n/b_n| \to 0$. If (a_n) is Benford, then by Theorem 4.2, $(\log |a_n|)$ is u.d. mod 1, so Proposition 4.3(i) implies that $(\log |b_n|)$ is also u.d. mod 1. By Theorem 4.2 again, (b_n) is Benford. As the roles of (a_n) and (b_n) can be interchanged, this establishes (i). The proof of (ii) is analogous.   ∎

EXAMPLE 4.13. (i) The sequence $(\lfloor \pi^n \rfloor) = (3, 9, 31, 97, 306, \ldots)$ is Benford [117, A001672]. This follows from Theorem 4.12(i), since $|\lfloor \pi^n \rfloor - \pi^n| < 1$ for all n, and since (π^n) is Benford, as was seen in Example 4.7(i). Alternatively, the conclusion follows directly from Weyl's criterion and an easy approximation argument.

(ii) The hyperbolic cosine function $f(t) = \cosh t = \frac{1}{2}(e^t + e^{-t})$ is Benford. Indeed, e^t is Benford, so by Theorem 4.4, $\frac{1}{2}e^t$ is Benford. Since $|\cosh t - \frac{1}{2}e^t| \leq \frac{1}{2}$ for all $t \geq 0$, Theorem 4.12(ii) implies that $\cosh t$ is a Benford function.

(iii) By Example 4.9(ii), the function $f(t) = t$ is not Benford, so by Theorem 4.4, $g(t) = t^k$ is not Benford for any $k \in \mathbb{Z}$. Thus, by Theorem 4.12(ii), no polynomial or rational function is Benford.   ✠

A third very useful tool from the basic theory of uniform distribution is Koksma's metric theorem [93, Thm. I.4.3]. For its formulation, recall that a property of real numbers is said to hold for (Lebesgue) *almost all* (*a.a.*) $x \in [a, b)$ if there exists a *nullset* N, i.e., $N \in \mathcal{B}[a, b)$ with $\lambda_{a,b}(N) = 0$, such that the property holds for all $x \notin N$. The probabilistic interpretation of a given property of real numbers holding for a.a. x is that this property holds *almost surely* (*a.s.*), i.e., with probability one, *for every random variable that has a density* (i.e., is absolutely continuous).

PROPOSITION 4.14. *Let $J \subset \mathbb{R}$ be an interval and, for every $n \in \mathbb{N}$, let $f_n : J \to \mathbb{R}$ be continuously differentiable. If $f'_m - f'_n$ is monotone on J, and $|f'_m(x) - f'_n(x)| \geq a > 0$ for all $m \neq n$, where a does not depend on $x \in J$, m, and n, then $\big(f_n(x)\big)$ is u.d. mod 1 for almost all $x \in J$.*

EXAMPLE 4.15. Fix a real number $b > 1$ and consider the sequence (b^n). Clearly, if b is an integer then $\langle b^n \rangle \equiv 0$, and so (b^n) is not u.d. mod 1. On the other hand, for non-integer values of b it is generally not known whether (b^n) is u.d. mod 1, even if b is rational. For example, it is a famous open problem whether or not $\big((3/2)^n\big)$ is u.d. mod 1; see [50, p. 137]. However, Proposition 4.14 implies that $(b^n x)$ is u.d. mod 1 for *almost all* $x \in \mathbb{R}$. Indeed, with $f_n(x) := b^n x$, clearly $f'_m - f'_n$ is constant (hence monotone), and for all $m \neq n$,

$$|f'_m(x) - f'_n(x)| = |b^m - b^n| \geq b(b-1) > 0 \quad \text{for all } x \in \mathbb{R}.$$

By Proposition 4.14, therefore, the set

$$U_b := \big\{ x \in \mathbb{R} : (b^n x) \text{ is u.d. mod 1} \big\}$$

has full measure, so $\mathbb{R} \setminus U_b$ is a nullset. That U_b is not trivial, i.e., $U_b \neq \mathbb{R} \setminus \{0\}$, is easy to see for integral b (in which case $\mathbb{Q} \cap U_b = \varnothing$), and in fact is true for every $b > 1$. The proof of Theorem 6.46 below shows that, despite being a nullset, the set $\mathbb{R} \setminus U_b$ is uncountable and dense in $\mathbb{R}$. Note that, even for integer b, it may be very hard to decide whether $x \in U_b$ for any specific $x \in \mathbb{R}$. For example, for no integer $q \geq 2$ is it known whether $(q^n \sqrt{2})$ is u.d. mod 1, i.e., whether $\sqrt{2} \in U_q$; see [42]. ✠

The next theorem establishes a necessary and sufficient condition for an asymptotically exponential sequence to be Benford.

THEOREM 4.16 ([22]). *Let (b_n) be a sequence of real numbers such that $\lim_{n \to \infty} |b_n/a^n|$ exists and is positive for some $a > 0$. Then (b_n) is Benford if and only if $\log |a|$ is irrational.*

PROOF. Observe that

$$(b_n) \text{ is Benford} \iff (a^n) \text{ is Benford}$$
$$\iff (n \log a) \text{ is u.d. mod 1} \iff \log a \text{ is irrational},$$

where the first equivalence follows from Theorem 4.12(i), the second from Theorem 4.2, and the third from Proposition 4.6(ii). ∎

EXAMPLE 4.17. **(i)** By Theorem 4.16 the sequence (2^n) is Benford since $\log 2$ is irrational, but (10^n) is not Benford since $\log 10 = 1 \in \mathbb{Q}$. Similarly, (0.2^n), (3^n), (0.3^n), $(0.01 \cdot 0.2^n + 0.2 \cdot 0.01^n)$ are Benford, whereas (0.1^n), $(\sqrt{10}^n)$, $(0.1 \cdot 0.02^n + 0.02 \cdot 0.1^n)$ are not.

(ii) The sequence $(0.2^n + (-0.2)^n)$ is not Benford, since all odd terms are zero, but $(0.2^n + (-0.2)^n + 0.03^n)$ is Benford — although this does not follow directly from Theorem 4.16.

(iii) By Proposition 4.14, the sequence $(nx) = (x, 2x, \ldots)$ is u.d. mod 1 for almost all real x, but clearly not for *all* x, as taking $x \in \mathbb{Q}$ shows. Consequently, by Theorem 4.2, (10^{nx}) is Benford for almost all, but not all, real x, e.g., (10^{nx}) is not Benford when x is rational.

(iv) By Proposition 4.6(iv) or Example 4.7(ii), the sequence $(\log n)$ is not u.d. mod 1, so by Theorem 4.16 the sequence (n) of positive integers is not Benford, and neither is (an) for any $a \in \mathbb{R}$; see Figure 4.3.

(v) Let $(p_n) = (2, 3, 5, 7, 11, 13, 17, \ldots)$ denote the sequence of prime numbers. By the Prime Number Theorem, $\lim_{n \to \infty} p_n/(n \ln n) = 1$; e.g., see [117, A000040]. Hence it follows from Proposition 4.6(iv) that (p_n) is not Benford; see Figure 4.3. ✠

EXAMPLE 4.18. The Fibonacci sequence $(F_n) = (1, 1, 2, 3, 5, 8, 13, \ldots)$ is given explicitly (i.e., non-recursively) by the well-known formula (e.g., see [117, A000045])

$$F_n = \frac{1}{\sqrt{5}} \left(\left(\tfrac{1}{2} + \tfrac{1}{2}\sqrt{5} \right)^n - \left(\tfrac{1}{2} - \tfrac{1}{2}\sqrt{5} \right)^n \right) = \frac{\varphi^n - (-\varphi^{-1})^n}{\sqrt{5}}, \quad n \in \mathbb{N},$$

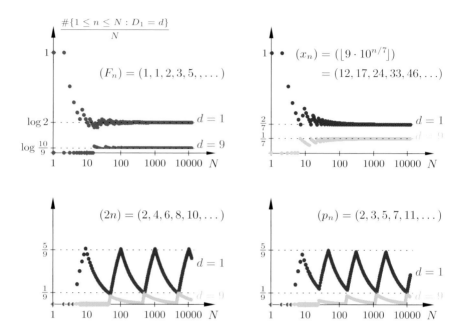

Figure 4.3: For every Benford sequence (x_n) and $d \in \{1, 2, \ldots, 9\}$, the sequence $\left(\#\{1 \leq n \leq N : D_1(x_n) = d\}/N\right)$ converges to $\log(1 + d^{-1})$ as $N \to \infty$. Thus if this sequence does not converge (bottom) or has a different limit (top right), then (x_n) is not Benford; see Example 4.17.

where $\varphi = \frac{1}{2}(1 + \sqrt{5}) = 1.618$. Since $\varphi > 1$ and, as is easily checked, $\log \varphi$ is irrational, Theorem 4.16 implies that (F_n) is Benford; see Figure 4.3. Sequences such as (F_n) which are generated by (linear) recurrence relations will be studied in detail in Chapters 6 and 7. ✠

4.3   UNIFORM DISTRIBUTION OF RANDOM VARIABLES

This section focuses on tools for the analysis of the Benford property of random variables. First, Theorem 4.2 alone is used to exhibit several Benford and non-Benford random variables.

EXAMPLE 4.19. **(i)** As a simple extension of Example 4.7(i), Theorem 4.2 implies that if U is uniform on $(0, a)$ for some $a > 0$, then for every $b \in \mathbb{R}^+$, b^U is Benford if and only if $a|\log b| \in \mathbb{N}$.

(ii) No exponential random variable is u.d. mod 1. Specifically, let X be an exponential random variable with mean $\sigma > 0$, i.e.,

$$F_X(x) = \max\{0, 1 - e^{-x/\sigma}\}, \quad x \in \mathbb{R},$$

so var $X = \sigma^2$. For every integer $k \geq 0$,

$$\mathbb{P}\big(k \leq X < k + \tfrac{1}{2}\big) = F_X\big(k + \tfrac{1}{2}\big) - F_X(k)$$
$$> F_X(k + 1) - F_X\big(k + \tfrac{1}{2}\big) = \mathbb{P}\big(k + \tfrac{1}{2} \leq X < k + 1\big),$$

and since $\sum_{k=0}^{\infty} \mathbb{P}(k \leq X < k + 1) = 1$, this implies that

$$\mathbb{P}\big(\langle X \rangle < \tfrac{1}{2}\big) = \sum_{k=0}^{\infty} \mathbb{P}\big(k \leq X < k + \tfrac{1}{2}\big) > \tfrac{1}{2},$$

showing that X is not u.d. mod 1. To obtain more quantitative information, observe that, for every $0 \leq s < 1$,

$$F_{\langle X \rangle}(s) = \mathbb{P}(\langle X \rangle \leq s) = \sum_{k=0}^{\infty} \big(F_X(k + s) - F_X(k)\big) = \frac{1 - e^{-s/\sigma}}{1 - e^{-1/\sigma}},$$

from which it follows via a straightforward calculation that

$$R(\sigma) := \max_{0 \leq s < 1} \big|F_{\langle X \rangle}(s) - s\big| = \frac{1}{e^{1/\sigma} - 1} - \sigma + \sigma \ln(\sigma e^{1/\sigma} - \sigma).$$

Note that $R(1) = \ln(e - 1) - \dfrac{e - 2}{e - 1} = 0.1233 < \tfrac{1}{8}$. Moreover,

$$8\sigma R(\sigma) - 1 = \mathcal{O}\big(\sigma^{-1}\big) \quad \text{as } \sigma \to +\infty,$$

which shows that even though X is not u.d. mod 1, the deviation of $\langle X \rangle$ from uniformity is small for large σ. As a consequence, 10^X resembles a Benford random variable ever more closely as $\sigma \to +\infty$.

(iii) If X is a normal random variable then X is not u.d. mod 1, and neither is $|X|$ nor $\max\{0, X\}$. While this is easily checked by a direct calculation as in (ii), it is again illuminating to obtain more quantitative information. To this end, assume that X is a normal variable with mean 0 and variance σ^2. By means of Fourier series [124], it can be shown that, for every $0 \leq s < 1$,

$$F_{\langle X \rangle}(s) - s = \sum_{n=1}^{\infty} \frac{\sin(2\pi n s)}{\pi n} e^{-2\pi^2 \sigma^2 n^2};$$

see Examples 8.10 and 8.25 below. From this, it follows that

$$R(\sigma) = \max_{0 \leq s < 1} \big|F_{\langle X \rangle}(s) - s\big| \leq \frac{1}{\pi} \sum_{n=1}^{\infty} n^{-1} e^{-2\pi^2 \sigma^2 n^2},$$

and hence in particular

$$\pi e^{2\pi^2 \sigma^2} R(\sigma) - 1 = \mathcal{O}\big(e^{-6\pi^2 \sigma^2}\big) \quad \text{as } \sigma \to +\infty,$$

which shows that $R(\sigma)$, the deviation of $\langle X \rangle$ from uniformity, goes to zero very rapidly as $\sigma \to +\infty$. Note that $R(1) < \frac{1}{\pi} \sum_{n=1}^{\infty} n^{-1} e^{-2\pi^2 n^2} = 8.515 \cdot 10^{-10}$. Thus even though a standard normal random variable X is not u.d. mod 1, the distribution of $\langle X \rangle$ is extremely close to uniform. In the same vein, a *log-normal* random variable with large variance is practically indistinguishable from a Benford random variable. ✠

The next theorem (Theorem 4.21 below) establishes two key facts about the uniform distribution modulo one of random variables that will lead, via Theorem 4.2, to general conclusions about Benford random variables, including the fact (Theorem 5.3 below) that the Benford distribution is the unique significand distribution that is scale-invariant. First, recall that for any random variable X with values in $[0,1)$, or, equivalently, for the associated probability measure P_X on $([0,1), \mathcal{B}[0,1))$, the *Fourier* (or *Fourier–Stieltjes*) *coefficients* of X and P_X are the bounded bi-infinite sequence $\left(\widehat{P_X}(k)\right)_{k \in \mathbb{Z}}$ of complex numbers, given by

$$
\widehat{P_X}(k) = \mathbb{E}\left[e^{2\pi \imath k X}\right] = \int_0^1 e^{2\pi \imath k s}\, \mathrm{d}P_X(s)
$$

$$
= \int_0^1 \cos(2\pi k s)\, \mathrm{d}P_X(s) + \imath \int_0^1 \sin(2\pi k s)\, \mathrm{d}P_X(s)\,, \quad k \in \mathbb{Z}\,.
$$

To facilitate the proof of Theorem 4.21, and various subsequent results, the next lemma records several of the most basic and important properties of Fourier coefficients that will be useful in this context; see [36] for an authoritative treatise on this material.

LEMMA 4.20. *Let X and X_n, $n \in \mathbb{N}$, be random variables with values in $[0,1)$, and with distributions P_X, P_{X_n}, and Fourier coefficients $\left(\widehat{P_X}(k)\right)_{k \in \mathbb{Z}}$, $\left(\widehat{P_{X_n}}(k)\right)_{k \in \mathbb{Z}}$, respectively. Then:*

(i) $\left|\widehat{P_X}(k)\right| \le 1$ *for all $k \in \mathbb{Z}$, and $\widehat{P_X}(0) = 1$;*

(ii) *If $\left|\widehat{P_X}(k)\right| = 1$ for some $k \ne 0$ then $\mathbb{P}(kX \in s + \mathbb{Z}) = 1$ for some $s \in [0,1)$;*

(iii) $\left(\widehat{P_X}(k)\right)_{k \in \mathbb{Z}}$ *uniquely determines P_X, i.e., $\widehat{P_{X_1}}(k) = \widehat{P_{X_2}}(k)$ for all $k \in \mathbb{Z}$ implies that $P_{X_1} = P_{X_2}$;*

(iv) *If X_1 and X_2 are independent then $\widehat{P_{\langle X_1 + X_2 \rangle}}(k) = \widehat{P_{X_1}}(k) \cdot \widehat{P_{X_2}}(k)$ for all $k \in \mathbb{Z}$;*

(v) $\lim_{n \to \infty} \widehat{P_{X_n}}(k) = \widehat{P_X}(k)$ *for every $k \in \mathbb{Z}$ if and only if*

$$
\lim_{n \to \infty} \mathbb{P}(X_n \le x) = \mathbb{P}(X \le x) \quad \text{for all } x \in \mathbb{R} \text{ with } \mathbb{P}(X = x) = 0;
$$
$$(4.1)$$

(vi) *X is uniform, i.e., $X = U(0,1)$, if and only if*

$$
\widehat{P_X}(k) = \widehat{\lambda_{0,1}}(k) = \begin{cases} 1 & \text{if } k = 0\,, \\ 0 & \text{otherwise.} \end{cases}
$$

Remark. In probabilistic terms, (4.1) says that the sequence (X_n) converges *in distribution* to X; see Section 8.1.

THEOREM 4.21. *Let X, Y be random variables. Then:*

(i) *If X is u.d. mod 1 and Y is independent of X then $X + Y$ is u.d. mod 1;*

(ii) *If $\langle X \rangle$ and $\langle X + a \rangle$ have the same distribution for some irrational $a \in \mathbb{R}$ then X is u.d. mod 1.*

PROOF. (i) By Lemma 4.20(vi), $\widehat{P_{\langle X \rangle}}(k) = 0$ for all $k \neq 0$, so by Lemma 4.20(iv),

$$\widehat{P_{\langle X+Y \rangle}}(k) = \widehat{P_{\langle X \rangle}}(k) \cdot \widehat{P_{\langle Y \rangle}}(k) = 0 \quad \text{for all } k \neq 0.$$

Thus, since $\widehat{P_{\langle X+Y \rangle}}(0) = 1$, Lemma 4.20(iii, vi) imply that $\langle X + Y \rangle = U(0,1)$, i.e., $X + Y$ is u.d. mod 1.

(ii) Note that if $X = a$ with probability one then $\widehat{P_{\langle X \rangle}}(k) = e^{2\pi \imath k a}$ for every $k \in \mathbb{Z}$. Consequently, if $\langle X \rangle$ and $\langle X + a \rangle$ have the same distribution then

$$\widehat{P_{\langle X \rangle}}(k) = \widehat{P_{\langle X+a \rangle}}(k) = e^{2\pi \imath k a} \widehat{P_{\langle X \rangle}}(k)$$

for every $k \in \mathbb{Z}$. If a is irrational then $e^{2\pi \imath k a} \neq 1$ for all $k \neq 0$, implying that $\widehat{P_{\langle X \rangle}}(k) = 0$. Thus by Lemma 4.20(iii, vi), $\widehat{P_{\langle X \rangle}} = \widehat{\lambda_{0,1}}$ and $P_{\langle X \rangle} = \lambda_{0,1}$, that is, $\langle X \rangle = U(0,1)$. ∎

Remark. Without independence, conclusion (i) of the theorem may fail, as the example of $X = U(0,1)$ and $Y = -X$ shows.

As will be recorded in Chapter 8, an immediate consequence of Theorems 4.2 and 4.21 is that the product of two independent positive random variables is Benford if either variable is Benford.

Random walks on the circle

The final two theorems in this section will be used in Chapter 8 to establish distributional and almost sure (probability one) Benford properties of limits of products of i.i.d. random variables. Since these two theorems are basic results for general random walks on the circle, they may also be of independent interest. The first theorem gives a necessary and sufficient condition for the *distribution of the paths* of a random walk on the unit circle (circle of circumference 1) to converge to a uniform distribution.

THEOREM 4.22. *For each $n \in \mathbb{N}$, let $S_n = X_1 + \ldots + X_n$, where $X_1, X_2, \ldots$ are i.i.d. random variables. Then the following are equivalent:*

(i) $\lim_{n \to \infty} \mathbb{P}(\langle S_n \rangle \leq s) = s$ *for all $0 \leq s < 1$;*

(ii) $\mathbb{P}\big(X_1 \in x + \frac{1}{m}\mathbb{Z}\big) < 1$ *for every $x \in \mathbb{R}$ and $m \in \mathbb{N}$.*

Proof. Assume (ii) holds. Then by Lemma 4.20(ii), $\left|\widehat{P_{\langle X_1 \rangle}}(k)\right| < 1$ for all $k \in \mathbb{Z} \setminus \{0\}$. Thus, by Lemma 4.20(iv)

$$\widehat{P_{\langle S_n \rangle}}(k) = P_{\langle \sum_{j=1}^{n} X_j \rangle}\widehat{}(k) = \widehat{P_{\langle X_1 \rangle}}(k)^n \xrightarrow{n \to \infty} 0 \quad \text{for all } k \in \mathbb{Z} \setminus \{0\},$$

which implies (i).

Conversely, suppose that (ii) does not hold, i.e., $\mathbb{P}\left(X_1 \in x + \frac{1}{m}\mathbb{Z}\right) = 1$ for some $x \in \mathbb{R}$ and $m \in \mathbb{N}$. In this case,

$$\widehat{P_{\langle X_1 \rangle}}(m) = \mathbb{E}\left[e^{2\pi \imath m X_1}\right] = e^{2\pi \imath m x},$$

and since, for every $n \in \mathbb{N}$,

$$\left|\widehat{P_{\langle S_n \rangle}}(m)\right| = \left|P_{\langle \sum_{j=1}^{n} X_j \rangle}\widehat{}(m)\right| = \left|\widehat{P_{\langle X_1 \rangle}}(m)^n\right| = \left|e^{2\pi \imath m n x}\right| = 1,$$

(i) clearly does not hold.   ∎

Note that if X_1 is continuous, i.e., $\mathbb{P}(X_1 = x) = 0$ for all $x \in \mathbb{R}$, then Theorem 4.22(ii) applies, and so the random walk on the unit circle ($\langle X_1 \rangle, \langle X_1 + X_2 \rangle, \ldots$) converges to uniformity in the distributional sense of (i). If X_1 has atoms, however, (i) and (ii) may or may not hold, even if X_1 is purely atomic with only two atoms.

EXAMPLE 4.23. **(i)** Assume that the random variables $X_1, X_2, \ldots$ are i.i.d. with $\mathbb{P}\left(X_1 = \sqrt{2}\right) = \mathbb{P}(X_1 = 1) = \frac{1}{2}$. Then X_1 satisfies Theorem 4.22(ii); hence the sequence of partial sums (S_n) is u.d. mod 1 in the sense of Theorem 4.22(i). **(ii)** If $\mathbb{P}\left(X_1 = \sqrt{2}\right) = \mathbb{P}\left(X_1 = 1 + \sqrt{2}\right) = \frac{1}{2}$ then X_1 is again purely atomic, but Theorem 4.22(ii) fails, so the partial sums (S_n) are not u.d. mod 1 in the distributional sense, even though, as the next theorem shows, the sequence (S_n) is u.d. mod 1 with probability one.   ✠

The second theorem for random walks on the unit circle establishes an analogous necessary and sufficient condition on the underlying distribution of the increments so that *almost all paths* are u.d. mod 1. Note that this result is a natural generalization, via the special case where $X_1 = \theta$ with probability one, of Weyl's Theorem which asserts that $(n\theta)$ is u.d. mod 1 if and only if θ is irrational; see Proposition 4.6(ii). The proof given here follows [18]; an alternative proof of a closely related result may be found in [131].

THEOREM 4.24 ([18]). *For each $n \in \mathbb{N}$, let $S_n = X_1 + \ldots + X_n$, where $X_1, X_2, \ldots$ are i.i.d. random variables. Then the following are equivalent:*

(i) *The sequence (S_n) is u.d. mod 1 with probability one;*

(ii) $\mathbb{P}\left(X_1 \in \frac{1}{m}\mathbb{Z}\right) < 1$ *for every $m \in \mathbb{N}$.*

PROOF. First, assume (ii) holds, and let X_0 be uniform on $(0,1)$ and independent of $X_1, X_2, \ldots$. Note that, as a consequence of (ii), the set

$$\operatorname{supp} P_{X_1} = \left\{ x \in \mathbb{R} : \mathbb{P}(|X_1 - x| < \varepsilon) > 0 \text{ for every } \varepsilon > 0 \right\}$$

contains either an irrational number, or rational numbers with arbitrarily large denominators. In either case, therefore, the set

$$\mathbb{S}_{X_1} := \left\{ \langle x_1 + x_2 + \ldots + x_n \rangle : n \in \mathbb{N}; \; x_1, x_2, \ldots, x_n \in \operatorname{supp} P_{X_1} \right\} \subset [0,1)$$

is *dense* in $[0,1)$.

Since the forthcoming main argument rests on an application of the ergodic theorem, some ergodic theory notation and terminology will be used. Specifically, endow the space of sequences in $[0,1)$,

$$\mathbb{S}_\infty := [0,1)^{\mathbb{N}_0} = \left\{ (s_k)_{k \in \mathbb{N}_0} : s_k \in [0,1) \text{ for all } k \right\},$$

with the (product) σ-algebra

$$\mathcal{B}_\infty := \bigotimes_{k \in \mathbb{N}_0} \mathcal{B}[0,1)$$
$$:= \sigma\big(\{ B_0 \times B_1 \times \ldots \times B_k \times [0,1) \times [0,1) \times \ldots : k \in \mathbb{N}_0; \; B_0, B_1, \ldots, B_k \in \mathcal{B}[0,1) \}\big).$$

A probability measure P_∞ is uniquely defined on $(\mathbb{S}_\infty, \mathcal{B}_\infty)$ by setting

$$P_\infty(B_0 \times B_1 \times \ldots \times B_k \times [0,1) \times [0,1) \times \ldots) = \mathbb{P}(X_0 \in B_0, X_1 \in B_1, \ldots, X_k \in B_k)$$
$$= \lambda_{0,1}(B_0) \prod_{j=1}^{k} \mathbb{P}(X_j \in B_j)$$

for all $k \in \mathbb{N}_0$ and all $B_0, B_1, \ldots, B_k \in \mathcal{B}[0,1)$. Define a map σ_∞ of $\mathbb{S}_\infty$ into itself by

$$\sigma_\infty\big((s_0, s_1, s_2, \ldots)\big) = \big(\langle s_0 + s_1 \rangle, s_2, s_3, \ldots\big), \quad (s_k)_{k \in \mathbb{N}_0} \in \mathbb{S}_\infty.$$

Clearly, σ_∞ is measurable, i.e., $\sigma_\infty^{-1}(A) \in \mathcal{B}_\infty$ for every $A \in \mathcal{B}_\infty$. Moreover, since $X_0 = U(0,1)$ is independent of the i.i.d. variables $X_1, X_2, \ldots$,

$$P_\infty\big(\sigma_\infty^{-1}(B_0 \times B_1 \times \ldots \times B_k \times [0,1) \times [0,1) \times \ldots)\big)$$
$$= \mathbb{P}(\langle X_0 + X_1 \rangle \in B_0, X_2 \in B_1, \ldots, X_{k+1} \in B_k)$$
$$= \lambda_{0,1}(B_0)\mathbb{P}(X_1 \in B_1, \ldots, X_k \in B_k)$$
$$= P_\infty(B_0 \times B_1 \times \ldots \times B_k \times [0,1) \times [0,1) \times \ldots),$$

which shows that the two probability measures P_∞ and $P_\infty \circ \sigma_\infty^{-1}$ on $(\mathbb{S}_\infty, \mathcal{B}_\infty)$ are identical, i.e., σ_∞ is P_∞-preserving. The crucial step in the argument now consists in showing that actually more is true, namely,

$$\sigma_\infty \text{ is ergodic}. \tag{4.2}$$

Recall that this simply means that every σ_∞-invariant set $A \in \mathcal{B}_\infty$ has P_∞-measure zero or one, or, more formally, $P_\infty(\sigma_\infty^{-1}(A)\Delta A) = 0$ for $A \in \mathcal{B}_\infty$ implies that $P_\infty(A) \in \{0,1\}$; here the symbol Δ denotes the symmetric difference of two sets, i.e., $A\Delta B = (A \setminus B) \cup (B \setminus A)$. It is straightforward to show (see [134, p. 300]) that ergodicity of σ_∞ follows from (and in fact is equivalent to) the following property:

$$\text{If } \lambda_{0,1}(\langle B + X_1 \rangle \Delta B) = 0 \text{ with probability one} \qquad (4.3)$$
$$\text{for some } B \in \mathcal{B}[0,1) \text{ then } \lambda_{0,1}(B) \in \{0,1\};$$

here and throughout the remainder of the proof, for any $B \in \mathcal{B}[0,1)$ and $x \in \mathbb{R}$, the notation $\langle B + x \rangle := \{\langle b + x \rangle : b \in B\}$ will be used. Thus, to prove (4.2) it is enough to verify (4.3), and this will now be done.

Assume, therefore, that $\lambda_{0,1}(B) > 0$, and $\lambda_{0,1}(\langle B + X_1 \rangle \Delta B) = 0$ with probability one. It follows that $\lambda_{0,1}(\langle B + X_1 + \ldots + X_n \rangle \Delta B) = 0$ with probability one for every $n \in \mathbb{N}$, and hence by Fubini's Theorem, $\lambda_{0,1}(\langle B + s \rangle \Delta B) = 0$ for every $s \in \mathbb{S}_{X_1}$. Define a probability measure Q on $\big([0,1), \mathcal{B}[0,1)\big)$ by

$$Q(C) = \frac{\lambda_{0,1}(C \cap B)}{\lambda_{0,1}(B)}, \quad C \in \mathcal{B}[0,1),$$

and observe that, for every $C \in \mathcal{B}[0,1)$ and $s \in \mathbb{S}_{X_1}$,

$$\big|Q(\langle C + s \rangle) - Q(C)\big| = \frac{|\lambda_{0,1}(\langle C + s \rangle \cap B) - \lambda_{0,1}(C \cap B)|}{\lambda_{0,1}(B)}$$

$$= \frac{|\lambda_{0,1}(C \cap \langle B - s \rangle) - \lambda_{0,1}(C \cap B)|}{\lambda_{0,1}(B)}$$

$$\leq \frac{\lambda_{0,1}(\langle B - s \rangle \Delta B)}{\lambda_{0,1}(B)} = \frac{\lambda_{0,1}(B \Delta \langle B + s \rangle)}{\lambda_{0,1}(B)} = 0.$$

Thus $Q(\langle C + s \rangle) = Q(C)$ for every $C \in \mathcal{B}[0,1)$ and $s \in \mathbb{S}_{X_1}$. Taking Fourier coefficients yields, for every $k \in \mathbb{Z}$,

$$e^{2\pi\imath k s}\widehat{Q}(k) = \widehat{Q}(k) \quad \text{for all } s \in \mathbb{S}_{X_1}. \qquad (4.4)$$

Since $\mathbb{S}_{X_1}$ is dense in $[0,1)$, given any $k \in \mathbb{Z} \setminus \{0\}$, there exists a sequence (s_n) in $\mathbb{S}_{X_1}$ with $\lim_{n\to\infty} |k|s_n = \frac{1}{2}$. Hence (4.4) implies that $\widehat{Q}(k) = 0$ for all $k \neq 0$, which in turn shows that $Q = \lambda_{0,1}$. Thus $\lambda_{0,1}(B)^2 = \lambda_{0,1}(B)$, and since $\lambda_{0,1}(B) > 0$, it follows that $\lambda_{0,1}(B) = 1$. This establishes (4.3) and, as explained in the preceding paragraph, also (4.2).

The scene is now set for an application of the Birkhoff Ergodic Theorem, which states that for every (Borel measurable) function $f : [0,1) \to \mathbb{C}$ with $\int_0^1 |f(x)|\,\mathrm{d}x < +\infty$,

$$\frac{1}{n}\sum_{j=0}^{n} f(\langle s_0 + s_1 + \ldots + s_j \rangle) \overset{n\to\infty}{\longrightarrow} \int_0^1 f(x)\,\mathrm{d}x$$

for all $(s_k)_{k \in \mathbb{N}_0} \in \mathbb{S}_\infty$, with the possible exception of a P_∞-nullset. In probabilistic terms, this means that

$$\lim_{n \to \infty} \frac{1}{n} \sum_{j=0}^{n} f(\langle X_0 + X_1 + \ldots + X_j \rangle) = \int_0^1 f(x) \, \mathrm{d}x \quad \text{a.s.} \qquad (4.5)$$

Assume henceforth that f is actually *continuous* with $\lim_{x \uparrow 1} f(x) = f(0)$, e.g., $f(x) = e^{2\pi \imath x}$. For any such f, as well as any $s \in [0,1)$ and $m \in \mathbb{N}$, denote the set

$$\left\{ \omega \in \Omega : \limsup_{n \to \infty} \left| \frac{1}{n} \sum_{j=1}^{n} f(\langle s + X_1(\omega) + \ldots + X_j(\omega) \rangle) - \int_0^1 f(x) \, \mathrm{d}x \right| < \frac{1}{m} \right\}$$

simply by $\Omega_{f,s,m}$. By (4.5), $1 = \int_0^1 \mathbb{P}(\Omega_{f,s,m}) \, \mathrm{d}s$, and hence $\mathbb{P}(\Omega_{f,s,m}) = 1$ for a.a. $s \in [0,1)$. Since f is uniformly continuous, for every $m \geq 2$ there exists $s_m > 0$ such that $\mathbb{P}(\Omega_{f,s_m,m}) = 1$ and $\Omega_{f,s_m,m} \subset \Omega_{f,0,\lfloor m/2 \rfloor}$. From

$$1 = \mathbb{P}\left(\bigcap_{m \geq 2} \Omega_{f,s_m,m} \right) \leq \mathbb{P}\left(\bigcap_{m \geq 2} \Omega_{f,0,\lfloor m/2 \rfloor} \right) \leq 1 \, ,$$

it is clear that

$$\lim_{n \to \infty} \frac{1}{n} \sum_{j=1}^{n} f(\langle X_1 + \ldots + X_j \rangle) = \int_0^1 f(x) \, \mathrm{d}x \quad \text{a.s.} \qquad (4.6)$$

Since the intersection of countably many sets of full measure itself has full measure, choosing $f(x) = e^{2\pi \imath k x}$, $k \in \mathbb{Z}$, in (4.6) shows that, with probability one,

$$\lim_{n \to \infty} \frac{1}{n} \sum_{j=1}^{n} e^{2\pi \imath k(X_1 + \ldots + X_j)} = \int_0^1 e^{2\pi \imath k x} \, \mathrm{d}x = 0 \quad \text{for all } k \in \mathbb{Z}, k \neq 0 \, .$$
$$(4.7)$$

By Weyl's criterion [93, Thm. I.2.1], (4.7) is equivalent to

$$\mathbb{P}\left(\left(\sum_{j=1}^{n} X_j \right) \text{ is u.d. mod } 1 \right) = 1 \, ,$$

which establishes (i).

Conversely, suppose (ii) does not hold, i.e., $\mathbb{P}\big(X_1 \in \frac{1}{m}\mathbb{Z}\big) = 1$ for some $m \in \mathbb{N}$. This means that mX_1 is an integer with probability one, and so is mS_n for all $n \in \mathbb{N}$. In other words, $(\langle mS_n \rangle) = (0,0,\ldots)$ with probability one. But since a sequence (x_n) is u.d. mod 1 if and only if (mx_n) is u.d. mod 1 for all $m \in \mathbb{N}$, by Proposition 4.3(i), this implies that (S_n) is not u.d. mod 1, i.e., (i) does not hold. ∎

EXAMPLE 4.25. **(i)** If $X_1, X_2, \ldots$ are i.i.d. and X_1 is continuous, or, more generally, if $\mathbb{P}(X_1 \in \mathbb{Q}) = 0$, then, by Theorem 4.24, the sequence of partial sums (S_n) is u.d. mod 1 with probability one.

(ii) Let $X_1, X_2, \ldots$ be i.i.d. with $\mathbb{P}(X_1 = j^{-1}) = 2^{-j}$ for all $j \in \mathbb{N}$. Then, by Theorem 4.24, (S_n) is u.d. mod 1 with probability one even though all the increments $S_{n+1} - S_n = X_{n+1}$ are rational with probability one, and so is S_n for every n. ✠

Clearly, if X_1 has a density, then property (ii) holds in both Theorems 4.22 and 4.24, so if $X_1, X_2, \ldots$ are i.i.d. and X_1 has a density, then the partial sums (S_n) are u.d. mod 1 in both the distributional sense and with probability one. As it turns out, these same two properties also hold for every random walk on the unit circle that has *identical* increments, i.e., if $X_n = X_1$ and hence $S_n = nX_1$ for all $n \in \mathbb{N}$ — provided that X_1 has a density.

PROPOSITION 4.26. *If the random variable X_1 has a density, then:*

(i) $\lim_{n \to \infty} \mathbb{P}(\langle nX_1 \rangle \leq s) = s$ *for all* $0 \leq s < 1$;

(ii) *The sequence* $(nX_1) = (X_1, 2X_1, \ldots)$ *is u.d.* mod 1 *with probability one.*

This last proposition, which will not be used anywhere in this book, follows immediately from Theorem 4.2 and Theorem 8.8 below; it may also be proved directly using the same tools as in the proof of Theorem 4.22, together with the identity $\widehat{P_{\langle nX_1 \rangle}}(k) = \widehat{P_{\langle X_1 \rangle}}(nk)$ and Proposition 4.14, respectively. Unlike properties (i) in Theorems 4.22 and 4.24, however, which both hold for all continuous X_1, the conclusion of Proposition 4.26(i) may fail if X_1 is continuous but does not have a density; see Example 8.9(ii) below.

Chapter Five

Scale-, Base-, and Sum-Invariance

The purpose of this chapter is to establish and illustrate three basic invariance properties of the Benford distribution that are instrumental in demonstrating whether or not certain datasets are Benford, and that also prove helpful for predicting which empirical data are likely to follow Benford's law closely.

5.1 THE SCALE-INVARIANCE PROPERTY

One popular hypothesis often related to Benford's law is that of *scale-invariance*. Informally put, scale-invariance captures the intuitively attractive notion that any universal law should be independent of units. For instance, if a sufficiently large aggregation of data is converted from meters to feet, US\$ to €, etc., then while the individual numbers may change, any statement about the overall distribution of significant digits should not be affected by this conversion. R. Pinkham [124] credits Hamming with the idea of scale-invariance, and attempts to prove that the Benford distribution is the only scale-invariant distribution. This idea has subsequently been used by numerous authors to explain the appearance of Benford's law in many real-life data sets, by arguing that the data in question should be invariant under changes of scale and thus must be Benford.

Although this scale-invariance conclusion is correct in the proper setting (see Theorem 5.3 below), Knuth [90] observed that Pinkham's argument fails because of the implicit assumption that there is a scale-invariant Borel probability measure on $\mathbb{R}^+$, when in fact no such probability measure exists; cf. [127]. Indeed, the only real-valued random variable X that is *scale-invariant*, i.e., X and aX have the same distribution for all scaling factors $a > 0$, is the random variable that is zero almost surely, that is, $\mathbb{P}(X = 0) = 1$. Clearly, any such random variable is scale-invariant since $X = aX$ with probability one. To see that this is the only scale-invariant random variable, simply observe that for any scale-invariant X,

$$\mathbb{P}(X \neq 0) = \lim_{a \to +\infty} \mathbb{P}(|X| > a^{-1}) = \lim_{a \to +\infty} \mathbb{P}(|a^2 X| > a)$$
$$= \lim_{a \to +\infty} \mathbb{P}(|X| > a) = 0,$$

where the third equality is due to scale-invariance. Thus no non-zero random variable is scale-invariant.

Note, however, that the measure on $(\mathbb{R}^+, \mathcal{B}^+)$ defined as

$$\mu([c,d]) = \int_c^d \frac{\log e}{x}\, \mathrm{d}x = \log \frac{d}{c} \quad \text{for every } 0 < c < d$$

is scale-invariant. To see this, for each $a > 0$ let μ_a denote the measure induced by μ and the scaling with factor a, that is, $\mu_a = \mu \circ f^{-1}$ with the linear function $f(x) = ax$. Then

$$\mu_a([c,d]) = \int_{c/a}^{d/a} \frac{\log e}{x}\, \mathrm{d}x = \log \frac{d}{c} = \mu([c,d]) \quad \text{for every } 0 < c < d\,,$$

and hence $\mu_a = \mu$. Note that μ is not finite, i.e., $\mu(\mathbb{R}^+) = +\infty$, but is σ-finite. (A measure μ on $(\Omega, \mathcal{A})$ is σ-finite if $\Omega = \bigcup_{n \in \mathbb{N}} A_n$ for some sequence (A_n) in $\mathcal{A}$, and $\mu(A_n) < +\infty$ for all n.)

In a similar spirit, a sequence (x_n) of real numbers may be called *scale-invariant* if

$$\lim_{N \to \infty} \frac{\#\{1 \le n \le N : ax_n \in [c,d]\}}{N} = \lim_{N \to \infty} \frac{\#\{1 \le n \le N : x_n \in [c,d]\}}{N}$$

for all $a > 0$ and $c < d$. For example, the sequences $(\sqrt{n})$, (n^{-1}), and

$$\left(2, 2^{-1}, 2, 3, 2^{-1}, 3^{-1}, 2, \ldots, (n-1)^{-1}, 2, 3, \ldots, n, 2^{-1}, 3^{-1}, \ldots, n^{-1}, 2 \ldots \right)$$

all are scale-invariant. As above, it is not hard to see that

$$\lim_{N \to \infty} \frac{\#\{1 \le n \le N : x_n \in [c,d]\}}{N} = 0 \quad \text{for every } c < d \text{ with } cd > 0\,,$$

whenever (x_n) is scale-invariant. Most entries of a scale-invariant sequence of real numbers, therefore, are close to either 0 or $\pm\infty$.

While no positive random variable X can be scale-invariant, as shown above, it may nevertheless have *scale-invariant significant digits*. For this, however, X has to be Benford. In fact, Theorem 5.3 below shows that being Benford is not only necessary but also sufficient for X to have scale-invariant significant digits. This result will first be stated in terms of probability distributions. Recall from Definition 2.9 that $\mathcal{S}$ denotes the significand σ-algebra on $\mathbb{R}^+$.

DEFINITION 5.1. Let $\mathcal{A} \supset \mathcal{S}$ be a σ-algebra on $\mathbb{R}^+$. A probability measure P on $(\mathbb{R}^+, \mathcal{A})$ has *scale-invariant significant digits* if

$$P(aA) = P(A) \quad \text{for all } a > 0 \text{ and } A \in \mathcal{S}\,,$$

or, equivalently, if for all $m \in \mathbb{N}$, all $d_1 \in \{1, 2, \ldots, 9\}$, all $d_j \in \{0, 1, \ldots, 9\}$, $j \ge 2$, and all $a > 0$,

$$P\big(\{x : D_j(ax) = d_j \text{ for } j = 1, 2, \ldots m\}\big) = P\big(\{x : D_j(x) = d_j \text{ for } j = 1, 2, \ldots, m\}\big)\,.$$

EXAMPLE 5.2. **(i)** The Benford probability measure $\mathbb{B}$ on $(\mathbb{R}^+, \mathcal{S})$ has scale-invariant significant digits. This follows from Theorem 5.3 below but can also be seen from a direct calculation. Indeed, if $A = \bigcup_{k \in \mathbb{Z}} 10^k [1, 10^s]$ with $0 \le s < 1$, then, given any $a > 0$,

$$aA = \bigcup_{k \in \mathbb{Z}} 10^{k + \log a} [1, 10^s] = \bigcup_{k \in \mathbb{Z}} 10^{k + \langle \log a \rangle} [1, 10^s] = \bigcup_{k \in \mathbb{Z}} 10^k B,$$

where the set B is given by

$$B = \begin{cases} \left[10^{\langle \log a \rangle}, 10^{\langle \log a \rangle + s} \right] & \text{if } \langle \log a \rangle < 1 - s, \\ \left[1, 10^{\langle \log a \rangle - 1 + s} \right] \cup \left[10^{\langle \log a \rangle}, 10 \right) & \text{if } \langle \log a \rangle \ge 1 - s. \end{cases}$$

From this, it follows that

$$\mathbb{B}(aA) = \left\{ \begin{matrix} \langle \log a \rangle + s - \langle \log a \rangle \\ \langle \log a \rangle - 1 + s + (1 - \langle \log a \rangle) \end{matrix} \right\} = s = \mathbb{B}(A),$$

showing that $\mathbb{B}$ has scale-invariant significant digits.

(ii) The Dirac probability measure δ_1 at $x = 1$ does not have scale-invariant significant digits, since $\delta_1\big(\{x : D_1(x) = 1\}\big) = 1$ but $\delta_1\big(\{x : D_1(2x) = 1\}\big) = 0$.

(iii) Let the random variable X be uniform on $(0, 1)$, i.e., $X = U(0, 1)$. Then, for example,

$$\mathbb{P}(D_1(X) = 1) = \tfrac{1}{9} < \tfrac{11}{27} = \mathbb{P}\left(D_1\left(\tfrac{3}{2}X\right) = 1\right),$$

so X does not have scale-invariant significant digits.   ✠

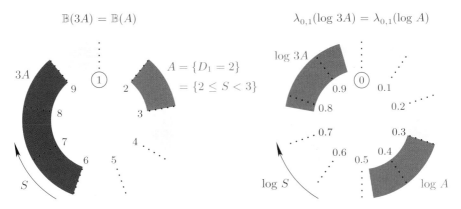

Figure 5.1: Visualizing the scale-invariant significant digits of the Benford distribution $\mathbb{B}$; see Example 5.2(i).

In fact, the Benford distribution is the only probability measure (on the significand σ-algebra $\mathcal{S}$) having scale-invariant significant digits.

THEOREM 5.3 (Scale-invariance characterization [73]). *A probability measure P on $(\mathbb{R}^+, \mathcal{A})$ with $\mathcal{A} \supset \mathcal{S}$ has scale-invariant significant digits if and only if $P(A) = \mathbb{B}(A)$ for every $A \in \mathcal{S}$, i.e., if and only if P is Benford.*

PROOF. Fix any probability measure P on $(\mathbb{R}^+, \mathcal{A})$, denote by P_0 its restriction to $(\mathbb{R}^+, \mathcal{S})$, and let $Q = P_0 \circ \ell^{-1}$ with ℓ as in Lemma 3.14. By Lemma 3.14, Q is a probability measure on $([0,1), \mathcal{B}[0,1))$. Moreover, under the correspondence established by ℓ, the statement

$$P_0(aA) = P_0(A) \quad \text{for all } a > 0, A \in \mathcal{S} \tag{5.1}$$

is equivalent to

$$Q(\langle B + x \rangle) = Q(B) \quad \text{for all } x \in \mathbb{R}, B \in \mathcal{B}[0,1), \tag{5.2}$$

where $\langle B + x \rangle = \{\langle b + x \rangle : b \in B\}$. Fix any random variable X for which the distribution of $\langle X \rangle$ is given by Q. Then (5.2) simply means that, for every $x \in \mathbb{R}$, the distributions of $\langle X \rangle$ and $\langle X + x \rangle$ coincide. By Theorem 4.21, this is the case if and only if X is u.d. mod 1, i.e., $Q = \lambda_{0,1}$. (For the "if" part, note that a *constant* random variable is independent of every other random variable.) Hence (5.1) is equivalent to $P_0 \circ \ell^{-1} = \lambda_{0,1} = \mathbb{B} \circ \ell^{-1}$, and so, by Lemma 3.14, (5.1) is equivalent to $P_0 = \mathbb{B}$. ∎

EXAMPLE 5.4. For every integer k, let

$$f_k(x) = \begin{cases} x^{-1} \log e & \text{if } 10^k \leq x < 10^{k+1}, \\ 0 & \text{otherwise}; \end{cases}$$

also, let $q_k \geq 0$. If $\sum_{k \in \mathbb{Z}} q_k = 1$ then, by Example 3.6(ii), $\sum_{k \in \mathbb{Z}} q_k f_k$ is the density of a Benford probability measure P on $(\mathbb{R}^+, \mathcal{B}^+)$. By Theorem 5.3, P has scale-invariant significant digits. Note that, in full agreement with earlier observations, P is not scale-invariant, as

$$q_k = P([10^k, 10^{k+1})) = P(10^{k-l}[10^l, 10^{l+1})) = P([10^l, 10^{l+1})) = q_l$$

cannot possibly hold for *all* pairs of integers (k, l). ✠

In analogy to Definition 5.1, a sequence (x_n) of real numbers is said to have *scale-invariant significant digits* if

$$\lim_{N \to \infty} \frac{\#\{1 \leq n \leq N : S(ax_n) \leq t\}}{N} = \lim_{N \to \infty} \frac{\#\{1 \leq n \leq N : S(x_n) \leq t\}}{N}$$

$$\text{for all } a > 0, t \in [1, 10). \tag{5.3}$$

Implicit in (5.3) is the assumption that the limits on both sides exist for all t. The definition of a *function* having scale-invariant significant digits is analogous. To formulate an analogue of Theorem 5.3 using this terminology, recall that a set $C \subset \mathbb{N}$ has *(natural) density* $\rho \in [0, 1]$ if $\lim_{N \to \infty} \#\{n \leq N : n \in C\}/N$ exists and equals ρ. For example, $\rho(\{n : n \text{ even}\}) = \frac{1}{2}$ and $\rho(\{n : n \text{ prime}\}) = 0$, whereas $\{n : D_1(n) = 1\}$ does not have a density.

THEOREM 5.5.   (i) *A sequence (x_n) of real numbers has scale-invariant significant digits if and only if the set $\{n : x_n \neq 0\}$ has density $\rho \in [0, 1]$, and either $\rho = 0$ or else $(x_{n_j})_{j \in \mathbb{N}}$ is Benford, where $n_1 < n_2 < \ldots$ and $\{n : x_n \neq 0\} = \{n_j : j \in \mathbb{N}\}$. In particular, if $\rho = 1$ then the sequence (x_n) has scale-invariant significant digits if and only if it is Benford.*

(ii) *Assume $\lambda(\{t \geq 0 : f(t) = 0\}) < +\infty$ for the (Borel measurable) function $f : [0, +\infty) \to \mathbb{R}$. Then f has scale-invariant significant digits if and only if f is Benford.*

PROOF. (i) Assume first that (x_n) has scale-invariant significant digits. By (5.3),

$$G(s) = \lim_{N \to \infty} \frac{\#\{n \leq N : S(x_n) \leq 10^s\}}{N}$$

exists for every $0 \leq s < 1$. Clearly, $0 \leq G(s) \leq 1$, and G is non-decreasing, and hence continuous except for at most countably many jump discontinuities. As will now be shown, G is in fact continuous on $[0, 1)$. To this end, suppose that G were discontinuous at $0 < s_0 < 1$. In this case, there exists $\delta > 0$ such that $G(s_0 + \varepsilon) - G(s_0 - \varepsilon) \geq \delta$ for all sufficiently small $\varepsilon > 0$. Pick any s_1 with $0 < s_1 < s_0$ and note that, for all $N \in \mathbb{N}$ and $\varepsilon < \min\{s_1, 1 - s_0\}$,

$$\#\{n \leq N : S(10^{s_1 - s_0} x_n) \leq 10^{s_1 + \varepsilon}\} - \#\{n \leq N : S(10^{s_1 - s_0} x_n) \leq 10^{s_1 - \varepsilon}\} =$$

$$= \#\{n \leq N : S(x_n) \leq 10^{s_0 + \varepsilon}\} - \#\{n \leq N : S(x_n) \leq 10^{s_0 - \varepsilon}\}.$$

With the assumption that the sequence (x_n) has scale-invariant significant digits, it follows that

$$G(s_1 + \varepsilon) - G(s_1 - \varepsilon) = G(s_0 + \varepsilon) - G(s_0 - \varepsilon) \geq \delta,$$

which in turn shows that G is discontinuous at s_1 as well. Since $0 < s_1 < s_0$ was arbitrary, this is impossible, and consequently G is continuous on $(0, 1)$. Moreover, for every $1 < t < 10$ and all sufficiently small $\varepsilon > 0$,

$$\limsup_{N \to \infty} \frac{\#\{n \leq N : S(x_n) = t\}}{N} \leq G(\log t + \varepsilon) - G(\log t - \varepsilon),$$

showing that $\rho(\{n \in \mathbb{N} : S(x_n) = t\}) = 0$. Since

$$G(0) = \lim_{N \to \infty} \frac{\#\{n \leq N : S(5x_n) \leq 1\}}{N}$$

$$= \lim_{N \to \infty} \frac{\#\{n \leq N : x_n = 0 \text{ or } S(x_n) = 2\}}{N},$$

again by scale-invariant significant digits, it is clear that $C := \{n \in \mathbb{N} : x_n \neq 0\}$ has density $\rho = \rho(C) = 1 - G(0)$. In addition, G is also (right-)continuous at $s = 0$.

To complete the proof of (i), note that there is nothing else to show in the case $G(0) = 1$. From now on, therefore, assume $G(0) < 1$. Define a continuous, non-decreasing function $H : [0, 1) \to \mathbb{R}$ as

$$H(s) = \frac{G(s) - G(0)}{1 - G(0)}, \quad 0 \le s < 1.$$

Note that C is infinite, and with $C = \{n_j : j \in \mathbb{N}\}$, where $n_1 < n_2 < \ldots$,

$$H(s) = \lim_{N \to \infty} \frac{\#\{n \le N : 1 \le S(x_n) \le 10^s\}}{\#\{n \le N : x_n \ne 0\}}$$

$$= \lim_{N \to \infty} \frac{\#\{j \le N : S(x_{n_j}) \le 10^s\}}{N};$$

so H only takes into account the non-zero entries of (x_n). Define $h : \mathbb{R} \to \mathbb{R}$ as

$$h(x) = H(\langle x \rangle) - \langle x \rangle \quad \text{for all } x \in \mathbb{R}.$$

Clearly, h is 1-periodic with $h(0) = 0$, and $|h(x)| \le 1$ for all $x \in \mathbb{R}$. Note that in terms of the function H, the invariance property (5.3) simply reads

$$H(s) = \begin{cases} H(1 + s - \langle \log a \rangle) - H(1 - \langle \log a \rangle) & \text{if } s < \langle \log a \rangle, \\ 1 - H(1 - \langle \log a \rangle) + H(s - \langle \log a \rangle) & \text{if } s \ge \langle \log a \rangle, \end{cases}$$

provided that $\log a \notin \mathbb{Z}$. In terms of h, this is equivalent to

$$h(x) = h(1 + x - \langle \log a \rangle) - h(1 - \langle \log a \rangle) \quad \text{for all } x \in \mathbb{R}, a > 0. \tag{5.4}$$

Thus, the function $h(1 + x - \langle \log a \rangle) - h(x)$ is constant for every $a > 0$. Since h is bounded and 1-periodic, it can be represented (at least in the L^2-sense) by a Fourier series $h(x) = \sum_{k \in \mathbb{Z}} c_k e^{2\pi \imath k x}$, from which it follows that

$$h(1 + x - \langle \log a \rangle) - h(x) = \sum_{k \in \mathbb{Z}} c_k \left(e^{2\pi \imath k (1 + x - \langle \log a \rangle)} - e^{2\pi \imath k x} \right)$$

$$= \sum_{k \in \mathbb{Z}} c_k \left(e^{-2\pi \imath k \langle \log a \rangle} - 1 \right) e^{2\pi \imath k x}.$$

Fix $a > 0$ such that $\log a$ is irrational, e.g., $a = 2$. Then $e^{-2\pi \imath k \langle \log a \rangle} \ne 1$ whenever $k \ne 0$, which in turn implies that $c_k = 0$ for all $k \ne 0$, i.e., h is constant almost everywhere. Thus $H(s) = s + c_0$ for a.a. $s \in [0, 1)$, and in fact $H(s) \equiv s$ because H is continuous with $H(0) = 0$. In summary, therefore,

$$\lim_{N \to \infty} \frac{\#\{j \le N : S(x_{n_j}) \le 10^s\}}{N} = s \quad \text{for all } s \in [0, 1),$$

showing that (x_{n_j}) is Benford.

For the converse in (i), note that if $\rho = 0$ then (5.3) holds with both sides equal to 1 for all $t \in [1, 10)$. Assume, therefore, that $\rho > 0$ and (x_{n_j}) is Benford.

In this case, $G(s) = s\rho + 1 - \rho$ for all $0 \leq s < 1$. Thus $h(x) \equiv 0$, so (5.4) and hence also (5.3) hold, i.e., (x_n) has scale-invariant significant digits.

The proof of (ii) is completely analogous, utilizing the function

$$G(s) = \lim_{T \to +\infty} \frac{\lambda(\{\tau \leq T : S(f(\tau)) \leq 10^s\})}{T}, \quad 0 \leq s < 1.$$

Note that the assumption $\lambda(\{t \geq 0 : f(t) = 0\}) < +\infty$ implies $G(0) = 0$ whenever f has scale-invariant significant digits. ∎

EXAMPLE 5.6. Let (x_n) be the sequence of either Fibonacci or prime numbers. In both cases, $x_n \neq 0$ for all n, and hence by Theorem 5.5(i) (x_n) has scale-invariant significant digits if and only if it is Benford. Thus by Examples 4.18 and 4.17(v), respectively, the sequence (F_n) has scale-invariant significant digits, and (p_n) does not. These facts are illustrated empirically in Figures 5.2 to 5.4, which show the relevant data for $a = 2$ and $a = 7$, respectively, and for the first 10^2 (Figures 5.2 and 5.3) and 10^4 (Figure 5.4) entries of both sequences. ✠

Remark. The notion of scale-invariant digits for sequences (x_n) in (5.3) is somewhat more general than that for probability measures P in Definition 5.1, in that the entries of (x_n) need not all be positive, whereas $P(\mathbb{R}^+) = 1$. It would be straightforward to extend Definition 5.1, and also Definition 5.10 below, to probability measures on $(\mathbb{R}, \mathcal{A})$ with $\mathcal{A} \supset \sigma(S)$. However, since the ensuing characterizations of Benford's law (Theorems 5.3 and 5.13, respectively) would in turn be more cumbersome to state, throughout the remainder of this chapter it will be assumed for simplicity that $P(\mathbb{R}^+) = 1$, or at the very least that $P(\{0\}) = 0$.

The next example is an elegant and entertaining application of the ideas underlying Theorems 5.3 and 5.5 to the mathematical theory of games. This game may be easily understood by a schoolchild, yet it has proven a challenge for game theorists not familiar with Benford's law, especially Theorem 5.3.

EXAMPLE 5.7 ([109]). Consider a two-person game where Player A and Player B each independently choose a (real) number greater than 1, and Player A wins if the product of their two numbers starts with a 1, 2, or 3; otherwise, Player B wins. Using the tools presented in this section, it may easily be seen that there is a strategy for Player A to choose her numbers so that she wins with probability at least $\log 4 \cong 60.2\%$, no matter what strategy Player B uses. Conversely, there is a strategy for Player B so that Player A wins no more than $\log 4$ of the time, no matter what strategy Player A uses.

The idea is simple, using the scale-invariance property of Benford's law established in Theorem 5.3: If Player A chooses her number X randomly according to Benford's law, then since Benford's law is scale-invariant, it follows from Theorem 4.21(i) and Example 5.2(i) that $X \cdot y$ is still Benford no matter what number y Player B chooses, so Player A will win with the probability that a

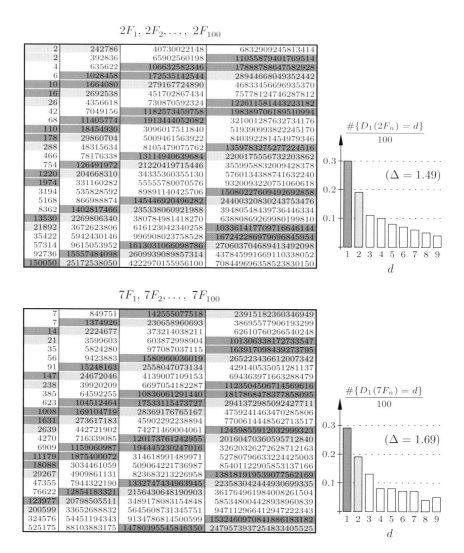

Figure 5.2: The first one hundred Fibonacci numbers have approximately scale-invariant significant digits: Multiplying $F_1, F_2, \ldots, F_{100}$ by 2 (top half) and by 7, respectively, leaves the distribution of first significant digits virtually unchanged; also see Figure 3.1.

Benford random variable has first significant digit less than 4, i.e., with probability exactly $\log 4$. Conversely, if Player B chooses his number Y according to Benford's law then, using scale-invariance again, $x \cdot Y$ is Benford, so Player A will again win with probability exactly $\log 4$. In game-theoretic parlance, this means that the game has a *value*, and the value is $\log 4$. Moreover, it can be

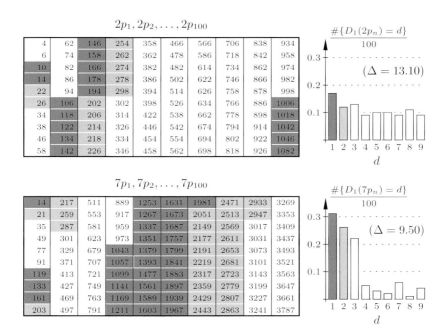

Figure 5.3: The first one hundred prime numbers do not have scale-invariant significant digits: Multiplying $p_1, p_2, \ldots, p_{100}$ by 2 (top half) and by 7, respectively, leads to very different distributions of first significant digits; also see Figure 3.2.

shown that Benford's law is the *only* optimal strategy for each player [20, pp. 48–50]. ✠

Theorem 5.3 shows that for a probability measure P on $(\mathbb{R}^+, \mathcal{B}^+)$ to have scale-invariant significant digits it is necessary (and sufficient) that P be Benford. In fact, as noted in [150], this conclusion already follows from a much weaker assumption: It is enough to require that the probability of a single significant digit remain unchanged under scaling.

THEOREM 5.8. *For every random variable X with $\mathbb{P}(X = 0) = 0$ the following statements are equivalent:*

(i) *X is Benford;*

(ii) *There exists a number $d \in \{1, 2, \ldots, 9\}$ such that*

$$\mathbb{P}(D_1(aX) = d) = \mathbb{P}(D_1(X) = d) \quad \text{for all } a > 0\,;$$

(iii) *There exists a number $d \in \{1, 2, \ldots, 9\}$ such that*

$$\mathbb{P}(D_1(aX) = d) = \log(1 + d^{-1}) \quad \text{for all } a > 0\,.$$

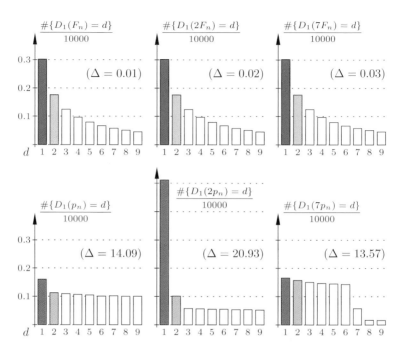

Figure 5.4: When the sample size is increased from $N = 10^2$ to $N = 10^4$, the Fibonacci numbers are even closer to scale-invariance (top half). For the primes, this is not the case; also see Figure 3.3.

PROOF. Assume first that X is Benford. By Theorem 5.3, X has scale-invariant significant digits. Thus for every $a > 0$,

$$\mathbb{P}(D_1(aX) = d) = \mathbb{P}(D_1(X) = d) = \log(1 + d^{-1}) \quad \text{for all } d = 1, 2, \ldots, 9,$$

so (i) implies (ii) and (iii). Clearly (ii) follows from (iii), so assume next that (ii) holds. For every $0 \le s < 1$ let $G(s) = \mathbb{P}(S(X) < 10^s)$. Hence G is non-decreasing and left-continuous, with $G(0) = 0$, and

$$\mathbb{P}(D_1(X) = d) = G(\log(1 + d)) - G(\log d).$$

Define a function $g : \mathbb{R} \to \mathbb{R}$ by setting $g(x) = G(\langle x \rangle) - \langle x \rangle$. Thus g is 1-periodic and Riemann-integrable, with $g(0) = 0$ and $|g(x)| \le 1$. Specifically,

$$\mathbb{P}(D_1(X) = d) = g(\log(1 + d)) - g(\log d) + \log(1 + d^{-1}),$$

and a straightforward calculation shows that, for all $a > 0$,

$$\mathbb{P}(D_1(aX) = d) = g(\log(1 + d) - \langle \log a \rangle) - g(\log d - \langle \log a \rangle) + \log(1 + d^{-1}).$$

With this, the assumption that $\mathbb{P}(D_1(aX) = d) = \mathbb{P}(D_1(X) = d)$ for all $a > 0$ simply means that the function $g(\log(1 + d) - x) - g(\log d - x)$ is constant. A

Fourier series argument similar to that in the proof of Theorem 5.5 now applies:
With $g(x) = \sum_{k \in \mathbb{Z}} c_k e^{2\pi \imath k x}$, it follows that

$$g(\log(1+d) - x) - g(\log d - x) = \sum_{k \in \mathbb{Z}} c_k \left(e^{2\pi \imath k \log(1+d)} - e^{2\pi \imath k \log d} \right) e^{-2\pi \imath k x}$$

$$= \sum_{k \in \mathbb{Z}} c_k e^{2\pi \imath k \log d} \left(e^{2\pi \imath k \log(1+d^{-1})} - 1 \right) e^{-2\pi \imath k x},$$

and since $\log(1 + d^{-1})$ is irrational for every $d \in \mathbb{N}$, necessarily $c_k = 0$ for all
$k \neq 0$, hence g is constant almost everywhere, and so $G(s) = s + c_0$ for a.a.
$s \in [0, 1)$. Since G is non-decreasing and left-continuous, $G(s) = s + c_0$ for *all*
$0 < s < 1$, and since $0 \leq G(s) \leq 1$, it follows that $c_0 = 0$, i.e., $G(s) \equiv s$, which
in turn implies (i). Thus the assertions (i), (ii), and (iii) are all equivalent. ∎

Close inspection of the above proof shows that Theorem 5.8 can be strength-
ened in various ways. On the one hand, other significant digits can be considered.
For example, the theorem and its proof remain largely unchanged if in (ii) it is
assumed that, for some $m \geq 2$ and some $d \in \{0, 1, \ldots, 9\}$,

$$\mathbb{P}(D_m(aX) = d) = \mathbb{P}(D_m(X) = d) \quad \text{for all } a > 0.$$

On the other hand, it is enough to assume in (ii) that, for some $d \in \{1, 2, \ldots, 9\}$,

$$\mathbb{P}(D_1(a_n X) = d) = \mathbb{P}(D_1(X) = d) \quad \text{for all } n \in \mathbb{N},$$

with the sequence (a_n) of positive numbers being such that $\{\langle \log a_n \rangle : n \in \mathbb{N}\}$
is *dense* in $[0, 1)$. Possible choices for such a sequence include (2^n), (n^2), and
the sequence of prime numbers. Thus, for example, X is Benford if and only if

$$\mathbb{P}(D_1(2^n X) = 1) = \mathbb{P}(D_1(X) = 1) \quad \text{for all } n \in \mathbb{N}.$$

EXAMPLE 5.9 ("Ones-scaling test" [150]). In view of Theorem 5.8, to informally test whether a sample of data comes from a Benford distribution, simply
compare the proportion of the sample that has first significant digit 1 with the
proportion after the data have been rescaled, i.e., multiplied by $a, a^2, a^3, \ldots$,
where $\log a$ is irrational, e.g., $a = 2$. In fact, it is enough to only consider re-
scalings by, for instance, a^{n^2} with $n = 1, 2, 3, \ldots$. On the other hand, note that
merely assuming

$$\mathbb{P}(D_1(2X) = d) = \mathbb{P}(D_1(X) = d) \quad \text{for all } d = 1, 2, \ldots, 9 \qquad (5.5)$$

is not sufficient to guarantee that X is Benford. For instance, (5.5) is satisfied
if X attains each of the four values $1, 2, 4, 8$ with equal probability $\frac{1}{4}$. ✠

Recall that a random variable X is *discrete* if $\mathbb{P}(X \in C) = 1$ for some
countable set $C \subset \mathbb{R}$. Clearly, no discrete random variable that is not identically
zero can have scale-invariant significant digits. On the other hand, even if C is
finite, X may be close to being scale-invariant. In this context, [23] introduces a
scale-distortion metric for n-point data sets, and shows that a sequence of real
numbers is Benford if and only if the scale-distortion of the first n entries (data
points) tends to zero as n goes to infinity.

5.2   THE BASE-INVARIANCE PROPERTY

One possible drawback to the hypothesis of scale-invariance in some tables is the special role played by the constant 1. For example, consider two physical laws, namely, Newton's *lex secunda* $F = ma$ and his universal law of gravitation $F = Gm_1m_2/r^2$. Both laws involve universal constants. In the lex secunda, the constant is usually made equal to 1 by the choice of units of measurement, and this 1 is then not recorded in most tables of universal constants. On the other hand, the constant G in the universal law of gravitation is typically recorded as a fundamental constant. If a "complete" list of universal physical constants also included the 1s, it would seem plausible that this special constant might occur with strictly positive frequency. But that would clearly violate scale-invariance, since then the constant 2, and in fact every other constant as well, would occur with this same positive probability, which is impossible.

Instead, suppose it is assumed that any reasonable universal significant-digit law should have *base-invariant significant digits*, that is, the law should be equally valid when rewritten in terms of bases other than 10. In fact, "every argument that applies to 10 applies to [other bases] b *mutatis mutandis*" [127, p. 536]. As will be seen shortly, a hypothesis of base-invariant significant digits characterizes mixtures of Benford's law and a Dirac probability measure concentrated at the special constant 1, which may occur with positive probability.

Just as the only scale-invariant real-valued random variable is one that is 0 with probability one, the only positive random variable X that is *base-invariant*, i.e., $X = 10^Y$ for some random variable Y such that $Y, 2Y, 3Y, \ldots$ all have the same distribution, is the random variable that almost surely equals 1, that is, $\mathbb{P}(X = 1) = 1$. This follows from the fact that nY has the same distribution for all $n \in \mathbb{N}$, and hence $\mathbb{P}(Y = 0) = 1$, as seen earlier.

On the other hand, a positive random variable (or sequence, function, distribution) can have *base-invariant significant digits*. The idea behind base-invariance of significant digits is simply this: The base-10 significand event A corresponds to the base-100 event $A^{1/2}$, since the new base $b = 100$ is the square of the original base $b = 10$. As a concrete example, denote by A the set of positive real numbers with first significant digit 1, i.e.,

$$A = \{x > 0 : D_1(x) = 1\} = \{x > 0 : S(x) \in [1, 2)\}.$$

It is easy to see that $A^{1/2}$ is the set

$$A^{1/2} = \left\{x > 0 : S(x) \in \left[1, \sqrt{2}\right) \cup \left[\sqrt{10}, \sqrt{20}\right)\right\}.$$

Now consider the base-100 significand function S_{100}, i.e., for any $x \neq 0$, $S_{100}(x)$ is the unique number in $[1, 100)$ such that $|x| = 100^k S_{100}(x)$ for some (necessarily unique) integer k. (To emphasize that the usual significand function S is taken relative to base 10, it will be denoted by S_{10} throughout this section.) Clearly,

$$A = \{x > 0 : S_{100}(x) \in [1, 2) \cup [10, 20)\}.$$

Hence, letting $s = \log 2$,

$$\{x > 0 : S_b(x) \in [1, b^{s/2}) \cup [b^{1/2}, b^{(1+s)/2})\} = \begin{cases} A^{1/2} & \text{if } b = 10, \\ A & \text{if } b = 100. \end{cases}$$

Thus, if a distribution P on the significand σ-algebra $\mathcal{S}$ has base-invariant significant digits, then $P(A)$ and $P(A^{1/2})$ should be the same, and similarly for other integral roots (corresponding to other integral powers of the original base $b = 10$). Thus $P(A) = P(A^{1/n})$ should hold for all $n \in \mathbb{N}$. (Recall from Lemma 2.15(iii) that $A^{1/n} \in \mathcal{S}$ for all $A \in \mathcal{S}$ and $n \in \mathbb{N}$, so those probabilities are well-defined.) This motivates the following definition.

DEFINITION 5.10. Let $\mathcal{A} \supset \mathcal{S}$ be a σ-algebra on $\mathbb{R}^+$. A probability measure P on $(\mathbb{R}^+, \mathcal{A})$ has *base-invariant significant digits* if $P(A) = P(A^{1/n})$ for all $A \in \mathcal{S}$ and all $n \in \mathbb{N}$.

EXAMPLE 5.11. **(i)** Recall that δ_a denotes the Dirac measure concentrated at the point a, that is, $\delta_a(A) = 1$ if $a \in A$, and $\delta_a(A) = 0$ if $a \notin A$. The probability measure δ_1 clearly has base-invariant significant digits since $1 \in A$ if and only if $1 \in A^{1/n}$. Similarly, δ_{10^k} has base-invariant significant digits for every $k \in \mathbb{Z}$. On the other hand, δ_2 does not have base-invariant significant digits since, with $A = \{x > 0 : S_{10}(x) \in [1, 3)\}$, $\delta_2(A) = 1$ but $\delta_2(A^{1/2}) = 0$.

(ii) It is easy to see that the Benford distribution $\mathbb{B}$ has base-invariant significant digits. Indeed, for any $0 \leq s < 1$, let

$$A = \{x > 0 : S_{10}(x) \in [1, 10^s]\} = \bigcup_{k \in \mathbb{Z}} 10^k [1, 10^s] \in \mathcal{S}.$$

Then, as seen in the proof of Lemma 2.15(iii),

$$A^{1/n} = \bigcup_{k \in \mathbb{Z}} 10^k \bigcup_{j=0}^{n-1} [10^{j/n}, 10^{(j+s)/n}]$$

and therefore

$$\mathbb{B}(A^{1/n}) = \sum_{j=0}^{n-1} \left(\log 10^{(j+s)/n} - \log 10^{j/n} \right) = \sum_{j=0}^{n-1} \left(\frac{j+s}{n} - \frac{j}{n} \right)$$
$$= s = \mathbb{B}(A).$$

(iii) The uniform distribution $\lambda_{0,1}$ on $(0, 1)$ does not have base-invariant significant digits. For instance, again taking $A = \{x > 0 : D_1(x) = 1\}$ yields

$$\lambda_{0,1}(A^{1/2}) = \sum_{n \in \mathbb{N}} 10^{-n} \left(\sqrt{2} - 1 + \sqrt{20} - \sqrt{10} \right) = \tfrac{1}{9} + \tfrac{1}{9} \left(\sqrt{5} - 1 \right) \left(2 - \sqrt{2} \right)$$
$$> \tfrac{1}{9} = \lambda_{0,1}(A).$$

(iv) The probability measure $\frac{1}{2}\delta_1 + \frac{1}{2}\mathbb{B}$ has base-invariant significant digits since both δ_1 and $\mathbb{B}$ do. ✠

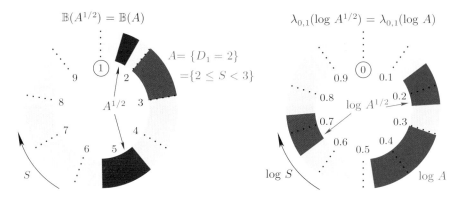

Figure 5.5: Visualizing the base-invariant significant digits of Benford's law; see Example 5.11(ii).

EXAMPLE 5.12. Completely analogously to the case of scale-invariance, it is possible to introduce a notion of a sequence or function having *base-invariant significant digits* and to formulate an analogue of Theorem 5.5 in the context of Theorem 5.13 below. With this, the sequence (F_n) has base-invariant significant digits, whereas the sequence (p_n) does not. This is illustrated empirically in Figures 5.6 to 5.8. ✠

The next theorem is the main result for base-invariant significant digits. It shows that convex combinations as in Example 5.11(iv) are the only probability distributions with base-invariant significant digits. To put the argument in perspective, recall that the proof of the scale-invariance theorem (Theorem 5.3) ultimately depended on Theorem 4.21, which in turn was proved analytically using Fourier analysis. The situation here is similar: An analytical result (Lemma 5.15 below) identifies all probability measures on $\big([0,1),\mathcal{B}[0,1)\big)$ that are invariant under *every* map $T_n(s) = \langle ns \rangle$ on $[0,1)$. Once this tool is available, it will be straightforward to prove the following analogue of Theorem 5.3.

THEOREM 5.13 (Base-invariance characterization [73]). *A probability measure P on $(\mathbb{R}^+, \mathcal{A})$ with $\mathcal{A} \supset \mathcal{S}$ has base-invariant significant digits if and only if, for some $q \in [0,1]$,*

$$P(A) = q\delta_1(A) + (1-q)\mathbb{B}(A) \quad \text{for every } A \in \mathcal{S}. \tag{5.6}$$

COROLLARY 5.14. *A continuous probability measure P on $\mathbb{R}^+$ has base-invariant significant digits if and only if $P(A) = \mathbb{B}(A)$ for all $A \in \mathcal{S}$, i.e., if and only if P is Benford.*

Recall that $\lambda_{0,1}$ is Lebesgue measure on $\big([0,1),\mathcal{B}[0,1)\big)$, and $T_n(s) = \langle ns \rangle$ for every $n \in \mathbb{N}$ and $s \in [0,1)$. Generally, if $T : [0,1) \to \mathbb{R}$ is measurable and

$S(F_1^2), S(F_2^2), \ldots, S(F_{100}^2)$

1.000	1.473	4.147	1.167
1.000	3.858	1.085	3.055
4.000	1.010	2.842	8.000
9.000	2.644	7.442	2.094
2.500	6.922	1.948	5.483
6.400	1.812	5.100	1.435
1.690	4.745	1.335	3.758
4.410	1.242	3.496	9.839
1.156	3.252	9.153	2.576
3.025	8.514	2.396	6.744
7.921	2.229	6.273	1.765
2.073	5.836	1.642	4.622
5.428	1.527	4.300	1.210
1.421	4.000	1.125	3.168
3.721	1.047	2.947	8.294
9.741	2.741	7.716	2.171
2.550	7.177	2.020	5.685
6.677	1.879	5.288	1.488
1.748	4.919	1.384	3.896
4.576	1.288	3.624	1.020
1.198	3.372	9.490	2.670
3.136	8.828	2.484	6.992
8.212	2.311	6.504	1.830
2.149	6.050	1.702	4.792
5.628	1.584	4.458	1.254

$S(F_1^7), S(F_2^7), \ldots, S(F_{100}^7)$

1.000	3.884	1.452	5.432
1.000	1.127	4.217	1.577
1.280	3.274	1.224	4.579
2.187	9.508	3.555	1.329
7.812	2.760	1.032	3.860
2.097	8.015	2.997	1.120
6.274	2.327	8.703	3.254
1.801	6.756	2.526	9.449
5.252	1.961	7.336	2.743
1.522	5.696	2.130	7.966
4.423	1.653	6.184	2.312
1.283	4.801	1.795	6.715
3.728	1.394	5.213	1.949
1.082	4.047	1.513	5.661
3.142	1.175	4.395	1.643
9.124	3.412	1.276	4.772
2.649	9.907	3.705	1.385
7.692	2.876	1.075	4.023
2.233	8.352	3.123	1.168
6.484	2.424	9.068	3.391
1.882	7.040	2.633	9.846
5.466	2.044	7.644	2.858
1.587	5.935	2.219	8.300
4.608	1.723	6.444	2.410
1.337	5.003	1.871	6.997

Figure 5.6: Illustrating the (approximate) base-invariance of the first one hundred Fibonacci numbers; see Example 5.12.

$T\big([0,1)\big) \subset [0,1)$, a probability measure P on $\big([0,1), \mathcal{B}[0,1)\big)$ is said to be T-invariant (or, T is P-preserving) if $P \circ T^{-1} = P$. Which probability measures are T_n-invariant for all $n \in \mathbb{N}$? A complete answer to this question is provided by the following lemma.

LEMMA 5.15 ([73]). *A probability measure P on $\big([0,1), \mathcal{B}[0,1)\big)$ is T_n-invariant for all $n \in \mathbb{N}$ if and only if $P = q\delta_0 + (1-q)\lambda_{0,1}$ for some $q \in [0,1]$.*

PROOF. Let $\big(\widehat{P}(k)\big)_{k \in \mathbb{Z}}$ denote the Fourier coefficients of P, and observe that

$$\widehat{P \circ T_n^{-1}}(k) = \widehat{P}(nk) \quad \text{for all } k \in \mathbb{Z}, n \in \mathbb{N}.$$

$$S(p_1^2), S(p_2^2), \ldots, S(p_{100}^2)$$

4.000	1.020	5.428	1.466
9.000	1.060	5.712	1.513
2.500	1.144	5.808	1.576
4.900	1.188	6.300	1.608
1.210	1.276	6.604	1.672
1.690	1.612	6.916	1.755
2.890	1.716	7.236	1.772
3.610	1.876	7.344	1.857
5.290	1.932	7.672	1.874
8.410	2.220	7.896	1.927
9.610	2.280	8.008	1.962
1.369	2.464	8.584	2.016
1.681	2.656	9.424	2.088
1.849	2.788	9.672	2.125
2.209	2.992	9.796	2.143
2.809	3.204	1.004	2.180
3.481	3.276	1.095	2.294
3.721	3.648	1.135	2.371
4.489	3.724	1.204	2.410
5.041	3.880	1.218	2.490
5.329	3.960	1.246	2.530
6.241	4.452	1.288	2.590
6.889	4.972	1.346	2.714
7.921	5.152	1.391	2.735
9.409	5.244	1.436	2.926

$$S(p_1^7), S(p_2^7), \ldots, S(p_{100}^7)$$

1.280	1.072	3.728	1.208
2.187	1.229	4.454	1.347
7.812	1.605	4.721	1.554
8.235	1.828	6.276	1.667
1.948	2.352	7.405	1.914
6.274	5.328	8.703	2.267
4.103	6.620	1.019	2.344
8.938	9.058	1.073	2.762
3.404	1.002	1.251	2.853
1.724	1.630	1.383	3.142
2.751	1.789	1.453	3.348
9.493	2.351	1.853	3.678
1.947	3.057	2.570	4.163
2.718	3.622	2.813	4.424
5.066	4.637	2.943	4.561
1.174	5.888	3.216	4.844
2.488	6.364	4.353	5.785
3.142	9.273	4.936	6.496
6.060	9.974	6.057	6.879
9.095	1.151	6.306	7.703
1.104	1.235	6.830	8.146
1.920	1.861	7.685	8.851
2.713	2.742	8.967	1.041
4.423	3.105	1.004	1.070
8.079	3.302	1.123	1.356

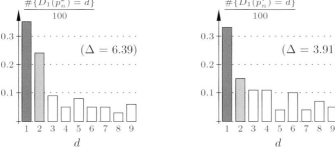

Figure 5.7: Illustrating the lack of base-invariance for the first one hundred prime numbers; see Example 5.12.

Assume first that $P = q\delta_0 + (1-q)\lambda_{0,1}$ for some $q \in [0,1]$. From $\widehat{\delta_0}(k) \equiv 1$ and $\widehat{\lambda_{0,1}}(k) = 0$ for all $k \neq 0$, it follows that

$$\widehat{P}(k) = \begin{cases} 1 & \text{if } k = 0\,, \\ q & \text{if } k \neq 0\,. \end{cases}$$

For every $n \in \mathbb{N}$ and $k \in \mathbb{Z}\setminus\{0\}$, therefore, $\widehat{P \circ T_n^{-1}}(k) = q$, and clearly $\widehat{P \circ T_n^{-1}}(0) = 1$. Thus $\widehat{P \circ T_n^{-1}}(k) = \widehat{P}(k)$ for all $k \in \mathbb{Z}$, and Lemma 4.20(iii) shows that $P \circ T_n^{-1} = P$ for all $n \in \mathbb{N}$.

Conversely, assume P is T_n-invariant for all $n \in \mathbb{N}$. In this case, for every $n \in \mathbb{N}$, $\widehat{P}(n) = \widehat{P \circ T_n^{-1}}(1) = \widehat{P}(1)$, and also $\widehat{P}(-n) = \widehat{P \circ T_n^{-1}}(-1) = \widehat{P}(-1)$.

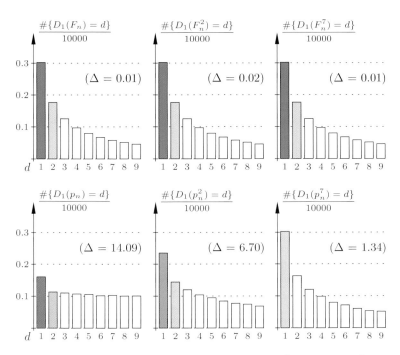

Figure 5.8: Increasing the sample size from $N = 10^2$ to $N = 10^4$ makes the sample of Fibonacci numbers' leading digits even closer to being base-invariant (top half). As in the case of scale-invariance, this is not at all true for the primes; see Figures 3.3 and 5.4. Note, however, that (p_n^7) conforms to the first-digit law considerably better than (p_n); cf. Theorem 8.8 below.

Since $\widehat{P}(-k) = \overline{\widehat{P}(k)}$, there exists $q \in \mathbb{C}$ such that

$$\widehat{P}(k) = \begin{cases} q & \text{if } k > 0\,, \\ 1 & \text{if } k = 0\,, \\ \overline{q} & \text{if } k < 0\,. \end{cases}$$

Also, observe that, for every $x \in \mathbb{R}$,

$$\lim_{n\to\infty} \frac{1}{n} \sum_{j=1}^{n} e^{2\pi i j x} = \begin{cases} 1 & \text{if } x \in \mathbb{Z}\,, \\ 0 & \text{if } x \notin \mathbb{Z}\,. \end{cases}$$

Using this and the Dominated Convergence Theorem, it follows from

$$P(\{0\}) = \int_0^1 \lim_{n\to\infty} \frac{1}{n} \sum_{j=1}^{n} e^{2\pi i j s} \, dP(s) = \lim_{n\to\infty} \frac{1}{n} \sum_{j=1}^{n} \widehat{P}(j) = q$$

that q is real, and in fact $q \in [0, 1]$. Hence the Fourier coefficients of P are exactly the same as those of the probability measure $q\delta_0 + (1-q)\lambda_{0,1}$. By uniqueness of Fourier coefficients (Lemma 4.20(iii)), therefore, $P = q\delta_0 + (1-q)\lambda_{0,1}$. ∎

Note that P is T_{mn}-invariant if it is both T_m- and T_n-invariant. Thus, in Lemma 5.15 it is enough to require only that P be T_n-invariant whenever n is a prime number. It is natural to ask how small the set $\mathbb{M}$ of natural numbers n can be so that T_n-invariance for all $n \in \mathbb{M}$ suffices in the "only if" part of Lemma 5.15. By the observation just made, it can be assumed that $\mathbb{M}$ is closed under multiplication, and hence is a (multiplicative) semi-group. If $\mathbb{M}$ is *lacunary*, i.e., $\mathbb{M} \subset \{p^m : m \in \mathbb{N}\}$ for some $p \in \mathbb{N}$, then probability measures P satisfying $P \circ T_n^{-1} = P$ for all $n \in \mathbb{M}$ exist in abundance, so $\mathbb{M}$ is too small to imply the "only if" part of Lemma 5.15. If, on the other hand, $\mathbb{M}$ is not lacunary, then in general it is not known whether an appropriate analogue of Lemma 5.15 may hold. For example, if $\mathbb{M} = \{2^{m_1}3^{m_2} : m_1, m_2 \in \mathbb{N}_0\}$ then the probability measure $P = \frac{1}{4}\sum_{j=1}^4 \delta_{j/5}$ is T_n-invariant for every $n \in \mathbb{M}$, but it is a famous open question of Furstenberg [54] whether any *continuous* probability measure with this property exists — except, of course, for $P = \lambda_{0,1}$. In the context of Benford's law, this famous question can equivalently be stated as follows: If X is a continuous random variable for which X, X^2, and X^3 all have the same distribution of significant digits, does it follow that X is Benford?

PROOF OF THEOREM 5.13. As in the proof of Theorem 5.3, fix any probability measure P on $(\mathbb{R}^+, \mathcal{A})$, denote by P_0 its restriction to $(\mathbb{R}^+, \mathcal{S})$, and consider the probability measure $Q = P_0 \circ \ell^{-1}$. Observe that P_0 has base-invariant significant digits if and only if Q is T_n-invariant for all $n \in \mathbb{N}$. Indeed, with $0 \le s < 1$ and $A = \{x > 0 : S_{10}(x) \le 10^s\}$,

$$Q \circ T_n^{-1}([0, s]) = Q\left(\bigcup_{j=0}^{n-1}\left[\frac{j}{n}, \frac{j+s}{n}\right]\right)$$
$$= P_0\left(\bigcup_{k \in \mathbb{Z}} 10^k \bigcup_{j=0}^{n-1}\left[10^{j/n}, 10^{(j+s)/n}\right]\right) = P_0\left(A^{1/n}\right),$$

and hence $Q \circ T_n^{-1} = Q$ for all n precisely if P_0 has base-invariant significant digits. In this case, by Lemma 5.15, $Q = q\delta_0 + (1-q)\lambda_{0,1}$ for some $q \in [0, 1]$, which in turn implies that $P_0(A) = q\delta_1(A) + (1-q)\mathbb{B}(A)$ for every $A \in \mathcal{S}$. ∎

5.3   THE SUM-INVARIANCE PROPERTY

No finite data set can obey Benford's law exactly, since the Benford probabilities of sets with m given significant digits become arbitrarily small as m goes to infinity, and no discrete probability measure with finitely many atoms can take arbitrarily small positive values. But, as first observed by Nigrini [112], if a table of real data approximately follows Benford's law, then the sum of the significands of all entries in the table with first significant digit 1 is very close to the sum of the significands of all entries with first significant digit 2, and to the sum of the significands of entries with the other possible first significant digits as well. This clearly implies that the table must contain more entries starting with 1 than with 2, more entries starting with 2 than with 3, and so forth. Similarly, the sums of significands of entries with $(D_1, D_2, \ldots, D_m) = (d_1, d_2, \ldots, d_m)$ are approximately equal for all tuples $(d_1, d_2, \ldots, d_m)$ of a fixed length m. In fact, even the sum-invariance of first or first and second digits yields a distribution close to Benford's law; see Figures 5.9 and 5.11. Nigrini conjectured, and partially proved, that this *sum-invariance* property also characterizes Benford's law. Note that it is the *significands* of the data, rather than the data themselves, that are added. Simply summing up the raw data will not lead to any meaningful conclusion, since the resulting sums may be dominated by a few very large numbers. It is only through considering significands that the magnitude of the individual numbers becomes irrelevant.

d	1	2	3	4	5	6	7	8	9
$N_d = \#\{x_n = d\}$	2520	1260	840	630	504	420	360	315	280
$100\, N_d/N$	35.34	17.67	11.78	8.83	7.06	5.89	5.04	4.41	3.92

$$N = \sum_d N_d = 7129 \qquad\qquad (\Delta = 5.24)$$

Figure 5.9: The smallest sample $x_1, x_2, \ldots, x_N$ from $\{1, 2, \ldots, 9\}$ that has *exact* sum-invariance for the first significant digit consists of $N = 7129$ elements; see Definition 5.16 below.

To motivate a precise definition of sum-invariance, let $A = \{x_1, x_2, \ldots, x_N\}$ be a finite data set (unordered tuple) of N different positive numbers, as in Nigrini's framework. Then the sum of the significands of all entries in A with first significant digit d is simply

$$\sum_{j=1}^{N} S(x_j) \mathbb{1}_{[d, d+1)}\big(D_1(x_j)\big) = \sum_{j=1}^{N} S(x_j) \mathbb{1}_{[d, d+1)}\big(S(x_j)\big).$$

(Recall that $\mathbb{1}_C$ denotes the indicator function of the set C, that is, $\mathbb{1}_C(x) = 1$ if $x \in C$, and $\mathbb{1}_C(x) = 0$ otherwise.) Thus if X is a random variable uniformly distributed on A, the expected value of the sum of the significands of the entries with first significant digit d is simply

$$\frac{1}{N} \sum_{j=1}^{N} S(x_j) \mathbb{1}_{[d, d+1)}\big(S(x_j)\big) = \mathbb{E}\big[S(X) \mathbb{1}_{[d, d+1)}\big(S(X)\big)\big],$$

so X has *sum-invariant first significant digits* if $\mathbb{E}\big[S(X) \mathbb{1}_{[d, d+1)}\big(S(X)\big)\big]$ is independent of $d \in \{1, 2, \ldots, 9\}$; see Figure 5.9 for such a finite data set A. Generalizing this to second and higher significant digits leads to the following natural notion of sum-invariance of significant digits for random variables. For convenience, for all $m \in \mathbb{N}$, all $d_1 \in \{1, 2, \ldots, 9\}$, and all $d_j \in \{0, 1, \ldots, 9\}$, $j \geq 2$, let

$$C(d_1, d_2, \ldots, d_m) = \big\{t \in [1, 10) : D_j(t) = d_j \text{ for } j = 1, 2, \ldots, m\big\}.$$

Note that $C(d_1, d_2, \ldots, d_m) \subset [1, 10)$ is an interval of length 10^{1-m}; in fact, $10^{m-1}C(d_1, d_2, \ldots, d_m) = [N, N+1)$ with $N = \sum_{j=1}^m 10^{m-j}d_j \in \mathbb{N}$. Also, clearly $S(x) \in C\big(D_1(x), D_2(x), \ldots, D_m(x)\big)$ for all $x \neq 0$ and $m \in \mathbb{N}$. For example,

$$C(4, 0, 3) = [4.03, 4.04) = 10^{-2}[403, 404).$$

DEFINITION 5.16. A random variable X has *sum-invariant significant digits* if, for every $m \in \mathbb{N}$, the value of $\mathbb{E}\big[S(X)\mathbb{1}_{C(d_1, d_2, \ldots, d_m)}\big(S(X)\big)\big]$ is independent of $d_1, d_2, \ldots, d_m$.

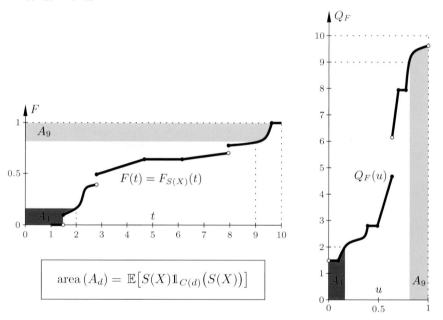

Figure 5.10: The area of the red region A_1 left of the distribution function F (and below the quantile function Q_F, as defined in (5.7) below) is the expected sum of the (significands of the) numbers with first significant digit 1; that of the green region A_9 is the expected sum of the numbers with first significant digit 9. Since these areas are unequal, any random variable X with $F = F_{S(X)}$ does not have sum-invariant significant digits; see Definition 5.16, Theorem 5.18, and Lemma 5.19.

EXAMPLE 5.17. **(i)** If X is uniformly distributed on $(0, 1)$, then X does not have sum-invariant significant digits. This follows from Theorem 5.18 below but can also be seen by a simple direct calculation: For every $d_1 \in \{1, 2, \ldots, 9\}$,

$$\mathbb{E}\big[S(X)\mathbb{1}_{C(d_1)}\big(S(X)\big)\big] = \sum_{n=1}^{\infty} 10^n \int_{10^{-n}d_1}^{10^{-n}(d_1+1)} x\, dx = \tfrac{1}{9}d_1 + \tfrac{1}{18},$$

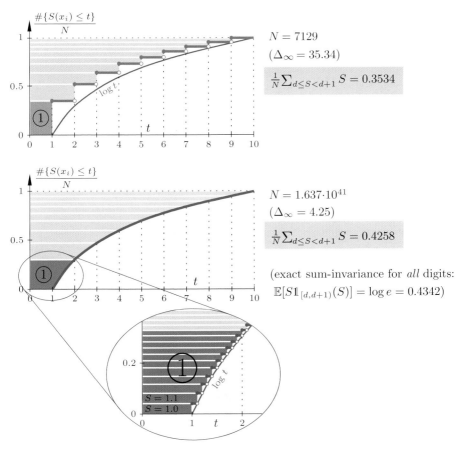

Figure 5.11: Empirical distribution functions $\#\{S(x_i) \leq t\}/N$ for two samples $x_1, x_2, \ldots, x_N$ from $\{1, 2, \ldots, 9\}$ and $\{10, 11, \ldots, 99\}$, respectively: The first sample is the smallest possible that exhibits exact sum-invariance for the *first* significant digit (top; see also Figure 5.9), and the second sample is the smallest possible that exhibits exact sum-invariance for the *first two* significant digits.

which clearly depends on d_1.

(ii) Similarly, if $\mathbb{P}(X = 1) = 1$ then X does not have sum-invariant significant digits, since

$$\mathbb{E}\big[S(X)\mathbb{1}_{C(d_1)}\big(S(X)\big)\big] = \begin{cases} 1 & \text{if } d_1 = 1, \\ 0 & \text{if } d_1 \geq 2. \end{cases}$$

(iii) If X is Benford then for every $m \in \mathbb{N}$, all $d_1 \in \{1, 2, \ldots, 9\}$, and all

$d_j \in \{0, 1, \ldots, 9\}$, $j \geq 2$,

$$\mathbb{E}\big[S(X)\mathbb{1}_{C(d_1,d_2,\ldots,d_m)}\big(S(X)\big)\big] = \int_{d_1+10^{-1}d_2+\ldots+10^{1-m}d_m}^{d_1+10^{-1}d_2+\ldots+10^{1-m}(d_m+1)} x \cdot x^{-1} \log e \, \mathrm{d}x$$
$$= 10^{1-m} \log e \,.$$

Thus X has sum-invariant significant digits. Note, however, that even in this example the higher moments generally depend on $d_1, d_2, \ldots, d_m$, since, e.g.,

$$\mathbb{E}\big[S(X)^2\mathbb{1}_{C(d_1)}\big(S(X)\big)\big] = \big(d_1 + \tfrac{1}{2}\big) \log e \quad \text{for all } d_1 \in \{1, 2, \ldots, 9\}\,.$$

This example shows that it would be too restrictive to require in Definition 5.16 that the *distribution* of the random variable $S(X)\mathbb{1}_{C(d_1,d_2,\ldots,d_m)}\big(S(X)\big)$, rather than its expectation, be independent of $d_1, d_2, \ldots, d_m$. ✠

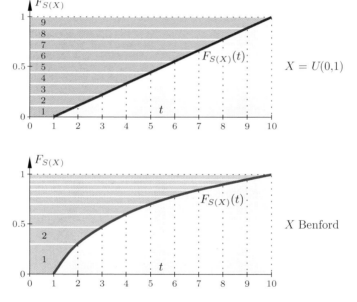

Figure 5.12: For a $U(0, 1)$ random variable X, the nine red trapezoids to the left of the distribution function of $S(X)$ do not have equal area (top), and hence X does not even have sum-invariant first significant digits. On the other hand, if X is Benford, then the nine blue regions all have equal area, showing that X has sum-invariant first significant digits; see Example 5.17.

The definitions of sum-invariance of significant digits for sequences, functions, and distributions are similar; for example, it follows from Theorem 5.18 below that the sequences (2^n) and (F_n) both have sum-invariant significant digits. Clearly, (10^n) does not have sum-invariant significant digits since all first digits are 1, i.e., for all N,

$$\frac{\sum_{n=1}^{N} S(10^n)\mathbb{1}_{C(d_1)}\big(S(10^n)\big)}{N} = \begin{cases} 1 & \text{if } d_1 = 1\,, \\ 0 & \text{if } d_1 \geq 2. \end{cases}$$

Not surprisingly, the sequence (p_n) of prime numbers does not have sum-invariant significant digits either; see Figure 5.13.

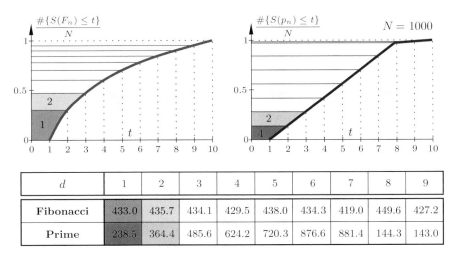

d	1	2	3	4	5	6	7	8	9
Fibonacci	433.0	435.7	434.1	429.5	438.0	434.3	419.0	449.6	427.2
Prime	238.5	364.4	485.6	624.2	720.3	876.6	881.4	144.3	143.0

Figure 5.13: Empirical distribution functions of the first $N = 10^3$ Fibonacci (top left) and prime numbers (top right). The value of $\sum_{D_1=d} S$ (bottom table) equals N times the area of the corresponding region to the left of the distribution function; it does not vary much (with d) for the Fibonaccis, but varies dramatically for the primes. Note that in the case of *exact* sum-invariance $N\mathbb{E}[S\mathbb{1}_{C(d)}(S)] = 434.2$ for all d.

By Example 5.17(iii), every Benford random variable has sum-invariant significant digits. As hinted at earlier, the converse is also true, i.e., sum-invariant significant digits characterize Benford's law.

THEOREM 5.18 (Sum-invariance characterization [4]). *A random variable X with $\mathbb{P}(X = 0) = 0$ has sum-invariant significant digits if and only if it is Benford.*

In direct analogy to the proofs of the scale- and base-invariance characterizations of Benford's law (Theorems 5.3 and 5.13), the proof of this theorem also relies on basic analytical facts. First, recall that the (*lower*) *quantile function* $Q_F : (0,1) \to \mathbb{R}$ of a distribution function F is given by

$$Q_F(s) = \inf\{x \in \mathbb{R} : F(x) \geq s\}\,. \tag{5.7}$$

So, for example, $Q_F\left(\frac{1}{2}\right)$ is the (smallest) median of F. If X is a random variable, for convenience write Q_{F_X} as Q_X. It is well known (and easy to check) that Q_F is non-decreasing and left-continuous, and that $Q_F \circ F(x) \leq x$ for all x with $0 < F(x) < 1$, with equality holding if and only if $F(x - \varepsilon) < F(x)$ for all $\varepsilon > 0$.

The first tool, Lemma 5.19 below, is an easy generalization of the known fact (e.g., [27, eq. (21.9)]; see Figure 5.14) that

$$\mathbb{E}[X] = \int_0^{+\infty} \big(1 - F_X(x)\big)\,\mathrm{d}x = \int_0^1 Q_X(s)\,\mathrm{d}s \tag{5.8}$$

for every nonnegative random variable X. Since no reference for this generalization is known to the authors, a proof is provided.

LEMMA 5.19. *Let X be a nonnegative random variable, with distribution function F and quantile function Q. Then*

$$\mathbb{E}\big[X\mathbb{1}_{(a,b]}(X)\big] = \int_{F(a)}^{F(b)} Q(s)\,\mathrm{d}s \quad \text{for all } 0 \leq a < b < +\infty. \tag{5.9}$$

PROOF. Assume first that the nonnegative random variable X is *simple*, i.e., there exist a positive integer N, real numbers $0 \leq x_1 < x_2 < \ldots < x_N$, and positive real numbers $p_1, p_2, \ldots, p_N$ with $\sum_{j=1}^N p_j = 1$ such that $\mathbb{P}(X = x_j) = p_j$ for all $j \in \{1, 2, \ldots, N\}$. Let $x_0 = 0$. It follows directly from the definitions of F and Q that $Q(s) = x_j$ for all $F(x_{j-1}) < s < F(x_j)$ and $j \in \{1, 2, \ldots, N\}$, and therefore

$$\int_{F(x_{j-1})}^{F(x_j)} Q(s)\,\mathrm{d}s = x_j\big(F(x_j) - F(x_{j-1})\big) = x_j p_j. \tag{5.10}$$

Fix $0 \leq a < b < +\infty$ and assume without loss of generality that $a < x_j \leq b$ for at least one $j \in \{1, 2, \ldots, N\}$; otherwise both sides of (5.9) are zero. Let $m = \min\{j : a < x_j \leq b\}$ and $M = \max\{j : a < x_j \leq b\}$. Then, by (5.10),

$$\mathbb{E}\big[X\mathbb{1}_{(a,b]}(X)\big] = \sum_{j=m}^{M} x_j p_j = \sum_{j=m}^{M} \int_{F(x_{j-1})}^{F(x_j)} Q(s)\,\mathrm{d}s = \int_{F(x_{m-1})}^{F(x_M)} Q(s)\,\mathrm{d}s$$

$$= \int_{F(a)}^{F(b)} Q(s)\,\mathrm{d}s,$$

which establishes (5.9) when X is simple; see also Figure 5.14.

Given an *arbitrary* nonnegative random variable X, there exists a sequence (X_n) of simple nonnegative random variables, with distribution and quantile functions F_n and Q_n, respectively, such that $X_n(\omega) \nearrow X(\omega)$ for all $\omega \in \Omega$, where the (abstract) probability space $(\Omega, \mathcal{A}, \mathbb{P})$ is the common domain of (X_n)

and X; see [27, eq. (20.1)]. Then

$$\mathbb{E}\big[X\mathbb{1}_{(a,b]}(X)\big] = \lim_{n\to\infty} \mathbb{E}\big[X_n\mathbb{1}_{(a,b]}(X_n)\big] = \lim_{n\to\infty} \int_{F_n(a)}^{F_n(b)} Q_n(s)\,\mathrm{d}s$$

$$= \int_{F(a)}^{F(b)} Q(s)\,\mathrm{d}s\,,$$

where the first equality is a consequence of the Monotone Convergence Theorem, the second has been established above since all X_n are simple, and the third follows from the fact that $Q_n(s) \nearrow Q(s)$ and $F_n(x) \searrow F(x)$ for all $s \in (0,1)$ and all $x \in \mathbb{R}$, respectively. (The latter fact is itself a consequence of the Monotone Convergence Theorem.) ∎

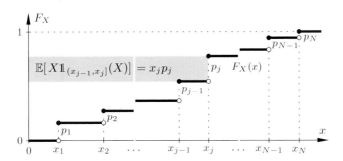

Figure 5.14: An illustration of Lemma 5.19 for a *simple* random variable X.

The second tool (Lemma 5.20 below) provides a useful link between integrals of quantile functions and logarithmic distribution functions.

LEMMA 5.20. *Let X be a random variable, with distribution function F and quantile function Q. Assume that $\mathbb{P}(a < X < b) = 1$ for some $a, b > 0$. Then the following are equivalent:*

(i) *There exists $C > 0$ with $\displaystyle\int_{F(x_1)}^{F(x_2)} Q(s)\,\mathrm{d}s = C(x_2 - x_1)$ for all $x_2 > x_1$, $x_1, x_2 \in (a,b)$;*

(ii) $F(x) = \dfrac{\log x - \log a}{\log b - \log a}$ *for all $x \in (a,b)$.*

Moreover, (i) *and* (ii) *both imply that $C = (b-a)^{-1}\mathbb{E}[X] = (\ln b - \ln a)^{-1}$.*

PROOF. Assume first that (i) holds. Then F is strictly increasing on (a,b), and hence $Q \circ F(x_1) = x_1$ for all $x_1 \in (a,b)$. Also, F is differentiable almost everywhere on (a,b); see [37, Thm. 1.3.1]. Moreover,

$$C(x_2 - x_1) \geq Q \circ F(x_1)\big(F(x_2) - F(x_1)\big) \geq a\big(F(x_2) - F(x_1)\big)$$

shows that F is absolutely continuous (in fact, Lipschitz). Since F is continuous and strictly increasing, so is Q. Pick any x_1 for which $F'(x_1)$ exists, and observe that by assumption (i),

$$
\begin{aligned}
C &= \lim_{x_2 \searrow x_1} \frac{1}{x_2 - x_1} \int_{F(x_1)}^{F(x_2)} Q(s)\,\mathrm{d}s \\
&= \lim_{x_2 \searrow x_1} \frac{F(x_2) - F(x_1)}{x_2 - x_1} \cdot \frac{1}{F(x_2) - F(x_1)} \int_{F(x_1)}^{F(x_2)} Q(s)\,\mathrm{d}s \\
&= F'(x_1) \cdot Q \circ F(x_1) = F'(x_1)\, x_1\,.
\end{aligned}
$$

Since F is absolutely continuous with $F(a) = 0$, for every $x_1 \in (a, b)$,

$$
F(x_1) = \int_a^{x_1} F'(x)\,\mathrm{d}x = \int_a^{x_1} x^{-1} C\,\mathrm{d}x = C(\ln x_1 - \ln a)\,,
$$

and since $\lim_{x \nearrow b} F(x) = 1$, $C = (\ln b - \ln a)^{-1}$. Thus $F(x) = \dfrac{\log x - \log a}{\log b - \log a}$, establishing (ii).

Conversely, (ii) implies (i) via a simple calculation, with $Q(s) = a^{1-s} b^s$, and (5.8) yields $\mathbb{E}[X] = \int_0^1 a^{1-s} b^s\,\mathrm{d}s = (b - a)(\ln b - \ln a)^{-1} = C(b - a)$. ∎

These two lemmas provide a simple proof of the sum-invariance characterization of Benford's law.

PROOF OF THEOREM 5.18. Since the definitions of a random variable X having sum-invariant significant digits and being Benford both involve only the significand of X, without loss of generality assume that $1 \le X < 10$ with probability one, so $S(X) = X$.

Assume first that X has sum-invariant significant digits. Note that this implies $\mathbb{E}[X \mathbb{1}_C(X)] = \frac{1}{9} 10^{1-m} \mathbb{E}[X] = \frac{1}{9} \lambda(C) \mathbb{E}[X]$ for every $C = C(d_1, d_2, \ldots, d_m)$. As a consequence, $\mathbb{P}(X = t) = 0$ for all $1 \le t < 10$; hence $\mathbb{P}(1 < X < 10) = 1$. Since the intervals $C(d_1, d_2, \ldots, d_m)$ generate the Borel σ-algebra $\mathcal{B}[1, 10)$, the equality $\frac{1}{9} \lambda(C) \mathbb{E}[X] = \mathbb{E}[X \mathbb{1}_C(X)]$ actually holds for all $C \in \mathcal{B}[1, 10)$. In particular, therefore, Lemma 5.19 shows that for all $1 < t_1 < t_2 < 10$,

$$
\frac{1}{9}(t_2 - t_1)\mathbb{E}[X] = \mathbb{E}\big[X \mathbb{1}_{(t_1, t_2]}(X)\big] = \int_{F_X(t_1)}^{F_X(t_2)} Q_X(s)\,\mathrm{d}s\,,
$$

and Lemma 5.20 with $a = 1$ and $b = 10$ yields $F_X(t) = \log t$ for all $t \in [1, 10)$, i.e., X is Benford.

The converse, i.e., that every Benford random variable has sum-invariant significant digits, has been shown in Example 5.17(iii). ∎

As an application, Theorem 5.18 provides another informal test for goodness-of-fit to Benford's law: Simply calculate the differences between the sums of the significands of the data corresponding to the same initial sequence of significant digits, and if these differences are large, the data are not a good fit to Benford's law; see [112]. Also, Theorem 5.18, together with Theorems 5.3 and 5.13, has the following immediate consequence.

THEOREM 5.21. *If a probability measure has scale-invariant or sum-invariant significant digits then it has base-invariant significant digits.*

Chapter Six

Real-valued Deterministic Processes

In many disciplines of science, one-dimensional deterministic (i.e., non-random) systems provide the simplest models for processes that evolve over time. Mathematically, these models take the form of one-dimensional difference or differential equations, and this chapter presents the basic theory of Benford's law for them. Specifically, conditions are studied under which these models conform to Benford's law by generating Benford sequences and functions, respectively. In the first seven sections, the focus is on discrete-time systems (i.e., difference equations) because they are somewhat easier to work with explicitly. Once the Benford properties of discrete-time systems are understood, it is straightforward to establish the analogous properties for continuous-time systems (i.e., differential equations), which is done in the chapter's eighth and final section. Throughout the chapter, recall that by Theorem 4.2 a sequence (x_n) of real numbers, or a real-valued function f, is Benford if and only if $(\log |x_n|)$, or $\log |f|$, is uniformly distributed modulo one.

6.1 ITERATION OF FUNCTIONS

In a one-dimensional *first-order difference equation* (or *one-step recursion*) the state of a process at time $n \in \mathbb{N}$, described by the real number x_n, is completely determined by the preceding state x_{n-1}. More formally,

$$x_n = f(x_{n-1}), \quad n \in \mathbb{N}, \tag{6.1}$$

where $f : C \to \mathbb{R}$ is a function that maps $C \subset \mathbb{R}$ into itself, i.e., $f(C) \subset C$. In what follows, the set C often equals $\mathbb{R}^+$ or $[c, +\infty)$ for some (large) $c \geq 0$. In accordance with standard terminology in dynamics [45, 88], the function f is also referred to as a *map*. Given any $x_0 \in C$, the difference equation (6.1) recursively defines a sequence (x_n) in C, often called the *orbit* of x_0 (under f). With the nth iterate of f denoted by f^n, i.e., $f^0 = \mathrm{id}_C$ and $f^n = f^{n-1} \circ f$ for every $n \in \mathbb{N}$, the orbit of x_0 is simply $(x_n) = (f^n(x_0))$. Note that this interpretation of the orbit as a *sequence* differs from terminology sometimes used in dynamics [88] where the orbit of $x_0 \in C$ is the mere *set* $\{f^n(x_0) : n \in \mathbb{N}\}$. Notice also that, here and throughout, a clear notational distinction is made between the iterates of a real-valued map f, written as $f^n(x)$, and its (usually different) integer powers, written as $f(x)^n$; thus, for instance with $f(x) = e^x$,

$$f^2(x) = e^{e^x} \neq e^{2x} = f(x)^2.$$

Despite its modest appearance, (6.1) can exhibit an enormous variety of different phenomena, depending on the particular properties of the map f. The simplest scenario occurs if $f^p(x_0) = x_0$ for some $p \in \mathbb{N}$. In this case, the orbit of x_0 is a periodic sequence

$$(x_n) = \left(f(x_0), f^2(x_0), \ldots, f^{p-1}(x_0), x_0, f(x_0), \ldots\right)$$

of period p, i.e., $x_{n+p} = x_n$ for all n, and x_0 is a *periodic point*, or a *fixed point* if $p = 1$. Furthermore, call x_0 and its orbit *attracting* if, in addition, $\lim_{n \to \infty} |f^{np+j}(x) - f^j(x_0)| = 0$ for every $j = 1, 2, \ldots, p$ whenever $|x - x_0|$ is sufficiently small. Clearly, (x_n) is not Benford in this situation, except possibly if $p = 1$ and $x_0 = 0$. Notice that the sequence $\left(S(x_n)\right)$ may be periodic even when (x_n) is not: E.g., for $f(x) = x\sqrt{10}$ and $x_0 = 1$ the generated orbit $(x_n) = (10^{n/2})$ tends to $+\infty$, but $\left(S(x_n)\right) = \left(10^{\langle n/2 \rangle}\right) = \left(\sqrt{10}, 1, \sqrt{10}, \ldots\right)$ is 2-periodic. In a slight abuse of terminology, call $+\infty$ a fixed point of f whenever $\lim_{x \to +\infty} f(x) = +\infty$, and call it an attracting fixed point if, in addition, $\lim_{n \to \infty} f^n(x) = +\infty$ for all sufficiently large x. For example, $f(x) = x - 1 + e^{-x}$ has 0 and $+\infty$ as an attracting and a non-attracting (repelling) fixed point, respectively, whereas $f(x) = e^x$ has $+\infty$ as its unique attracting fixed point.

The following examples illustrate the wide scope of (6.1) with respect to Benford's law, and also motivate many of the results presented throughout this chapter.

EXAMPLE 6.1. **(i)** Consider the map $f(x) = x + 2\sqrt{x} + 1$. The orbit of any $x_0 \geq 0$ under f can be computed explicitly as $(x_n) = \left((\sqrt{x_0} + n)^2\right)$. In particular, if $x_0 = 0$ then $(x_n) = (n^2)$. Since $(\log x_n) = \left(2 \log(\sqrt{x_0} + n)\right)$ is increasing but $(\log x_n / \log n)$ is bounded, (x_n) is not Benford, by Proposition 4.6(iv). Theorem 6.4 below yields this conclusion directly from simple properties of the map f; see Example 6.5(ii).

(ii) As seen in Example 4.17(v), the sequence $(p_n) = (2, 3, 5, 7, 11, \ldots)$ of prime numbers is not Benford. To see that (p_n) fits into the framework of (6.1), let $f(x) = x + 1$ whenever $x \leq 2 = p_1$, and on the interval $[p_n, p_{n+1}]$ let f be linear with $f(p_n) = p_{n+1}$ and $f(p_{n+1}) = p_{n+2}$. Then $(p_n) = \left(f^n(1)\right)$. Note that f is continuous, and

$$|f(x) - x| \leq \max\{p_{n+2} - p_{n+1}, p_{n+1} - p_n\} \quad \text{for all } x \in [p_n, p_{n+1}].$$

Again, Theorem 6.4 below yields directly that (p_n) is not Benford; see Example 6.6. ✠

EXAMPLE 6.2. **(i)** By Theorem 4.16, the sequence $(a^n x_0)$ is Benford for every $x_0 \neq 0$ or for none, depending on whether $\log |a|$ is irrational or not. Note that this sequence is simply the orbit of x_0 under the map $f(x) = ax$. The Benford properties of (essentially) linear maps are clarified by Theorem 6.13 below.

(ii) The Fibonacci sequence $(F_n) = (1, 1, 2, 3, 5, \ldots)$ is Benford; see Example 4.18. By means of the (linear) maps $f_1(x) \equiv x$ and $f_n(x) := \dfrac{F_n}{F_{n-1}} x$ for $n \geq 2$, the sequence (F_n) can be considered the unique solution, with $x_0 = 1$, of the difference equation

$$x_n = f_n(x_{n-1}), \quad n \in \mathbb{N}. \tag{6.2}$$

Note that, unlike the map f in (6.1), the maps f_n in (6.2) explicitly depend on n. In dynamical systems parlance, (6.2) is *nonautonomous* (i.e., explicitly time-varying) whereas (6.1) is *autonomous* (i.e., not explicitly time-varying). Nonautonomous difference equations will be studied in some detail in Section 6.6. Specifically, Theorem 6.40 and Lemma 6.44 directly apply to (6.2), showing that the sequence $\big(f_n \circ \cdots f_1(x_0)\big)$ is Benford for all $x_0 \neq 0$. Note also that while f_n does depend on n, this dependence becomes negligible rather quickly as $n \to \infty$: With $\varphi = \frac{1}{2}(1 + \sqrt{5}) > 1$,

$$\varphi^{2n}|f_n(x) - \varphi x| \leq (2 + 3\varphi)|x| \quad \text{for all } n \in \mathbb{N},\, x \in \mathbb{R};$$

in particular, $\lim_{n \to \infty} f_n(x) = \varphi x$ uniformly on every compact interval.

(iii) Let $f_n(x) = nx$, so again f_n explicitly depends on n. Unlike in (ii), the sequence (f_n) does not converge in any reasonable sense. Nevertheless, results for nonautonomous systems presented in Section 6.6 apply and show that $\big(f_n \circ \cdots \circ f_1(x_0)\big) = (n!x_0)$ is Benford for all $x_0 \neq 0$; in particular, $(n!)$ is Benford [15, 46]. ✠

EXAMPLE 6.3. Consider the map $f(x) = 10x^2$. Note that $f(0) = 0$ and $f(\pm\frac{1}{10}) = \frac{1}{10}$; otherwise $\big(f^n(x_0)\big)$ tends to $+\infty$ or to 0, depending on whether $|x_0| > \frac{1}{10}$ or $|x_0| < \frac{1}{10}$. Informally put, every orbit with $|x_0| \neq \frac{1}{10}$ is attracted to one of the two fixed points at $+\infty$ and 0. Unlike the case in Example 6.2 above, not every orbit is Benford. For instance, if $x_0 = 1$ then $S(x_n) = S\big(f^n(1)\big) = 1$ for all n. For another example, let $x_0 = 10^{1/3}$ and note that $\big(S(x_n)\big) = \big(10^{\langle(-1)^n/3\rangle}\big) = (10^{2/3}, 10^{1/3}, 10^{2/3}, \ldots)$ is 2-periodic. In fact, the explicit formula $f^n(x) = 10^{2^n-1}x^{2^n}$ shows that $\big(S(x_n)\big)$ is eventually periodic whenever $\log|x_0|$ is rational, so clearly $\big(f^n(x_0)\big)$ is not Benford in this case. Nevertheless, Proposition 6.20 below asserts that $\big(f^n(x_0)\big)$ is Benford for *most* $x_0 \in \mathbb{R}$. ✠

From the above examples the reader may have gained the following general impressions (see also Figure 6.1): Sequences with polynomial growth or decay like (n^2) or (p_n) are not Benford; sequences with exponential growth or decay, generated in the simplest case as orbits of linear maps like $f(x) = ax$, are either all Benford or none, depending on a; and finally, sequences with super-exponential growth or decay, generated as orbits of maps like $f(x) = 10x^2$, may or may not be Benford, and such maps exhibit a delicate mixture of Benford and non-Benford orbits. A main goal of this chapter is to demonstrate rigorously that, by and large, these impressions are correct. For the sake of notational clarity, results are often formulated only for maps $f : \mathbb{R}^+ \to \mathbb{R}^+$ even though they can easily be extended to maps $f : \mathbb{R} \to \mathbb{R}$. The details of such extensions are occasionally hinted at in examples, but more often are left to the interested reader.

	1	2	3	4	5	6	7	8	9	Δ
(n^2)	19.19	14.69	12.37	10.95	9.84	9.08	8.45	7.91	7.52	10.91
(p_n)	16.01	11.29	10.97	10.55	10.13	10.13	10.27	10.03	10.06	14.09
(2^n)	30.10	17.61	12.49	9.70	7.91	6.70	5.79	5.12	4.58	0.00
(F_n)	30.11	17.62	12.50	9.68	7.92	6.68	5.80	5.13	4.56	0.01
$(n!)$	29.56	17.89	12.76	9.63	7.94	7.15	5.71	5.10	4.26	0.54
$x_n = 10x_{n-1}^2$ $x_0 = 2$	30.19	17.66	12.68	9.56	7.83	6.97	5.45	5.13	4.53	0.34

Figure 6.1: Relative frequencies (in percent) of the leading significant (decimal) digit for the first 10^4 terms of the sequences (x_n) discussed in Examples 6.1 to 6.3.

Throughout, the following standard terminology is used: For any two functions $f, g : \mathbb{R}^+ \to \mathbb{R}$, the statement

$$f = \mathcal{O}(g) \quad \text{as } x \to +\infty$$

means that $\limsup_{x \to +\infty} |f(x)/g(x)| < +\infty$, whereas

$$f = o(g) \quad \text{as } x \to +\infty$$

means that $\lim_{x \to +\infty} f(x)/g(x) = 0$. Note that, for every $a \in \mathbb{R}$, $f = o(x^a)$ implies $f = \mathcal{O}(x^a)$, which in turn implies $f = o(x^{a+\varepsilon})$ for every $\varepsilon > 0$. For *sequences* of functions $f_n : \mathbb{R}^+ \to \mathbb{R}$, a uniform version of this terminology is employed that reduces to the above in case $f_n \equiv f$: Specifically,

$$f_n = \mathcal{O}(g) \quad \textit{uniformly as } x \to +\infty,$$

if there exists $c > 0$ such that $|f_n(x)/g(x)| < c$ for all $n \in \mathbb{N}$ and $x \geq c$; and

$$f_n = o(g) \quad \textit{uniformly as } x \to +\infty,$$

if, for every $\varepsilon > 0$, there exists $c > 0$ such that $|f_n(x)/g(x)| < \varepsilon$ for all $n \in \mathbb{N}$ and $x \geq c$.

6.2   SEQUENCES WITH POLYNOMIAL GROWTH

Recall from Example 4.7(ii) that neither of the sequences (n) and (n^2) is Benford. Clearly, both sequences are orbits $\big(f^n(0)\big)$ under the maps $f(x) = x + 1$ and

$f(x) = x + 2\sqrt{x} + 1$, respectively. As the following theorem shows, for these maps the orbit $\big(f^n(x_0)\big)$ is not Benford for *any* x_0. The simple reason is that in both cases the map f is close to the identity map, and no orbit under the latter is Benford.

THEOREM 6.4. *Let* $f : \mathbb{R}^+ \to \mathbb{R}^+$ *be a map such that* $f(x) = x + g(x)$ *with* $g(x) \geq 0$ *for all sufficiently large* x. *If* $g = o(x^{1-\varepsilon})$ *as* $x \to +\infty$ *with some* $\varepsilon > 0$, *then for all sufficiently large* x_0, *the orbit* $\big(f^n(x_0)\big)$ *is not Benford.*

PROOF. Pick $\xi > 0$ such that $g(x) \geq 0$ whenever $x \geq \xi$. For every $x_0 \geq \xi$, the non-decreasing sequence $(x_n) = \big(f^n(x_0)\big)$ is clearly not Benford whenever bounded. Assume therefore that $\lim_{n\to\infty} x_n = +\infty$, and let $y_n = \log x_n$ for every $n \in \mathbb{N}$. Then (y_n) is non-decreasing as well, with $\lim_{n\to\infty} y_n = +\infty$, and $y_n = y_{n-1} + h(y_{n-1})$, where

$$h(y) = \log\left(1 + \frac{g(10^y)}{10^y}\right).$$

Recall that $g = o(x^{1-\varepsilon})$ as $x \to +\infty$, by assumption. Fix $0 < \delta < \varepsilon$ and consider the auxiliary function $H : \mathbb{R}^+ \to \mathbb{R}^+$ given by

$$H(y) = \frac{1}{\delta} \log\left(1 + 10^{-\delta y}\right).$$

Since $h = o(10^{-\varepsilon y})$ as $y \to +\infty$, but $\lim_{y\to+\infty} 10^{\delta y} H(y) = \delta^{-1} \log e$, it follows that $h(y) \leq H(y)$ for the appropriate $\eta > 0$ and all $y \geq \eta$. Fix any $Y_0 \geq \eta$ and define a sequence (Y_n) recursively as $Y_n = Y_{n-1} + H(Y_{n-1})$, $n \geq 1$. Given any $y_0 \geq \max\{\eta, \log \xi\}$, note that

$$y_1 - Y_1 = y_0 - Y_0 + h(y_0) - H(Y_0) \leq y_0 - Y_0 + H(y_0) - H(Y_0).$$

If $y_0 \leq Y_0$ then $y_1 - Y_1 \leq H(y_0) \leq H(\eta)$. On the other hand, if $y_0 > Y_0$ then $y_1 - Y_1 \leq y_0 - Y_0$ since H is decreasing. In either case, therefore,

$$\eta \leq y_0 \leq y_1 \leq Y_1 + \max\{H(\eta), |y_0 - Y_0|\},$$

and iterating the same argument shows that

$$\eta \leq y_n \leq Y_n + \max\{H(\eta), |y_0 - Y_0|\} \quad \text{for all } n. \tag{6.3}$$

Next note that Y_n is explicitly, i.e., non-recursively, given by

$$Y_n = \frac{1}{\delta} \log\left(1 + 10^{\delta Y_{n-1}}\right) = \frac{1}{\delta} \log\left(2 + 10^{\delta Y_{n-2}}\right) = \ldots = \frac{1}{\delta} \log\left(n + 10^{\delta Y_0}\right). \tag{6.4}$$

From (6.3) and (6.4), it follows that $(y_n/\log n)$ is bounded, and hence by Proposition 4.6(iv), the sequence (y_n) is not u.d. mod 1 for $y_0 \geq \max\{\eta, \log \xi\}$. Whenever $x_0 \geq \max\{10^\eta, \xi\}$, therefore, the orbit $\big(f^n(x_0)\big)$ is not Benford. ∎

EXAMPLE 6.5. **(i)** For the map $f(x) = x + 1$, $g(x) \equiv 1 = \mathcal{O}(1) = \mathcal{O}(x^0)$ as $x \to +\infty$. Consequently, $\big(f^n(x_0)\big) = (x_0 + n)$ is not Benford for any sufficiently large x_0, nor in fact for any $x_0 \in \mathbb{R}$.
(ii) Similarly, for the map $f(x) = x + 2\sqrt{x} + 1$ in Example 6.1(i),

$$0 < g(x) = 2\sqrt{x} + 1 = \mathcal{O}(x^{1/2}) \quad \text{as } x \to +\infty.$$

For every $x_0 \in \mathbb{R}^+$, therefore, $\big(f^n(x_0)\big)$ is not Benford either. ✠

EXAMPLE 6.6. Consider the map introduced in Example 6.1(ii), namely,

$$f(x) = \begin{cases} x + 1 & \text{if } x \leq p_1 = 2, \\ \dfrac{p_{n+2} - p_{n+1}}{p_{n+1} - p_n}(x - p_n) + p_{n+1} & \text{if } p_n \leq x < p_{n+1}, \end{cases}$$

where $(p_n) = (2, 3, 5, 7, 11, \ldots)$ is the sequence of prime numbers. As observed in Example 6.1(ii),

$$1 \leq g(x) = f(x) - x \leq \max\{p_{n+2} - p_{n+1}, p_{n+1} - p_n\} \quad \text{for all } x \in [p_n, p_{n+1}].$$

A classical result due to G. Hoheisl asserts that, for some $\varepsilon > 0$,

$$p_{n+1} - p_n = o(p_n^{1-\varepsilon}) \quad \text{as } n \to \infty; \tag{6.5}$$

see [5] where it is shown that ε can be chosen as large as $\varepsilon = \frac{19}{40} = 0.475$. From (6.5) it is clear that $g = o(x^{1-\varepsilon})$ as $x \to +\infty$. Thus Theorem 6.4 implies that $\big(f^n(x_0)\big)$ is not Benford for any $x_0 \in \mathbb{R}$. In particular, $\big(f^n(1)\big) = (p_n)$ is not Benford, as already seen in Example 4.17(v). ✠

The next example demonstrates that the two key assumptions on g in Theorem 6.4 are, in a sense, best possible.

EXAMPLE 6.7. **(i)** Clearly $g(x) \geq 0$ for all sufficiently large x whenever $\liminf_{x \to +\infty} g(x) > 0$. However, the conclusion of Theorem 6.4 may fail under the assumption $\liminf_{x \to +\infty} g(x) \geq 0$, which, on first sight, may seem only insignificantly weaker. To see this, consider

$$f(x) = x - \frac{x}{x^2 + 2},$$

for which

$$g(x) = -\frac{x}{x^2 + 2} = \mathcal{O}(x^{-1}) \quad \text{as } x \to +\infty;$$

hence $\lim_{x \to +\infty} g(x) = 0$. Nevertheless, $f^n(x_0) \to 0$ for every $x_0 \in \mathbb{R}^+$ (in fact, even for every $x_0 \in \mathbb{R}$), and since $f'(0) = \frac{1}{2}$, by Corollary 6.18 below, $\big(f^n(x_0)\big)$ is Benford for every $x_0 \neq 0$.
(ii) The assumption in Theorem 6.4 that $g = o(x^{1-\varepsilon})$ for some $\varepsilon > 0$ is sharp in that it cannot be weakened to $g = o(x)$. To see this, consider the map $f : \mathbb{R}^+ \to \mathbb{R}^+$ with

$$f(x) = 10^{\sqrt{(\log x)^2 + 1}} > x,$$

for which

$$g(x) = f(x) - x = x\left(10^{\sqrt{(\log x)^2+1}-\log x} - 1\right) = x\left(10^{(\sqrt{(\log x)^2+1}+\log x)^{-1}} - 1\right)$$

shows that $g = \mathcal{O}(x/\log x)$ as $x \to +\infty$. Moreover, for every $x_0 > 0$ and $n \in \mathbb{N}$,

$$f^n(x_0) = 10^{\sqrt{(\log f^{n-1}(x_0))^2+1}} = 10^{\sqrt{(\log f^{n-2}(x_0))^2+2}} = \ldots = 10^{\sqrt{(\log x_0)^2+n}},$$

and since $(\sqrt{a+n})$ is u.d. mod 1 whenever $a \geq 0$, by Example 4.7(iii), $(f^n(x_0))$ is Benford for every $x_0 > 0$ — despite the fact that $g(x) > 0$ for all $x > 0$, and $g = o(x)$ as $x \to +\infty$.   ✠

EXAMPLE 6.8. Theorem 6.4 carries over to maps $f : \mathbb{R} \to \mathbb{R}$ more or less verbatim, with the proviso that $g(x) = |f(x)| - |x| \geq 0$ whenever $|x|$ is sufficiently large, and $g = o(|x|^{1-\varepsilon})$ as $|x| \to +\infty$. Thus, for instance, no orbit $(f^n(x_0))$ under $f(x) = -2x + \sqrt[3]{x^3 - x - 1}$ is Benford whenever $|x_0|$ is sufficiently large, because

$$0 \leq |f(x)| - |x| = |x|\left(\left|2 - \sqrt[3]{1 - x^{-2} - x^{-3}}\right| - 1\right) = \mathcal{O}(|x|^{-1}) \quad \text{as } |x| \to +\infty.$$ ✠

Many results in this section have immediate counterparts for maps with 0 rather than $+\infty$ as an attracting fixed point. One only has to consider reciprocals and recall from Theorem 4.4 that a real sequence (x_n) is Benford if and only if (x_n^{-1}) is Benford.

COROLLARY 6.9. Let $f : \mathbb{R} \to \mathbb{R}$ be C^2, and assume that $|f(x)| \leq |x|$ for some $\delta > 0$ and all $|x| \leq \delta$. If $|f'(0)| = 1$ then $(f^n(x_0))$ is not Benford whenever $|x_0| \leq \delta$.

EXAMPLE 6.10. (i) The smooth map $f(x) = \sin x$ satisfies $|f(x)| < |x|$ for all $x \neq 0$, and $\lim_{n\to\infty} f^n(x_0) = 0$ for every x_0. Since $f'(0) = 1$, Corollary 6.9 applies and shows that $(f^n(x_0))$ is not Benford for any $x_0 \in \mathbb{R}$. Using Proposition 4.6(iv), this could also be seen directly by observing that the sequence $(f^n(x_0))$ is monotone and $\lim_{n\to\infty} nf^n(x_0)^2 = 3$ for all $x_0 \neq 0$.
(ii) The smooth map $f : \mathbb{R} \to \mathbb{R}$ with

$$f(x) = x - \frac{x^3}{x^4 + 1}$$

also satisfies $|f(x)| < |x|$ for all $x \neq 0$, and $\lim_{n\to\infty} f^n(x_0) = 0$ for every x_0. Again, $f'(0) = 1$, and $(f^n(x_0))$ is not Benford for any $x_0 \in \mathbb{R}$, by Corollary 6.9.
(iii) For the smooth map $f : \mathbb{R} \to \mathbb{R}$ given by $f(x) = x - \frac{1}{6}x^3$,

$$|f(x)| = |x|\left|1 - \tfrac{1}{6}x^2\right| < |x| \quad \text{for all } x \in \left(-2\sqrt{3}, 2\sqrt{3}\right) \setminus \{0\}.$$

Note that f is simply the third-order Taylor polynomial of $\sin x$ at $x = 0$. Hence it is plausible that near the fixed point at $x_0 = 0$, the behavior of f is very similar to that of the sine function in (i). Indeed, $(f^n(x_0))$ is not Benford whenever

$|x_0| < 2\sqrt{3}$, by Corollary 6.9. Also, $f(\pm 2\sqrt{3}) = \mp 2\sqrt{3}$, and hence $\left(f^n(\pm 2\sqrt{3})\right)$ is 2-periodic and thus not Benford either. On the other hand, if $|x_0| > 2\sqrt{3}$ then $|f^n(x_0)| \to +\infty$ as $n \to \infty$, but this scenario is not covered by Theorem 6.4 or its extension mentioned in Example 6.8, simply because the difference $|f(x)| - |x|$ grows as $\frac{1}{6}|x|^3$ as $|x| \to +\infty$, and hence clearly is not $o(|x|^{1-\varepsilon})$ for any $\varepsilon > 0$. However, it follows from Theorem 6.23 below that $\left(f^n(x_0)\right)$ is Benford for *most but not all* $|x_0| > 2\sqrt{3}$; see also Example 6.25.   ✠

EXAMPLE 6.11. In Corollary 6.9, the smoothness assumption on f can be weakened somewhat, but the conclusion is not generally true if f is merely C^1. For instance, the function

$$f(x) = \begin{cases} 10^{-\sqrt{(\log x)^2+1}} & \text{if } x > 0, \\ 0 & \text{if } x = 0, \\ -10^{-\sqrt{(\log |x|)^2+1}} & \text{if } x < 0, \end{cases}$$

satisfies $|f(x)| < |x|$ for all $x \neq 0$, and $f'(0) = 1$. But f is only C^1, and for every $x_0 \neq 0$ and $n \in \mathbb{N}$,

$$\log|f^n(x_0)| = -\sqrt{\left(\log|f^{n-1}(x_0)|\right)^2 + 1} = \ldots = -\sqrt{(\log |x_0|)^2 + n}\,.$$

By Example 4.7(iii), the sequence $(-\sqrt{a+n})$ is u.d. mod 1 for $a \geq 0$, and so the orbit $\left(f^n(x_0)\right)$ is Benford whenever $x_0 \neq 0$.   ✠

6.3   SEQUENCES WITH EXPONENTIAL GROWTH

Recall from Theorem 4.16 that, for any real number a and any $x_0 \neq 0$, the sequence $(a^n x_0)$ is Benford if and only if $\log |a|$ is irrational. In the dynamical systems terminology of the present chapter, for the map $f(x) = ax$, every orbit $\left(f^n(x_0)\right)$ with $x_0 \neq 0$ is Benford, or none is, depending on whether $\log |a|$ is irrational or not. The following example provides an alternative, graphical description of the dynamics of f with regard to Benford's law.

EXAMPLE 6.12. Given $a > 0$, note that the map $f(x) = ax$ has the property that, for every $x, \widetilde{x} \in \mathbb{R}^+$,

$$S(x) = S(\widetilde{x}) \quad \Longrightarrow \quad S\big(f(x)\big) = S\big(f(\widetilde{x})\big)\,. \tag{6.6}$$

As a consequence of (6.6), there exists a (uniquely determined) map f_0 on $[1, 10)$ such that $S\big(f(x)\big) = f_0\big(S(x)\big)$ for all $x > 0$. (Note that the same map f_0 is obtained if a is replaced by $10^k a$ for any integer k.) For every $n \in \mathbb{N}$ and $x_1 > 0$, therefore, $S\big(f^n(x_0)\big) = f_0^n\big(S(x_0)\big)$, and by means of the map $h : [0, 1) \to [0, 1)$ with $h(s) = \log f_0(10^s)$, it follows that

$$S\big(f^n(x_0)\big) = f_0^n\big(S(x_0)\big) = 10^{h^n(\langle \log x_0 \rangle)}\,.$$

Hence, the orbit $\left(f^n(x_0)\right)$ of x_0 under f is Benford if and only if the orbit $\left(h^n(s_0)\right)$ of s_0 under h is u.d. in $[0,1)$, where $s_0 = \log S(x_0) = \langle \log x_0 \rangle$. The latter is the case precisely when $\log a$ is irrational. For a concrete example, choose $a = 2$. Then

$$f_0(t) = \begin{cases} 2t & \text{if } 1 \le t < 5 \,, \\ t/5 & \text{if } 5 \le t < 10 \,, \end{cases}$$

and

$$h(s) = \langle s + \log 2 \rangle = \begin{cases} s + \log 2 & \text{if } 0 \le s < \log 5 \,, \\ s - \log 5 & \text{if } \log 5 \le s < 1 \,; \end{cases}$$

see Figure 6.2.   ✠

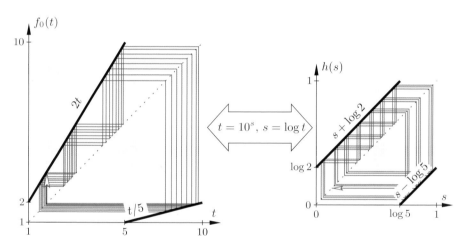

Figure 6.2: For the map $f(x) = 2x$, the sequence $\left(f^n(x_0)\right)$ is Benford for all $x_0 > 0$; see Example 6.12.

In general, if a map $f : \mathbb{R}^+ \to \mathbb{R}^+$ has the form $f(x) = ax + g(x)$ with a function g that is small in some sense but not identically zero, then (6.6) will usually not hold, and so the simple graphical analysis of Example 6.12 fails. However, as the following theorem, the main result of the present section, shows, the all-or-nothing behavior with regard to Benford's law does persist nevertheless, provided that g is small enough.

THEOREM 6.13. *Let $f : \mathbb{R}^+ \to \mathbb{R}^+$ be a map such that $f(x) = ax + g(x)$ with $a > 1$ and $g = o(x)$ as $x \to +\infty$.*

(i) *If $\log a \notin \mathbb{Q}$ then $\left(f^n(x_0)\right)$ is Benford for every sufficiently large x_0.*

(ii) *If $\log a \in \mathbb{Q}$ and $g = o(x/\log x)$ as $x \to +\infty$ then, for every sufficiently large x_0, $\left(f^n(x_0)\right)$ is not Benford.*

PROOF. (i) Since $g(x) = o(x)$ as $x \to +\infty$, there exists $\xi > 0$ such that $|g(x)| \leq \frac{1}{2}(a-1)x$ for all $x \geq \xi$, and consequently

$$x_n = f(x_{n-1}) = ax_{n-1} + g(x_{n-1}) \geq \tfrac{1}{2}(a+1)\,x_{n-1}\,,$$

provided that $x_{n-1} \geq \xi$. Thus $f^n(x_0) \geq \xi$ for all n, and $f^n(x_0) \to +\infty$ as $n \to \infty$ whenever $x_0 \geq \xi$; moreover, $g(x_n)/x_n \to 0$. Letting $y_n = \log x_n$ for every n,

$$y_n - y_{n-1} = \log \frac{f(x_{n-1})}{x_{n-1}} = \log a + \log\left(1 + \frac{g(x_{n-1})}{ax_{n-1}}\right) \overset{n\to\infty}{\longrightarrow} \log a\,,$$

and Proposition 4.6(i) implies that (y_n) is u.d. mod 1 if $\log a$ is irrational. In that case, therefore, $\big(f^n(x_0)\big)$ is Benford for every $x_0 \geq \xi$.

(ii) Assume now that $g = o(x/\log x)$ as $x \to +\infty$. As in (i), pick $\xi > 0$ such that $x_n = f^n(x_0) \geq \xi$ for every n, and $f^n(x_0) \to +\infty$ as $n \to \infty$ whenever $x_0 \geq \xi$. Note that, given any $\varepsilon > 0$,

$$\left|\log\left(1 + \frac{g(x_n)}{ax_n}\right)\right| \leq \frac{|g(x_n)|}{2ax_n} < \frac{\varepsilon \log a}{\log x_n}$$

for all sufficiently large n, where the first inequality holds since $|\log(1+x)| \leq \frac{1}{2}|x|$ for all $x \geq -\frac{1}{4}$. Letting $z_n = y_n - n \log a$ yields $\lim_{n\to\infty} z_n/n = 0$ and

$$n|z_{n+1} - z_n| = n\left|\log\left(1 + \frac{g(x_n)}{ax_n}\right)\right| \leq \frac{\varepsilon n \log a}{\log x_n} = \frac{\varepsilon \log a}{\log a + z_n/n}\,,$$

so $\limsup_{n\to\infty} n|z_{n+1} - z_n| \leq \varepsilon$. In fact, $\lim_{n\to\infty} n(z_{n+1} - z_n) = 0$ since $\varepsilon > 0$ was arbitrary. By Proposition 4.6(ii, vi), the sequence (y_n) is u.d. mod 1 (if and) only if $(n \log a)$ is, and thus $\big(f^n(x_0)\big)$ is not Benford whenever $x_0 \geq \xi$ and $\log a \in \mathbb{Q}$. ∎

COROLLARY 6.14. *Let $f : \mathbb{R}^+ \to \mathbb{R}^+$ be a map such that, with some $a > 1$, $f(x) - ax = o(x/\log x)$ as $x \to +\infty$. Then, for every sufficiently large x_0, $\big(f^n(x_0)\big)$ is Benford if and only if $\log a$ is irrational.*

Remark. In Theorem 6.13 and Corollary 6.14, and also in Theorem 6.4 of the previous section, the map f is not even required to be (Borel) measurable, let alone continuous.

EXAMPLE 6.15. **(i)** By Corollary 6.14, every orbit under $f(x) = 2x + e^{-x}$ is Benford. Note that no simple explicit formula is available for f^n.

(ii) The map $f(x) = 2x - e^{-x}$ has a unique (repelling) fixed point, namely, $x^* = 0.5671$. If $x_0 > x^*$ then $f^n(x_0) \to +\infty$, so Corollary 6.14 implies that $\big(f^n(x_0)\big)$ is Benford. On the other hand, if $x_0 < x^*$ then $f^n(x_0) \to -\infty$ super-exponentially fast. This scenario is not covered by any result so far. However, Proposition 6.31 below implies that $\big(f^n(x_0)\big)$ is Benford for *most but not all* $x_0 < x^*$. ✠

Just as with Theorem 6.4, Corollary 6.14 carries over to maps $f : \mathbb{R} \to \mathbb{R}$: If $|f(x)| - a|x| = o(|x|/\log|x|)$ as $|x| \to +\infty$ for some $a > 1$ then, for every sufficiently large $|x_0|$, the sequence $(f^n(x_0))$ is Benford if and only $\log a$ is irrational.

EXAMPLE 6.16. Consider the map $f(x) = -2x + 3$. Since

$$\big||f(x)| - 2|x|\big| = \big||2x - 3| - |2x|\big| \le 3 \quad \text{for all } x \in \mathbb{R},$$

clearly $|f(x)| - 2|x| = \mathcal{O}(1)$ as $|x| \to +\infty$, so $(f^n(x_0))$ is Benford provided that $|x_0|$ is sufficiently large, and in fact is Benford for every $x_0 \ne 1$. Note that, for every $n \ge 2$,

$$x_n = f^n(x_0) = f(x_{n-1}) = -2x_{n-1} + 3 = -x_{n-1} - (-2x_{n-2} + 3) + 3$$
$$= -x_{n-1} + 2x_{n-2}.$$

Thus $(f^n(x_0))$ solves a linear *second*-order difference equation (or *two*-step recursion). The Benford properties of (solutions of) linear difference equations are studied systematically in Chapter 7. For instance, Theorem 7.41 implies that not *every* solution of $x_n = -x_{n-1} + 2x_{n-2}$, and correspondingly not *every* orbit $(f^n(x_0))$, can be Benford. In the present example, however, the only exception occurs for $x_0 = 1$. ✠

As the following example shows, the requirement that $g = o(x/\log x)$ as $x \to +\infty$ in Theorem 6.13(ii) is sharp in the sense that the latter result, as well as the "only if" part in Corollary 6.14, may fail if $g = o(x/(\log x)^\delta)$ for some $0 \le \delta < 1$. Under the latter, weaker assumption, $(f^n(x_0))$ may be Benford for all x_0 even if $\log a$ is rational.

EXAMPLE 6.17. Given any $0 \le \delta < 1$, fix $0 < \varepsilon < 1 - \delta$, define a homeomorphism $h_\varepsilon : \mathbb{R} \to \mathbb{R}$ as

$$h_\varepsilon(y) = \begin{cases} y + y^\varepsilon & \text{if } y \ge 0, \\ y - |y|^\varepsilon & \text{if } y < 0, \end{cases}$$

and observe that

$$\frac{h_\varepsilon^{\pm 1}(y)}{y} - 1 = o\big(|y|^{-\delta}\big) \quad \text{as } |y| \to +\infty,$$

as well as that $|h_\varepsilon^{-1}(y)| \ge 1$ for every $|y| \ge 2$. Next define a map $f : \mathbb{R}^+ \to \mathbb{R}^+$ by setting

$$f(x) = 10^{h_\varepsilon(1 + h_\varepsilon^{-1}(\log x))}.$$

It is straightforward to check that $f(x) - 10x = o(x/(\log x)^\delta)$ as $x \to +\infty$. Thus $a = 10$, and $\log a = 1$ is rational. On the other hand, for every $n \in \mathbb{N}$,

$$f^n(x) = 10^{h_\varepsilon(n + h_\varepsilon^{-1}(\log x))}, \quad x > 0.$$

Hence, given any $x_0 > 0$, with $c = h_\varepsilon^{-1}(\log x_0) \in \mathbb{R}$,

$$\langle \log f^n(x_0) \rangle = \langle n + c + (n+c)^\varepsilon \rangle = \langle c + (n+c)^\varepsilon \rangle = \langle H(n) \rangle \quad \text{for all } n \geq |c|,$$

with $H(t) = c + (t+c)^\varepsilon$. By Proposition 4.6(v), the sequence $(H(n))$ is u.d. mod 1, and so $(f^n(x_0))$ is Benford for every $x_0 \in \mathbb{R}^+$. ✠

Theorem 6.13 yields the following straightforward corollary when reciprocals are considered.

COROLLARY 6.18. *Let $f : \mathbb{R} \to \mathbb{R}$ be C^2 with $f(0) = 0$ and $0 < |f'(0)| < 1$. Then, for every $x_0 \neq 0$ sufficiently close to 0, $(f^n(x_0))$ is Benford if and only if $\log|f'(0)|$ is irrational.*

EXAMPLE 6.19. The map $f(x) = x - \frac{1}{3} + \frac{1}{3}e^{-x}$ is smooth, with $f(0) = 0$ and $0 < f'(0) = \frac{2}{3} < 1$. Since $\log f'(0) = \log 2 - \log 3$ is irrational, the sequence $(f^n(x_0))$ is Benford for every $x_0 \neq 0$ sufficiently close to 0. In fact, it is easy to see that $\lim_{n \to \infty} f^n(x_0) = 0$ for every $x_0 \in \mathbb{R}$, and so $(f^n(x_0))$ is Benford unless $f^n(x_0) = 0$ for some $n \in \mathbb{N}$. Since f is strictly convex, the latter happens only if $f(x_0) = 0$, which in turn implies that either $x_0 = 0$ or $x_0 = -1.903$. For every $x_0 \notin \{0, -1.903\}$, therefore, $(f^n(x_0))$ is Benford. ✠

Remark. The "if" part of Corollary 6.18 remains valid if f is merely assumed to be C^1. However, examples in the spirit of Example 6.11 show that the "only if" part may fail under this weaker assumption.

6.4 SEQUENCES WITH SUPER-EXPONENTIAL GROWTH

To appreciate how substantially different the Benford properties of iterations of maps like $f(x) = x^2 + 1$ or $f(x) = e^x$ are from those of the maps in the previous two sections, consider two very simple examples: the linear map $f_1(x) = 10x$ and the non-linear (quadratic) map $f_2(x) = 10x^2$. Applying the map f_1 does not affect the significand at all; more formally, for every $x_0 \in \mathbb{R}$,

$$S(f_1^n(x_0)) = S(10^n x_0) = S(x_0) \quad \text{for all } n \in \mathbb{N}.$$

This, of course, corresponds to the fact that $(f_1^n(x_0))$ is not Benford for any x_0. For the map f_2 on the other hand, it is also clear that $(f_2^n(x_0))$ may not be Benford for *some* x_0. For example, $x_0 = 1$ yields $S(f_2^n(1)) = 1$ for all n. However, if $x_0 = 2$ or $x_0 = \pi$ then the numbers $S(f_2^n(x_0))$, $n = 1, 2, \ldots$, are all different, and so $(f_2^n(x_0))$ *might* be Benford; see Figure 6.3, which suggests that this is indeed the case.

The following result shows that $(f_2^n(x_0))$ is Benford for (Lebesgue) almost all $x \in \mathbb{R}$, but the set of *exceptional points*, i.e., points x_0 for which $(f_2^n(x_0))$ is *not* Benford, is also rather large. The result follows easily using Theorem 6.23 below, which handles maps sufficiently close to monomials. (Formally, the result is a special case of Theorem 6.46, for which a proof is given in Section 6.6.)

	1	2	3	4	5	6	7	8	9	Δ
$x_0 = 2$	30.19	17.66	12.68	9.56	7.83	6.97	5.45	5.13	4.53	0.34
$x_0 = 3$	30.46	16.99	12.93	9.33	7.54	6.87	5.84	5.28	4.76	0.61
$x_0 = \pi$	30.38	17.07	12.62	9.14	8.15	6.32	6.17	5.10	5.05	0.55

Figure 6.3: Relative frequencies of the leading significant digit for the first 10^4 terms of the sequence (x_n) with $x_n = 10x_{n-1}^2$, for three different values of x_0; see also Example 6.21.

PROPOSITION 6.20. *Let $f(x) = ax^b$ with $a > 0$, $b > 1$. Then $(f^n(x_0))$ is Benford for almost all $x_0 > 0$, but every non-empty open interval in $\mathbb{R}^+$ contains uncountably many x_0 for which $(f^n(x_0))$ is not Benford.*

EXAMPLE 6.21. Let $f(x) = 10x^2$. By Proposition 6.20, $(f^n(x_0))$ is Benford for almost all $x_0 > 0$, and in fact even for almost all $x_0 \in \mathbb{R}$ because $f(x_0) \geq 0$, but not for *all* x_0. For instance, $f^n(x_0) = 10^{2^n-1}x_0^{2^n}$ always has first digit $D_1 = 1$ if $x_0 = 10^k$ for some $k \in \mathbb{Z}$. As in the case of the exactly linear map in Example 6.12, $f(x) = 10x^2$ has the property (6.6) and hence induces a map f_0 on $[1, 10)$, according to $S(f(x)) = f_0(S(x))$. Concretely,

$$ f_0(t) = S(f(t)) = \begin{cases} t^2 & \text{if } 1 \leq t < \sqrt{10}, \\ t^2/10 & \text{if } \sqrt{10} \leq t < 10. \end{cases} $$

The associated map h on $[0, 1)$ is simply $h(s) = \log f_0(10^s) = \langle 2s \rangle$. Again, $(f^n(x_0))$ is Benford if and only if $(h^n(s_0))$ is u.d. in $[0, 1)$, where s_0 is given by $s_0 = \log S(x_0) = \langle \log |x_0| \rangle$; see Figure 6.4. ✠

To put Proposition 6.20 into perspective, given any map $f : \mathbb{R}^+ \to \mathbb{R}^+$, let

$$ B = \{x \in \mathbb{R}^+ : (f^n(x)) \text{ is Benford}\}. \tag{6.7} $$

With this, if f has the special form $f(x) = ax^b$, then Proposition 6.20 asserts that $\mathbb{R}^+ \setminus B = \{x \in \mathbb{R}^+ : x \notin B\}$, the set of exceptional points, is a (Lebesgue) nullset. Thus *most* $x > 0$ belong to B. However, $\mathbb{R}^+ \setminus B$ is also uncountable and everywhere dense in $\mathbb{R}^+$. In fact, the proof of Theorem 6.46 (a generalization of Proposition 6.20) given in Section 6.6 shows that the set B is of *first category*, i.e., a countable union of nowhere dense sets. From a topological point of view, therefore, most $x \in \mathbb{R}^+$ do *not* belong to B. This discrepancy between the measure-theoretic and the topological point of view is not uncommon in ergodic theory and may explain why it is difficult to find even a single point x_0 for which $(f^n(x_0))$ is Benford for, say, $f(x) = 10x^2$ — despite the fact that Proposition 6.20 guarantees the existence of such points in abundance. The sets B and

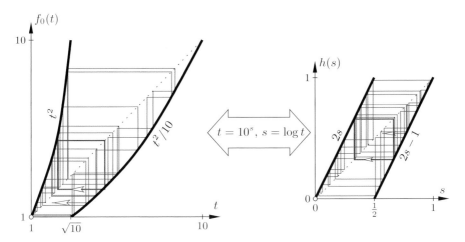

Figure 6.4: For the map $f(x) = 10x^2$, the sequence $\left(f^n(x_0)\right)$ is Benford for almost all, but not all, x_0. An exceptional orbit, corresponding to $S(x_0) = 10^{1/3}$, is indicated in red; see Examples 6.3 and 6.21.

$\mathbb{R}^+ \setminus B$ are intertwined in a rather complicated way, and the reader may want to compare this to the much more clear-cut situation prevailing for the maps of Corollary 6.14 where, for every $c > 0$ sufficiently large, either $[c, +\infty) \cap B$ or $[c, +\infty) \setminus B$ is empty, depending on whether $\log a$ is rational or not.

The maps of Proposition 6.20 will now be used as basic models for studying the Benford properties of more general maps. More concretely, consider maps $f : \mathbb{R}^+ \to \mathbb{R}^+$ with the property that

$$f(x) - ax^b = o(x^b) \quad \text{as } x \to +\infty, \tag{6.8}$$

for some $a > 0$ and $b > 1$. Recall that in order to understand the Benford properties of the *essentially linear* maps f of Section 6.3, all that was needed was a growth condition on the deviation $f(x) - ax$; see Theorem 6.13. For the *power-like* maps given by (6.8), the situation is more complicated in that even if $f(x) - ax^b$ decays very rapidly, the Benford properties of the orbits $\left(f^n(x_0)\right)$ may be quite unlike those in Proposition 6.20.

EXAMPLE 6.22. The goal of this example is to demonstrate that, given any (decreasing) function $g : \mathbb{R}^+ \to \mathbb{R}^+$, the deviation $f(x) - ax^b$ in (6.8) may decay faster than g, i.e., $f(x) - ax^b = o(g)$ as $x \to +\infty$, and yet, in stark contrast to Proposition 6.20, $\left(f^n(x_0)\right)$ may be Benford for every (large) $x_0 \in \mathbb{R}^+$, or for none at all. Throughout, let (N_n) be any increasing sequence of positive integers, to be further specified shortly.

(i) Fix $0 < \eta < 1$ such that $(2^n \eta)$ is u.d. mod 1. (Such η exist; see Example 4.15.) Define a map $h : [\eta, +\infty) \to \mathbb{R}$ as follows: Given any $y \geq \eta$, there exists a unique integer $l(y)$ with $2^{l(y)}\eta \leq y < 2^{l(y)+1}\eta$. With this, define h as

$$h(y) = 2^{1+l(y)}\eta + 2^{-N_{l(y)}} \left\lfloor 2^{1+N_{l(y)}}\left(y - 2^{l(y)}\eta\right)\right\rfloor, \quad y \geq \eta,$$

and note that

$$2y - 2^{-N_{l(y)}} < h(y) \leq 2y \quad \text{for all } y \geq \eta.$$

Moreover, it is readily confirmed that $h^2(y) = 2h(y)$ for every $y \geq \eta$. It follows that $h^n(y) - 2^{n+l(y)}\eta$ is an integer whenever $n > N_{l(y)}$, showing in turn that $\left(h^n(y)\right)$ is u.d. mod 1 for all $y \geq \eta$.

Now consider the map $f : [10^\eta, +\infty) \to \mathbb{R}^+$ given by $f(x) = 10^{h(\log x)}$, for which

$$f(x) - x^2 = \mathcal{O}\left(x^2 2^{-N_{l(\log x)}}\right) \quad \text{as } x \to +\infty. \tag{6.9}$$

From (6.9), it is clear that $f(x) - x^2 = o(g)$ as $x \to +\infty$, provided that (N_n) is growing sufficiently fast. Since $\log f^n(x_0) = h^n(\log x_0)$, it follows that $\left(f^n(x_0)\right)$ is Benford for every $x_0 \geq 10^\eta$. In other words, $[10^\eta, +\infty) \setminus B$ is empty.

(ii) Define a map $h : [0, +\infty) \to \mathbb{R}$ as

$$h(y) = 2^{-N_{\lfloor y \rfloor}} \left\lfloor 2^{1+N_{\lfloor y \rfloor}} y \right\rfloor + 1, \quad y \geq 0.$$

Note that

$$2y \leq 2y + 1 - 2^{-N_{\lfloor y \rfloor}} < h(y) \leq 2y + 1 \quad \text{for all } y \geq 0,$$

and $h^n(y)$ is an integer whenever n is sufficiently large. Thus $\left(h^n(y)\right)$ is not u.d. mod 1 for any $y \geq 0$. Analogously to (i), consider the map $f : [1, +\infty) \to \mathbb{R}^+$ given by $f(x) = 10^{h(\log x)}$, for which

$$f(x) - 10x^2 = \mathcal{O}\left(x^2 2^{-N_{\lfloor \log x \rfloor}}\right) \quad \text{as } x \to +\infty.$$

Again, it is clear that $f(x) - 10x^2 = o(g)$ as $x \to +\infty$, provided that (N_n) grows fast enough. As in (i), $\log f^n(x_0) = h^n(\log x_0)$, and so $\left(f^n(x_0)\right)$ is not Benford for any $x_0 \geq 1$, i.e., $[1, +\infty) \cap B$ is empty. ✠

As evidenced by Example 6.22, in order to relate the Benford properties of maps satisfying (6.8) to those of the reference map $f(x) = ax^b$, merely prescribing the order of magnitude as $x \to +\infty$ of the deviation $g(x) = f(x) - ax^b$ is not enough. Rather, the analytic properties of f, such as continuity or differentiability, also play a role. As it turns out, if f is *continuous* then there always exist many points x_0 for which $\left(f^n(x_0)\right)$ is Benford, but also many exceptional points, i.e., neither of the sets B and $\mathbb{R}^+ \setminus B$ in (6.7) is empty. In fact, both sets contain many points. (Note that neither of the two maps f considered in Example 6.22 is continuous.)

THEOREM 6.23. *Let $f : \mathbb{R}^+ \to \mathbb{R}^+$ be a map such that $f(x) = ax^b + g(x)$ with $a > 0$, $b > 1$, and $g = o(x^b)$ as $x \to +\infty$.*

(i) *If f is continuous then, for every $c > 0$, there exist uncountably many $x_0 \geq c$ for which $(f^n(x_0))$ is Benford, but also uncountably many $x_0 \geq c$ for which $(f^n(x_0))$ is not Benford, i.e., $[c, +\infty) \cap B$ and $[c, +\infty) \setminus B$ are both uncountable.*

(ii) *If f is continuously differentiable and $g' = o(x^{b-1}/\log x)$ as $x \to +\infty$, then there exists $c > 0$ such that $(f^n(x_0))$ is Benford for almost all $x_0 \geq c$, i.e., $[c, +\infty) \setminus B$ is a (Lebesgue) nullset.*

For the proof of Theorem 6.23, it is convenient to utilize (a simple variant of) the technique of *shadowing*. While the argument employed here is elementary, note that in dynamical systems theory, shadowing is a powerful and sophisticated tool; e.g., see [121, 123]. To appreciate the usefulness of a shadowing argument in the context of Benford's law, consider any map within the scope of Theorem 6.23, that is, let $f(x) = ax^b + g(x)$ with $a > 0$, $b > 1$, and $g(x) = o(x^b)$ as $x \to +\infty$. Assume $a = 1$ for convenience and notice that $(x_n) = (f^n(x_0))$ is strictly increasing, and that $\lim_{n \to \infty} f^n(x_0) = +\infty$ whenever x_0 is sufficiently large. With $y_n = \log x_n$, therefore,

$$y_n = \log f(10^{y_{n-1}}) = by_{n-1} + \log\left(1 + \frac{g(10^{y_{n-1}})}{10^{by_{n-1}}}\right) =: h(y_{n-1}), \qquad n \in \mathbb{N},$$

and the crucial question is whether or not $(y_n) = (h^n(y_0))$ is u.d. mod 1 for some, or even most, $y_0 = \log x_0$.

Recall from Example 4.15 that $(b^n y)$ is u.d. mod 1 for almost all, but not all, $y \in \mathbb{R}$. However, given any y_0, the two sequences $(h^n(y_0))$ and $(b^n y_0)$ may have very little in common. In fact, $(h^n(y_0) - b^n y_0)$ will typically be unbounded, even if $h(y) - by$ is bounded. For example, with the very simple map $h(y) = 2y + 1$, the orbit $(h^n(y_0)) = (2^n y_0 + 2^n - 1)$ is bounded if and only if $y_0 = -1$, whereas $(2^n y_0)$ is bounded only for $y_0 = 0$. For every y_0, therefore, $(h^n(y_0) - 2^n y_0)$ is unbounded. In this situation, the key observation of shadowing, made precise by Lemma 6.24 below, is that while $(h^n(y_0) - b^n y_0)$ is unbounded for every given y_0, there nevertheless exists a *unique* point $\widehat{s}(y_0) \in \mathbb{R}$, informally referred to as the *shadow* of y_0, with the property that $(h^n(y_0) - b^n \widehat{s}(y_0))$ remains bounded. (In the case of $h(y) = 2y+1$ considered earlier, $h^n(y_0) = 2^n y_0 + 2^n - 1 = 2^n(y_0+1) - 1$ shows that $\widehat{s}(y_0) = y_0 + 1$.) Thus $(h^n(y_0))$ resembles $(b^n y)$ for *some* y — in fact $y = \widehat{s}(y_0)$ — and a closer inspection of the map $\widehat{s}$ and its analytic properties is virtually all that is needed in order to establish Theorem 6.23.

LEMMA 6.24. (Shadowing Lemma) *Let $h : \mathbb{R} \to \mathbb{R}$ be a map such that $h(y) = by + \Gamma(y)$ with $b > 1$. If $\limsup_{y \to +\infty} |\Gamma(y)| < +\infty$ then there exist $\eta \in \mathbb{R}$ and a function $\widehat{s} : [\eta, +\infty) \to \mathbb{R}$ such that the sequence $(h^n(y) - b^n \widehat{s}(y))$ is bounded for every $y \geq \eta$. The function $\widehat{s}$ is continuous whenever h is continuous. Moreover, if $\lim_{y \to +\infty} \Gamma(y) = 0$ then $\lim_{y \to +\infty}(\widehat{s}(y) - y) = 0$ and, for every $y \geq \eta$, $\lim_{n \to \infty}(h^n(y) - b^n \widehat{s}(y)) = 0$.*

PROOF. Since $\limsup_{y\to+\infty} |\Gamma(y)|$ is finite, there exist $\eta_0 \in \mathbb{R}$ and $\gamma > 0$ such that $|\Gamma(y)| \le \gamma$ for all $y \ge \eta_0$. Thus if $y \ge \eta := \max\{\eta_0, 2\gamma/(b-1)\}$ then

$$h(y) = by + \Gamma(y) \ge by - \gamma = (b-1)y - \gamma + y \ge y + \gamma\,,$$

and hence $h^n(y) \ge y \ge \eta$ for all n, and $\lim_{n\to\infty} h^n(y) = +\infty$. In this case, therefore,

$$h^n(y) = bh^{n-1}(y) + \Gamma \circ h^{n-1}(y) = \ldots = b^n y + \sum_{j=1}^{n} b^{n-j}\Gamma \circ h^{j-1}(y)\,.$$

Since $b > 1$, the number

$$\widehat{s}(y) = y + \sum_{j=1}^{\infty} b^{-j}\Gamma \circ h^{j-1}(y) \tag{6.10}$$

is well-defined for every $y \ge \eta$. Moreover, for every n,

$$|h^n(y) - b^n\widehat{s}(y)| = \left|\sum_{j=n+1}^{\infty} b^{n-1}\Gamma \circ h^{j-1}(y)\right| = \left|\sum_{j=1}^{\infty} b^{-j}\Gamma \circ h^{j+n-1}(y)\right|$$
$$\le \frac{\gamma}{b-1}\,. \tag{6.11}$$

Thus the sequence $\big(h^n(y) - b^n\widehat{s}(y)\big)$ is bounded. (Since $b > 1$, given any $y \ge \eta$, it is clear that $\widehat{s}(y)$ is the *only* point with this property.) If h is continuous on $[\eta, +\infty)$ then so is $\Gamma \circ h^{j-1}$ for every $j \in \mathbb{N}$. In this case, the sum on the right in (6.10) converges uniformly, by the Weierstrass M-test, and hence the function $\widehat{s}$ is continuous also.

Finally, assume that $\lim_{y\to+\infty} \Gamma(y) = 0$. Given any $\varepsilon > 0$, there exists $y_\varepsilon \ge \eta$ such that $|\Gamma(y)| \le \varepsilon(b-1)$ whenever $y \ge y_\varepsilon$. Thus, if $y \ge y_\varepsilon$ then $h^{j-1}(y) \ge y \ge y_\varepsilon \ge \eta$ for all $j \in \mathbb{N}$, which in turn implies that

$$|\widehat{s}(y) - y| < \sum_{j=1}^{\infty} b^{-j}\varepsilon(b-1) = \varepsilon\,,$$

and hence $\lim_{y\to+\infty}(\widehat{s}(y) - y) = 0$. Similarly, if $y \ge \eta$ then $|\Gamma \circ h^{j+n-1}(y)| \le \gamma$ for all j and n, and $\lim_{n\to\infty} \Gamma \circ h^{j+n-1}(y) = 0$. Thus (6.11) and the Dominated Convergence Theorem yield $\lim_{n\to\infty}\big(h^n(y) - b^n\widehat{s}(y)\big) = 0$. ∎

PROOF OF THEOREM 6.23. Assume that $f(x) = ax^b + g(x)$, with $a > 0$, $b > 1$, and $g = o(x^b)$ as $x \to +\infty$. Without loss of generality, it can be assumed that $a = 1$. (Otherwise replace $f(x)$ by $\alpha f(x/\alpha)$ with $\alpha = a^{(b-1)^{-1}}$.) Given any $x_0 > 0$, let $y_n = \log x_n = \log f^n(x_0)$ and observe that $y_n = h(y_{n-1})$, where $h : \mathbb{R} \to \mathbb{R}$ is given by

$$h(y) = by + \log\left(1 + \frac{g(10^y)}{10^{by}}\right)\,. \tag{6.12}$$

Note that $g = o(x^b)$ as $x \to +\infty$ implies $\lim_{y\to+\infty}(h(y) - by) = 0$.

To prove assertion (i) in Theorem 6.23, assume that f is continuous. From (6.12), it is clear that h is also continuous, and by Lemma 6.24, there exist $\eta \in \mathbb{R}$ and a continuous function $\widehat{s} : [\eta, +\infty) \to \mathbb{R}$ with $\widehat{s}(y) - y \to 0$ as $y \to +\infty$ such that $\lim_{n\to\infty}\left(h^n(y) - b^n\widehat{s}(y)\right) = 0$ for all $y \geq \eta$. Thus, by Proposition 4.3(i), for $y \geq \eta$, $\left(h^n(y)\right)$ is u.d. mod 1 if and only if $\left(b^n\widehat{s}(y)\right)$ is. Fix any $c > 0$ with $\log c \geq \eta$. By the Intermediate Value Theorem, $\widehat{s}\big([\log c, +\infty)\big) \supset [\widehat{s}(\log c), +\infty)$, and by Proposition 6.20, the sets

$$\{y \geq \widehat{s}(\log c) : (b^n y) \text{ is u.d. mod 1}\,\}$$

and

$$\{y \geq \widehat{s}(\log c) : (b^n y) \text{ is not u.d. mod 1}\,\}$$

are both uncountable. Hence the set

$$U_c = \left\{y \geq \log c : \left(h^n(y)\right) \text{ is u.d. mod 1}\right\} = \left\{y \geq \log c : \left(b^n\widehat{s}(y)\right) \text{ is u.d. mod 1}\right\}$$

is uncountable, and so is $[\log c, +\infty) \setminus U_c$. In other words, the sets

$$[c, +\infty) \cap B = \left\{x_0 \geq c : \left(f^n(x_0)\right) \text{ is Benford}\right\}$$

and

$$[c, +\infty) \setminus B = \left\{x_0 \geq c : \left(f^n(x_0)\right) \text{ is not Benford}\right\}$$

are both uncountable.

To prove (ii), assume that f is C^1 and $g' = o(x^{b-1}/\log x)$ as $x \to +\infty$. With $h : \mathbb{R} \to \mathbb{R}$ given by (6.12), recall that $\widehat{s}(y) = y + \sum_{j=1}^{\infty} b^{-j}\Gamma \circ h^{j-1}(y)$, where

$$\Gamma(y) = \log\left(1 + \frac{g(10^y)}{10^{by}}\right), \quad y \in \mathbb{R},$$

and consequently

$$\Gamma'(y) = \frac{10^y g'(10^y) - bg(10^y)}{10^{by} + g(10^y)}.$$

The assumption $g' = o(x^{b-1}/\log x)$ as $x \to +\infty$ implies that $\Gamma'(y) = o(y^{-1})$ as $y \to +\infty$. Deduce from

$$\frac{\mathrm{d}}{\mathrm{d}y}\left(b^{-j}\Gamma \circ h^{j-1}(y)\right) = b^{-j}\Gamma' \circ h^{j-1}(y)(h^{j-1})'(y)$$

$$= b^{-1}\Gamma' \circ h^{j-1}(y)\prod_{\ell=1}^{j-1}\left(1 + \frac{\Gamma' \circ h^{\ell-1}(y)}{b}\right)$$

that $\lim_{y\to+\infty}\left(b^{-j}\Gamma \circ h^{j-1}(y)\right)' = 0$, and also, with the appropriate constant $\gamma_0 > 0$, that

$$\left|\frac{\mathrm{d}}{\mathrm{d}y}\left(b^{-j}\Gamma \circ h^{j-1}(y)\right)\right| \leq \gamma_0 b^{-j} \quad \text{for all } j \in \mathbb{N}, y \geq \eta.$$

Thus the function $\widehat{s}$ defined by (6.10) is C^1, with its derivative $\widehat{s}'$ given by termwise differentiation, and the Dominated Convergence Theorem implies

$$\lim_{y \to +\infty} \widehat{s}'(y) = \lim_{y \to +\infty} \left(1 + \sum_{j=1}^{\infty} \left(b^{-j} \Gamma \circ h^{j-1}(y) \right)' \right) = 1 \,.$$

For every sufficiently large $c > 0$, therefore, $\widehat{s}$ is a diffeomorphism of $[\log c, +\infty)$ onto $[\widehat{s}(\log c), +\infty)$, and so maps nullsets onto nullsets. In particular, the set $[\log c, +\infty) \setminus U_c$ is a nullset. In other words, $[c, +\infty) \setminus B$ is a nullset. ∎

Remark. (i) Under the assumptions of Theorem 6.23(ii), the above proof shows that, for every sufficiently large $c > 0$, the set $[c, +\infty) \setminus B$ is dense in $[c, +\infty)$. By (i) of the theorem, it is also uncountable. Again, it can be shown that $[c, +\infty) \cap B$ is of first category; see [13].

(ii) The assumption in Theorem 6.23(ii) that $g' = o(x^{b-1}/\log x)$ as $x \to +\infty$ can be relaxed to $g' = o\big(x^{b-1}/(\log \log x)^{1+\varepsilon}\big)$ for some $\varepsilon > 0$. On the other hand, as demonstrated by Example 6.27 below, the conclusion of that theorem may fail if $g' = \mathcal{O}(x^{b-1})$.

EXAMPLE 6.25. Let the map f be polynomial of degree at least two, i.e.,

$$f(x) = a_p x^p + a_{p-1} x^{p-1} + \ldots + a_1 x + a_0 \,,$$

where $p \in \mathbb{N} \setminus \{1\}$ and $a_0, a_1, \ldots, a_{p-1}, a_p \in \mathbb{R}$ with $a_p \neq 0$. Assume without loss of generality that $a_p > 0$. (Otherwise replace $f(x)$ by $-f(-x)$ if p is even, or by $f^2(x)$ if p is odd.) The map f is continuously differentiable and $(f(x) - a_p x^p)' = \mathcal{O}(x^{p-2})$ as $x \to +\infty$. By Theorem 6.23, $\big(f^n(x_0)\big)$ is Benford for almost all, but not all, x_0 with $|x_0|$ sufficiently large.

For a simple concrete example, let $f(x) = x^2 + 1$. Since $f^n(x) \geq n$ for every $n \in \mathbb{N}$ and $x \in \mathbb{R}$, therefore, $\big(f^n(x_0)\big)$ is Benford for almost all $x_0 \in \mathbb{R}$, but the set $\big\{x_0 : \big(f^n(x_0)\big) \text{ is not Benford}\big\}$ is also uncountable and dense in $\mathbb{R}$. At the time of this writing, it is not known whether $\big(f^n(0)\big) = (1, 2, 5, 26, 677, \ldots)$, or in fact $\big(f^n(k)\big)$ for any integer k, is Benford; note that $f^n(0)$ equals the number of binary trees of height less than n [117, A003095].

For another concrete example, consider the (polynomial) map $f(x) = x - \frac{1}{6}x^3$ of Example 6.10(iii). As seen earlier, $|f^n(x_0)| \to +\infty$ whenever $|x_0| > 2\sqrt{3}$, and it is clear from the above that $\big(f^n(x_0)\big)$ is Benford for almost all, but not all, x_0 with $|x_0| > 2\sqrt{3}$. ✠

Careful inspection of the proof of Theorem 6.23(ii) shows that it suffices to require that the map $f : \mathbb{R}^+ \to \mathbb{R}^+$ be *absolutely continuous*. Recall that this means that f is an indefinite integral of its derivative f', which exists almost everywhere on $\mathbb{R}^+$. For example, every convex or continuously differentiable function is absolutely continuous. (This notion of absolute continuity of functions is consistent with the standard notion of absolute continuity of random variables as follows: A real-valued random variable X is absolutely continuous if and only if its distribution function F_X is absolutely continuous.) If Theorem

6.23(ii) is thus generalized, the symbol g' has to be understood as the (measurable) function that coincides with $f'(x) - abx^{b-1}$ whenever $f'(x)$ exists, and $g'(x) := 0$ otherwise.

EXAMPLE 6.26. Consider the continuous map $f : \mathbb{R}^+ \to \mathbb{R}^+$ with

$$f(x) = (2n - 1)x - n(n - 1) \quad \text{for all } n \in \mathbb{N}, x \in (n - 1, n].$$

Clearly, the map f, a piecewise linear interpolation of the map $f_0(x) = x^2$, is not continuously differentiable. However,

$$0 \le f(x) - x^2 = -x^2 + (2n - 1)x - n(n - 1) = \tfrac{1}{4} - \tfrac{1}{4}\big(2x - (2n - 1)\big)^2 \le \tfrac{1}{4},$$

and so Theorem 6.23(i) applies, since $g(x) = f(x) - x^2 = \mathcal{O}(1)$ as $x \to +\infty$. Moreover, f is convex, hence absolutely continuous, and

$$|g'(x)| = |f'(x) - 2x| = |2n - 1 - 2x| < 1 \quad \text{for all } x \in (n - 1, n).$$

Thus $g' = \mathcal{O}(1)$ as $x \to +\infty$, where $g'(n) := 0$ for every $n \in \mathbb{N}$. Theorem 6.23(ii), generalized to absolutely continuous maps f as outlined above, then shows that $\big(f^n(x_0)\big)$ is Benford for almost all sufficiently large x_0 — in fact for almost all $x_0 > 1$ — but there are also uncountably many exceptional points. ✠

EXAMPLE 6.27. This example shows that the condition $g' = o(x^{b-1}/\log x)$ in Theorem 6.23(ii) is needed even if f is very smooth. Let $h : \mathbb{R} \to \mathbb{R}$ be the (real-analytic) function with $h(0) = 0$ and

$$h(y) = 2y - \frac{\sin(2\pi y^2)}{2\pi y}, \quad y \ne 0.$$

Since $h(k) = 2k \in \mathbb{Z}$ and $h'(k) = 0$ for all $k \in \mathbb{Z} \setminus \{0\}$, clearly $\operatorname{dist}(h^n(y), \mathbb{Z}) \to 0$ as $n \to \infty$ whenever $|y| \ge 1$ and $\operatorname{dist}(y, \mathbb{Z})$ is sufficiently small; here, as usual, $\operatorname{dist}(y, \mathbb{Z}) = \min\{|y - k| : k \in \mathbb{Z}\}$. Specifically, it is straightforward to show that $h(J) \subset J$, with the countable union of intervals

$$J = \bigcup_{k \in \mathbb{Z} \setminus \{0\}} \left[k - \frac{1}{25|k|}, k + \frac{1}{25|k|}\right],$$

and $\lim_{n \to \infty}(h^n(y) - 2^n k) = 0$ for every $y \in J$ and the appropriate integer k. For every $y \in J$, therefore, $\big(h^n(y)\big)$ is not u.d. mod 1. (It is tempting to speculate whether the set $\{y \in \mathbb{R} : \big(h^n(y)\big) \text{ is u.d. mod 1}\}$ is actually a nullset. If it were, then the sequence $(2^n \sqrt{2})$, for instance, would not be u.d. mod 1 or, in number-theoretic parlance, the number $\sqrt{2}$ would not be 2-*normal*. This in turn would settle a well-known open problem; cf. Example 4.15.)

Next define a (real-analytic) map $f : \mathbb{R}^+ \to \mathbb{R}^+$ as

$$f(x) = 10^{h(\log x)} = x^2 10^{-\frac{\sin(2\pi(\log x)^2)}{2\pi \log x}} = x^2 + g(x),$$

with the deviation g given by

$$g(x) = f(x) - x^2 = x^2 \left(10^{-\frac{\sin(2\pi(\log x)^2)}{2\pi \log x}} - 1 \right) = \mathcal{O} \left(\frac{x^2}{\log x} \right) \quad \text{as } x \to +\infty .$$

Hence Theorem 6.23(i) applies, with $a = 1$, $b = 2$. Also, it is readily confirmed that $g' = \mathcal{O}(x)$ as $x \to +\infty$. However, it is clear from the above that $\big(f^n(x_0)\big)$ is not Benford whenever $\log x_0 \in J$ or, equivalently, whenever

$$x_0 \in 10^J := \{10^y : y \in J\} = \bigcup\nolimits_{k \in \mathbb{Z}\setminus\{0\}} 10^k \left[10^{-\frac{1}{25|k|}}, 10^{\frac{1}{25|k|}} \right] ,$$

and the set 10^J has positive (in fact, infinite) Lebesgue measure. Thus the conclusion of Theorem 6.23(ii) does not hold.   ✠

As usual, Theorem 6.23 has a simple corollary via taking reciprocals.

COROLLARY 6.28. [15, Thm. 4.1] *Let* $f : \mathbb{R} \to \mathbb{R}$ *be a smooth map with* $f(0) = 0$, $f'(0) = 0$, *and* $f^{(p)}(0) \neq 0$ *for some* $p \in \mathbb{N} \setminus \{1\}$. *Then* $\big(f^n(x_0)\big)$ *is Benford for almost all* x_0 *sufficiently close to* 0, *but there are also uncountably many exceptional points.*

EXAMPLE 6.29. For the smooth map $f(x) = x - 1 + e^{-x}$, $f(0) = f'(0) = 0$ but $f''(0) = 1 \neq 0$. For almost all x_0 sufficiently close to 0, therefore, $\big(f^n(x_0)\big)$ is Benford. Since $\lim_{n \to \infty} f^n(x_0) = 0$ for every x_0, and f maps nullsets to nullsets, it is clear from Corollary 6.28 that in fact $\big(f^n(x_0)\big)$ is Benford for almost all, but not all, $x_0 \in \mathbb{R}$.   ✠

EXAMPLE 6.30. In Corollary 6.28, the assumption $f^{(p)}(0) \neq 0$ for some $p \geq 2$ is crucial in that the conclusion may fail if $f^{(p)}(0) = 0$ for all $p \in \mathbb{N}$. For a simple example, let $h : \mathbb{R} \to \mathbb{R}$ be any smooth non-decreasing map with $h(y) = (2k-1)^3$ for every $k \in \mathbb{Z}$ and all $y \in [2k - 1, 2k]$. The map $f : \mathbb{R} \to \mathbb{R}$ with $f(0) = 0$ and

$$f(x) = 10^{h(\log |x|)} , \quad x \neq 0 ,$$

is smooth, and $f^{(p)}(0) = 0$ for every $p \in \mathbb{N}$. If $\log |x_0| \in \bigcup_{k \in \mathbb{Z}}[2k - 1, 2k]$ or, equivalently, if

$$x_0 \in \bigcup\nolimits_{k \in \mathbb{Z}} 10^{2k-1}\big([-10, -1] \cup [1, 10]\big) =: J ,$$

then $\big(f^n(x_0)\big)$ is not Benford since $S\big(f^n(x_0)\big) \equiv 1$. Note that $J \cap (-\varepsilon, \varepsilon)$ has positive measure for every $\varepsilon > 0$. Thus the conclusion of Corollary 6.28 fails for the smooth map f.   ✠

To conclude this section, note that maps such as $f(x) = e^x$ that have an even faster growth than $f(x) = ax^b$ are not covered by any of the results discussed so far. The following proposition addresses some such maps; the result is a special case of Theorem 6.49 in Section 6.6, and a proof is given there.

PROPOSITION 6.31. *Let $f : \mathbb{R}^+ \to \mathbb{R}^+$ be a map such that, for some $c \geq 0$, both of the following conditions hold:*

(i) *The function $\log f(10^x)$ is convex on $(c, +\infty)$;*

(ii) $\dfrac{\log f(10^x) - \log f(10^c)}{x - c} > 1$ *for all $x > c$.*

Then $(f^n(x_0))$ is Benford for almost all sufficiently large x_0, but there also exist uncountably many $x_0 > c$ for which $(f^n(x_0))$ is not Benford.

Remark. When applied to $f(x) = ax^b$ with $a > 0$, $b > 1$, the assertion of Proposition 6.31 is weaker than that of Proposition 6.20 in that the latter does not require x_0 to be "sufficiently large" (a property that may depend on a and b) and also guarantees denseness of exceptional points.

EXAMPLE 6.32. For $f(x) = e^x$, the map $h(x) = \log f(10^x) = 10^x \log e$ is convex on $\mathbb{R}$, and $h(x)/x > h'(0) = 1$ for all $x > 0$. Hence Proposition 6.31 applies with $c = 0$ and shows that $(f^n(x_0))$ is Benford for almost all sufficiently large x_0. In fact, since $f^n(x_0) > 2^{n-2}$ for every $x_0 \in \mathbb{R}$ and $n \geq 2$, the sequence $(f^n(x_0))$ is Benford for almost all, but not all, $x_0 \in \mathbb{R}$. Again, it can be shown that the set of exceptional points is dense in $\mathbb{R}$, besides being uncountable by Proposition 6.31. ✠

EXAMPLE 6.33. The (polynomial) map $f(x) = x^2 - 2x + 2$ satisfies (ii) in Proposition 6.31, provided that $c > \frac{1}{2}\log 2$. However, $\log f(10^x)$ is *concave* on $\left[\log(2 + \sqrt{2}), +\infty\right)$, and consequently Proposition 6.31 does not apply. Recall that, on the other hand, Theorem 6.23 does apply, showing that $(f^n(x_0))$ is Benford for almost all, but not all, $x_0 \notin [0, 2]$; see also Example 6.25. ✠

Notice that Proposition 6.31 does not impose any restriction on the order of magnitude as $x \to +\infty$ of the deviation of f from some reference map f_0, such as $f_0(x) = x$, $f_0(x) = ax$, and $f_0(x) = ax^b$ used in Theorems 6.4, 6.13, and 6.23, respectively.

EXAMPLE 6.34. The map $f : \mathbb{R}^+ \to \mathbb{R}^+$ with

$$f(x) = \frac{x^m}{(m-1)!} \quad \text{for every } m \in \mathbb{N}, \ x \in (m-1, m],$$

is neither continuously differentiable nor $\mathcal{O}(x^b)$ as $x \to +\infty$, for any $b > 1$. Nevertheless, Proposition 6.31 applies directly with $c = 0$, showing that $(f^n(x_0))$ is Benford for almost all, but not all, $x_0 > 1$. ✠

6.5   AN APPLICATION TO NEWTON'S METHOD

In scientific calculations using digital computers and floating-point arithmetic, roundoff errors are inevitable, and as Knuth points out in his classic text *The Art of Computer Programming* [90, pp. 253–255],

> [I]n order to analyze the average behavior of floating-point arith-
> metic algorithms (and in particular to determine their average run-
> ning time), we need some statistical information that allows us to
> determine how often various cases arise ... [If, for example, the] lead-
> ing digits tend to be small [, that] makes the most obvious techniques
> of "average error" estimation for floating-point calculations invalid.
> The relative error due to rounding is usually ... more than expected.

One of the most widely used floating-point algorithms is *Newton's method* for finding the roots of a given (differentiable) function numerically. Thus when using Newton's method, it is important to keep track of the distribution of significant digits (or significands) of the approximations generated by the method. As will be seen shortly, the differences between successive Newton approximations, and the differences between the successive approximations and the unknown root, often exhibit exactly the type of non-uniformity of significant digits alluded to by Knuth — they typically follow Benford's law.

Throughout this section, let $g : I \to \mathbb{R}$ be a differentiable function defined on some open interval $I \subset \mathbb{R}$, and denote by N_g the map associated with g by Newton's method, that is,

$$N_g(x) = x - \frac{g(x)}{g'(x)} \quad \text{for all } x \in I \text{ with } g'(x) \neq 0 \,.$$

For N_g to be defined wherever g is, set $N_g(x) = x$ if $g'(x) = 0$. Using Newton's method for finding roots of g (i.e., real numbers x^* with $g(x^*) = 0$) amounts to picking an initial point $x_0 \in I$ and iterating N_g. Henceforth in this section, (x_n) denotes the resulting sequence of approximations starting at x_0, that is, $(x_n) = \left(N_g^n(x_0) \right)$.

Clearly, if (x_n) converges to x^*, say, and if N_g is continuous at x^*, then $N_g(x^*) = x^*$, so x^* is a fixed point of N_g, and $g(x^*) = 0$. (Note that according to the definition of N_g used here, $N_g(x^*) = x^*$ could also mean that $g'(x^*) = 0$. If, however, $g'(x^*) = 0$ but $g(x^*) \neq 0$, then N_g is not continuous at x^* unless g is constant.) It is this correspondence between the roots of g and the fixed points of N_g that makes Newton's method work locally. Often, every fixed point x^* of N_g is attracting, i.e., $\lim_{n \to \infty} N_g^n(x_0) = x^*$ for all x_0 sufficiently close to x^*. (Observe that if g is *linear* near x^*, i.e., $g(x) = a(x - x^*)$ for some $a \neq 0$ and all x near x^*, then $N_g(x) = x^*$ for all x near x^*.)

To formulate a result about Benford's law for Newton's method, it will be assumed that the function $g : I \to \mathbb{R}$ is *real-analytic*. Recall that this means that g can be represented by its Taylor series in a neighborhood of each point of I. Although real-analyticity is a strong assumption indeed, the class of real-analytic functions covers most practically relevant cases, including all polynomials, and all rational, exponential, and trigonometric functions, and compositions thereof.

If $g : I \to \mathbb{R}$ is real-analytic and $x^* \in I$ is a root of g, i.e., if $g(x^*) = 0$, then $g(x) = (x - x^*)^m h(x)$ for some $m \in \mathbb{N}$ and some real-analytic $h : I \to \mathbb{R}$ with $h(x^*) \neq 0$. The number m is the *multiplicity* of the root x^*; if $m = 1$

then x^* is referred to as a *simple* root. The following theorem becomes plausible upon observing that $g(x) = (x - x^*)^m h(x)$ implies that N_g is real-analytic in a neighborhood of x^*, and

$$
\begin{aligned}
N_g'(x) &= \frac{g(x)g''(x)}{g'(x)^2} \\
&= \frac{m(m-1)h(x)^2 + 2m(x-x^*)h'(x)h(x) + (x-x^*)^2 h''(x)h(x)}{m^2 h(x)^2 + 2m(x-x^*)h'(x)h(x) + (x-x^*)^2 h'(x)^2},
\end{aligned}
$$

so that in particular $N_g'(x^*) = 1 - m^{-1}$. For the following main result on Benford's law for Newton's method, recall that $(x_n) = \big(N_g^n(x_0)\big)$ is the sequence of Newton's method approximations for the root x^* of g, starting at x_0.

THEOREM 6.35 ([19]). *Let the function* $g : I \to \mathbb{R}$ *be real-analytic with* $g(x^*) = 0$, *and assume that* g *is not linear.*

(i) *If* x^* *is a simple root, then* $(x_n - x^*)$ *and* $(x_{n+1} - x_n)$ *are both Benford for (Lebesgue) almost all, but not all, x_0 in a neighborhood of* x^*.

(ii) *If* x^* *is a root of multiplicity at least two, then* $(x_n - x^*)$ *and* $(x_{n+1} - x_n)$ *are Benford for all $x_0 \neq x^*$ sufficiently close to* x^*.

The proof of Theorem 6.35 given in [19] uses the following lemma, which may be of independent interest for studying Benford's law in other numerical approximation procedures. Part (i) is an analogue of Theorem 4.12, and (ii) and (iii) follow directly from Corollaries 6.28 and 6.18, respectively.

LEMMA 6.36. *Let* $f : I \to I$ *be* C^∞, *and assume that* $f(x^*) = x^*$ *for some* $x^* \in I$.

(i) *If* $f'(x^*) \neq 1$, *then for every x_0 with* $\lim_{n\to\infty} f^n(x_0) = x^*$, *the sequence* $(f^n(x_0) - x^*)$ *is Benford precisely when* $\big(f^{n+1}(x_0) - f^n(x_0)\big)$ *is Benford.*

(ii) *If* $f'(x^*) = 0$ *but* $f^{(p)}(x^*) \neq 0$ *for some* $p \in \mathbb{N}\backslash\{1\}$, *then* $(f^n(x_0) - x^*)$ *is Benford for (Lebesgue) almost all, but not all, x_0 in a neighborhood of* x^*.

(iii) *Suppose* $0 < |f'(x^*)| < 1$. *Then, for every $x_0 \neq x^*$ sufficiently close to* x^*, $(f^n(x_0) - x^*)$ *is Benford if and only if* $\log|f'(x^*)|$ *is irrational.*

In order to relate Theorem 6.35 to the results of the previous sections, suppose the real-analytic function $g : I \to \mathbb{R}$ is not linear, and assume $g(x^*) = 0$. For convenience, define a (real-analytic) auxiliary map f as $f(x) = N_g(x + x^*) - x^*$. Then $f(0) = 0$, $f'(0) = N_g'(x^*) = 1 - m^{-1}$, and $f^n(x - x^*) = N_g^n(x) - x^*$ for all $n \in \mathbb{N}$ and $x \in I$. It follows that $(x_n - x^*) = \big(f^n(x_0 - x^*)\big)$ and $(x_{n+1} - x_n) = \big(f^{n+1}(x_0 - x^*) - f^n(x_0 - x^*)\big)$; hence the Benford properties of $(x_n - x^*)$ and $(x_{n+1} - x_n)$ can also be studied directly using Corollaries 6.18 and 6.28, together with Lemma 6.36(i).

EXAMPLE 6.37. **(i)** Let $g(x) = e^x - 2$. Then g has a unique simple root, namely, $x^* = \ln 2 = 0.6931$, and $N_g(x) = x - 1 + 2e^{-x}$. By Theorem 6.35(i), the sequences $(x_n - x^*)$ and $(x_{n+1} - x_n)$ are both Benford for almost all x_0 near x^*. In fact, the auxiliary map f defined above simply takes the form $f(x) = x - 1 + e^{-x}$, and Example 6.29, together with Lemma 6.36(i), shows that $(x_n - x^*)$ and $(x_{n+1} - x_n)$ are Benford for almost all, but not all, $x_0 \in \mathbb{R}$; see Figure 6.5.

(ii) Let $g(x) = (e^x - 2)^3$. Then g has a triple root at $x^* = \ln 2$, and $N_g(x) = x - \frac{1}{3} + \frac{2}{3}e^{-x}$. By Theorem 6.35(ii), the sequences $(x_n - x^*)$ and $(x_{n+1} - x_n)$ are both Benford for every $x_0 \neq 1$ near x^*; see Figure 6.5. The auxiliary map f here is $f(x) = x - \frac{1}{3} + \frac{1}{3}e^{-x}$, and again it follows directly from Example 6.19 and Lemma 6.36(i) that $(x_n - x^*)$ and $(x_{n+1} - x_n)$ are Benford unless $N_g(x_0) = x^*$, that is, unless $x_0 = x^*$ or $x_0 = x^* - 1.903 = -1.210$; in the latter case, clearly, both sequences are identically zero.   ✠

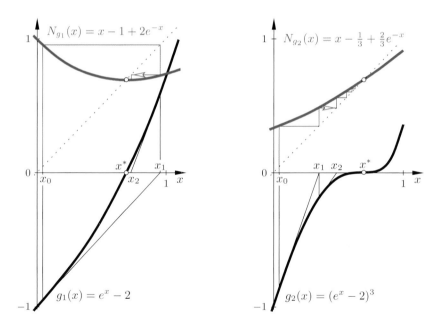

Figure 6.5: Visualizing Newton's method and Theorem 6.35: For the function $g_1(x) = e^x - 2$, the sequence $\left(N_{g_1}^n(x_0)\right)$ converges to the unique simple root $x^* = \ln 2$ super-exponentially (left); for $g_2(x) = g_1(x)^3 = (e^x - 2)^3$, convergence of $\left(N_{g_2}^n(x_0)\right)$ to the triple root x^* is only exponential; see Example 6.37.

Utilizing Lemma 6.36, an analogue of Theorem 6.35 can be established for other root-finding algorithms as well.

EXAMPLE 6.38. Let $g(x) = x + x^3$ and consider the successive approximations (ξ_n) for the root $x^* = 0$ of g, generated iteratively from $\xi_0 \in \mathbb{R}$ by the Jacobi–

Steffensen method,

$$\xi_n = \xi_{n-1} - \frac{g(\xi_{n-1})^2}{g(\xi_{n-1}) - g(\xi_{n-1} - g(\xi_{n-1}))}, \quad n \geq 1 .$$

For almost all, but not all, ξ_0 near $x^* = 0$, the sequence (ξ_n) is Benford. This follows from Lemma 6.36(ii), since $\xi_n = J_g^n(\xi_0)$, with the Jacobi–Steffensen transformation

$$J_g(x) = -x^5 \frac{1 - x^2}{1 + x^2 - x^4 + x^6} ,$$

and $J_g(0) = J_g'(0) = \ldots = J_g^{(4)}(0) = 0$, yet $J_g^{(5)}(0) \neq 0$. Alternatively, $J_g = N_h$ with the real-analytic function $h(x) = (x + x^3)e^{\frac{1}{4}x^4 - x^2}$, so Theorem 6.35(i) applies directly as well. ✠

If g fails to be real-analytic, then N_g may not be well-behaved analytically. For instance, N_g may have discontinuities even if g is C^∞ and $g' > 0$. Pathologies like this can cause Newton's method to fail for a variety of reasons, of which the reader can get an impression from [19, Sec. 4]. Even if N_g is smooth, (x_n) may not be Benford.

EXAMPLE 6.39. (i) For the C^∞-function g given by

$$g(x) = \begin{cases} e^{\frac{1}{2}(x^2 - x^{-2})} & \text{if } x \neq 0 , \\ 0 & \text{if } x = 0 , \end{cases}$$

the associated Newton map

$$N_g(x) = x - \frac{x^3}{x^4 + 1}$$

is smooth (in fact, real-analytic), and $\lim_{n\to\infty} x_n = \lim_{n\to\infty} N_g^n(x_0) = 0 = x^*$ for every $x_0 \in \mathbb{R}$. However, $N_g'(0) = 1$ and, as seen already in Example 6.10(ii), the sequence (x_n) is not Benford for any x_0.

(ii) Similarly, if g is the C^∞-function

$$g(x) = \begin{cases} e^{-3x^{-2}} & \text{if } x \neq 0 , \\ 0 & \text{if } x = 0 , \end{cases}$$

then the associated map $N_g(x) = x - \frac{1}{6}x^3$ is smooth. Here $\lim_{n\to\infty} x_n = 0 = x^*$ only if $|x_0| < 2\sqrt{3}$, and (x_n) is not Benford in this case; see Example 6.10(iii). On the other hand, recall from Example 6.25 that (x_n) is Benford for almost all, but not all, x_0 with $|x_0| > 2\sqrt{3}$. (Note, however, that $\lim_{n\to\infty} |x_n| = +\infty$ for $|x_0| > 2\sqrt{3}$, and so, as a root-*finding* algorithm, Newton's method fails completely in this case.) ✠

As will be seen in Chapter 10, the appearance of Benford's law in Newton's method and other root-finding methods has important implications for estimating errors in scientific calculations that rely on such algorithms.

6.6   TIME-VARYING SYSTEMS

Except for Example 6.2(ii, iii), the sequences considered in this chapter so far have all been generated by the iteration of a single map f or, in dynamical systems parlance, by an *autonomous* system (or recurrence relation). Autonomous systems constitute a classical and well-studied field. Beyond this field there has been, in the recent past, an increased interest in systems that are *nonautonomous*, i.e., explicitly time-varying in one way or another [89]. This development is motivated and driven by important practical applications as well as purely mathematical questions. In this context, it is interesting to study how the results discussed previously extend to systems with the map f changing with n. In full generality, this is a very wide topic with many open problems, both conceptual and computational. Only a small number of pertinent results (some with partial proofs or no proofs at all) and examples are presented here. The interested reader is referred to [13] for a fuller account and references, and to [91, 95] for an intriguing specific problem.

Throughout this section, let (f_n) be a sequence of functions that map $C \subset \mathbb{R}$ into itself. As in previous sections, most results and examples are for $C = \mathbb{R}^+$ but can easily be adjusted to other settings. Given any $x_0 \in C$, the sequence

$$(x_n) = \big(f_1(x_0), f_2\big(f_1(x_0)\big), \ldots\big) = \big(f_n \circ \ldots \circ f_1(x_0)\big)$$

is the *(nonautonomous) orbit* of x_0 (under $f_1, f_2, \ldots$). Equivalently, (x_n) is the unique solution of the (nonautonomous) recurrence relation

$$x_n = f_n(x_{n-1}), \quad n \in \mathbb{N}.$$

As an analogue of the maps leading to exponential growth (as studied in Section 6.3), consider a sequence of maps $f_n : \mathbb{R}^+ \to \mathbb{R}^+$ of the form

$$f_n(x) = a_n x + g_n(x), \quad n \in \mathbb{N}, \tag{6.13}$$

where $a_n > 0$ and the function g_n is small in some sense. Note that unlike in Theorem 6.13, a_n may be strictly less than 1 for some n in (6.13). The following result contains Corollary 6.14 as a special case.

THEOREM 6.40. *For every* $n \in \mathbb{N}$, *let* $f_n : \mathbb{R}^+ \to \mathbb{R}^+$ *be a map such that* $f_n(x) = a_n x + g_n(x)$ *with* $a_n > 0$. *Assume that* $\liminf_{n \to \infty} a_n > 1$ *and that* $g_n = o(x/\log x)$ *uniformly as* $x \to +\infty$. *Then, for every sufficiently large* x_0, $\big(f_n \circ \ldots \circ f_1(x_0)\big)$ *is Benford if and only if* $\big(\prod_{j=1}^{n} a_j\big) = (a_1, a_1 a_2, \ldots)$ *is Benford.*

PROOF. Pick $\delta > 0$ and $N_0 \in \mathbb{N}$ such that $a_n \geq 1 + 2\delta$ for all $n \geq N_0$. Also, choose $\xi_1 > 10$ large enough to ensure $|g_n(x) \log x| < \delta x$ for all $x \geq \xi_1$ and $n \in \mathbb{N}$. With this,

$$f_n(x) = x \left(a_n + \frac{g_n(x)}{x}\right) > x(a_n - \delta) \geq (1 + \delta)x \quad \text{for all } x \geq \xi_1, n \geq N_0.$$

Observe that $\lim_{x \to +\infty} f_{N_0} \circ \ldots \circ f_1(x) = +\infty$ for all $x \geq \xi_1$. Hence there exists $\xi_2 > 0$ such that $f_{N_0} \circ \ldots \circ f_1(x) \geq \xi_1$ whenever $x \geq \xi_2$. For every $x \geq \xi_2$ and $n \geq N_0$, therefore,

$$f_n \circ \ldots \circ f_1(x) = f_n \circ \ldots \circ f_{N_0+1}\big(f_{N_0} \circ \ldots \circ f_1(x)\big) \geq (1+\delta)^{n-N_0} \xi_1 \,.$$

Given any $x_0 \geq \xi_2$, define $y_n = \log x_n = \log f_n \circ \ldots \circ f_1(x_0)$. Moreover, let $z_n = y_n - \log \prod_{j=1}^n a_j$, and note that, for every $\varepsilon > 0$,

$$n|z_{n+1} - z_n| = n \left| \log \left(1 + \frac{g_{n+1}(x_n)}{a_{n+1} x_n} \right) \right| \leq \frac{n|g_{n+1}(x_n)|}{2 x_n} \leq \frac{\varepsilon n}{\log x_n}$$
$$\leq \frac{\varepsilon n}{(n - N_0)\log(1+\delta) + \log \xi_1}$$

for all sufficiently large n. Thus $\limsup_{n \to \infty} n|z_{n+1} - z_n| \leq \varepsilon / \log(1+\delta)$, and since $\varepsilon > 0$ was arbitrary, $\lim_{n \to \infty} n(z_{n+1} - z_n) = 0$. By Proposition 4.6(vi), the sequence (y_n) is u.d. mod 1 if and only if $\big(\log \prod_{j=1}^n a_j\big)$ is. In other words, (x_n) is Benford if and only if $\big(\prod_{j=1}^n a_j\big)$ is. ∎

EXAMPLE 6.41. **(i)** Let the (linear) maps $f_n : \mathbb{R} \to \mathbb{R}$ be $f_n(x) = \big(2 + \frac{1}{n}\big)x$. Thus $a_n = 2 + \frac{1}{n} > 2$, and Theorem 6.40 applies with $g_n \equiv 0$. It follows that $\big(f_n \circ \ldots \circ f_1(x_0)\big)$ is Benford for all $x_0 > 0$ (in fact, it is Benford for all $x_0 \neq 0$), provided that $\big(\prod_{j=1}^n a_j\big)$ is Benford. The latter clearly is the case since $\log \prod_{j=1}^n a_j - \log \prod_{j=1}^{n-1} a_j = \log a_n \overset{n \to \infty}{\longrightarrow} \log 2 \notin \mathbb{Q}$, and so $\big(\log \prod_{j=1}^n a_j\big)$ is u.d. mod 1, by Proposition 4.6(i).

(ii) Consider the (non-linear) maps $f_n : \mathbb{R}^+ \to \mathbb{R}^+$ given by

$$f_n(x) = \tfrac{1}{2}\big(x + \sqrt{|5x^2 - 4(-1)^n|}\big) = \begin{cases} \tfrac{1}{2}\big(x + \sqrt{5x^2 + 4}\big) & \text{if } n \text{ is odd}, \\ \tfrac{1}{2}\big(x + \sqrt{|5x^2 - 4|}\big) & \text{if } n \text{ is even}, \end{cases}$$

for which, with $\varphi = \frac{1}{2}(1 + \sqrt{5})$ again,

$$|f_n(x) - \varphi x| = \left| \frac{|5x^2 - 4(-1)^n| - 5x^2}{2\sqrt{|5x^2 - 4(-1)^n|} + 2x\sqrt{5}} \right| \leq \frac{2}{x\sqrt{5}}, \quad x > 0,$$

and hence $f_n(x) - \varphi x = \mathcal{O}(x^{-1})$ uniformly as $x \to +\infty$. Since $\log \varphi$ is irrational, Theorem 6.40 shows that $(x_n) = \big(f_n \circ \ldots \circ f_1(x_0)\big)$ is Benford for all sufficiently large $x_0 > 0$, and in fact (x_n) is Benford for all $x_0 \in \mathbb{R}$ because $f_1(x_0) > 0$ and $\lim_{n \to \infty} x_n = +\infty$ for every x_0. Specifically, the special case $x_0 = 0$ yields $(x_n) = (1, 1, 2, 3, 5, \ldots) = (F_n)$, and hence is another proof that the sequence (F_n) of Fibonacci numbers is Benford, as already seen in Example 4.18. ✠

The following two examples demonstrate that Theorem 6.40 is best possible in that its conclusion may fail if either the assumption $\liminf_{n \to \infty} a_n > 1$ or the *uniformity* in $g_n = o(x / \log x)$ are weakened. (That the *order of magnitude* in $g_n = o(x / \log x)$ is sharp as well has already been demonstrated in Example 6.17.)

EXAMPLE 6.42. In Theorem 6.40, the assumption $\liminf_{n\to\infty} a_n > 1$ cannot be replaced by the weaker requirement that $a_n > 1$ for all sufficiently large n, let alone by the still weaker assumption that $\liminf_{n\to\infty} a_n \geq 1$. To see this, let $0 < \delta < 1$ and

$$a_n = 10^{(n+1)^\delta - n^\delta}, \quad n \in \mathbb{N}.$$

Note that $a_n > 1$, and $a_n - (1 + \delta n^{\delta-1}/\log e) = \mathcal{O}(n^{2\delta-2})$ as $n \to \infty$. Also, the sequence $\left(\prod_{j=1}^n a_j\right) = \left(10^{(n+1)^\delta - 1}\right)$ is Benford because (n^δ) is u.d. mod 1 by Proposition 4.6(v). For every constant $c > 0$, the smooth map $f_\infty : \mathbb{R}^+ \to \mathbb{R}$ with

$$f_\infty(x) = x + c - \frac{x}{\log(2+x)^2}$$

is convex, with $f_\infty(0) = c$, $f_\infty'(0) < 0$, and $\lim_{x\to+\infty} f_\infty(x) = +\infty$. Thus it is possible to choose $c = c^*$ such that f_∞ has a unique fixed point $x^* > 0$ with $f_\infty'(x^*) = 0$. (Numerically, one finds $c^* = 5.155$ and $x^* = 1.584$.) With this, next consider the maps $f_n : \mathbb{R}^+ \to \mathbb{R}^+$ given by

$$f_n(x) = a_n x + c^* - \frac{x}{\log(2+x)^2}, \quad n \in \mathbb{N}.$$

With $g_n(x) \equiv c^* - x/\log(2+x)^2$, clearly $g_n = o(x/\log x)$ uniformly as $x \to +\infty$. Thus all assumptions of Theorem 6.40 are satisfied — except for the fact that $\lim_{n\to\infty} a_n = 1$. Since $f_n(x) \leq a_n x + c^*$ for all $n \in \mathbb{N}$ and $x \in \mathbb{R}^+$, it is clear that $\log x_n = \log f_n \circ \ldots \circ f_1(x_0) = \mathcal{O}(n^\delta)$ as $n \to \infty$. On the other hand, it is readily confirmed that $f_n(x) < x$ for all sufficiently large n and $x^* + 1 \leq x \leq 10^{n^{(1-\delta)/3}}$. Therefore, whenever $\delta < \frac{1}{3}(1 - \delta)$, or equivalently $\delta < \frac{1}{4}$, $\limsup_{n\to\infty} x_n < +\infty$. Since $f_n \to f_\infty$ uniformly on every compact subset of $\mathbb{R}^+$, it follows that $\lim_{n\to\infty} x_n = x^*$ for every $x^* \in \mathbb{R}^+$. Hence for $0 < \delta < \frac{1}{4}$ and any $x_0 \in \mathbb{R}^+$, the sequence $(x_n) = \left(f_n \circ \ldots \circ f_1(x_0)\right)$ is not Benford but $\left(\prod_{j=1}^n a_j\right)$ is. Thus the conclusion of Theorem 6.40 fails. ✠

EXAMPLE 6.43. To require that $g_n = o(x/\log x)$ as $x \to +\infty$ for every $n \in \mathbb{N}$ is not enough in Theorem 6.40, i.e., the uniformity assumption cannot in general be dropped from that theorem. For a simple example demonstrating this, consider the maps $f_n : \mathbb{R}^+ \to \mathbb{R}^+$ with

$$f_n(x) = 1 + \sqrt{4x^2 + 9^n} - 3^n = 2x + 1 + \frac{9^n}{\sqrt{4x^2 + 9^n} + 2x} - 3^n, \quad n \in \mathbb{N}.$$

Here $a_n \equiv 2$, and for every $n \in \mathbb{N}$,

$$g_n(x) = 1 + \frac{9^n}{\sqrt{4x^2 + 9^n} + 2x} - 3^n = \mathcal{O}(1) \quad \text{as } x \to +\infty.$$

Thus all assumptions of Theorem 6.40 are met — except that $g_n = o(x/\log x)$ does not hold *uniformly* as $x \to +\infty$ because, for instance,

$$g_n(x)\frac{\log x}{x}\bigg|_{x=3^n} = -n(3 - \sqrt{5})\log 3 + n3^{-n}\log 3.$$

Since $f_n(x) \leq 2x + 1$, it is clear that $x_n = \mathcal{O}(2^n)$ as $n \to \infty$. On the other hand, $f_n(x) < x$ for all sufficiently large n and $2 \leq x \leq \frac{1}{2}3^n$. Also, since

$$f_n(x) = 1 + \sqrt{4x^2 + 9^n} - 3^n = 1 + \frac{4x^2}{\sqrt{4x^2 + 9^n} + 3^n} \overset{n \to \infty}{\longrightarrow} 1$$

uniformly on every compact subset of $\mathbb{R}^+$, it follows that $\lim_{n \to \infty} x_n = 1$ for every $x_0 > 0$. Since $\left(\prod_{j=1}^n a_j\right) = (2^n)$ is Benford while $(x_n) = \left(f_n \circ \ldots \circ f_1(x_0)\right)$ is not, the conclusion of Theorem 6.40 again fails. ✠

As Theorem 6.40 shows, in order to understand the Benford property of nonautonomous orbits $(x_n) = \left(f_n \circ \ldots \circ f_1(x_0)\right)$ with f_n given by (6.13), one only has to decide whether the sequence $\left(\prod_{j=1}^n a_j\right)$ is Benford. The following simple observation helps with this task.

LEMMA 6.44. *Let (a_n) be a sequence of positive real numbers. Then the sequence $\left(\prod_{j=1}^n a_j\right) = (a_1, a_1 a_2, \ldots)$ is Benford if*

(i) $\lim_{n \to \infty} a_n = a_\infty$ *exists, $a_\infty > 0$, and $\log a_\infty$ is irrational; or*

(ii) $a_n = g(n)$ *for all $n \in \mathbb{N}$, where g is any non-constant polynomial.*

PROOF. If (i) holds then

$$\log \prod_{j=1}^{n+1} a_j - \log \prod_{j=1}^n a_j = \log a_{n+1} \overset{n \to \infty}{\longrightarrow} \log a_\infty,$$

and hence $\left(\prod_{j=1}^n a_j\right)$ is Benford whenever $\log a_\infty \notin \mathbb{Q}$, by Proposition 4.6(i).

Assume in turn that (ii) holds, i.e., $a_n = g(n)$ for all $n \in \mathbb{N}$ and some non-constant polynomial $g(t) = \alpha_p t^p + \alpha_{p-1} t^{p-1} + \ldots + \alpha_0$, where $p \in \mathbb{N}$ and $\alpha_0, \ldots, \alpha_{p-1}, \alpha_p \in \mathbb{R}$ with $\alpha_p > 0$. It is straightforward to see that, for all $n \in \mathbb{N}$,

$$\log \prod_{j=1}^n a_j = \sum_{j=1}^n \log g(j)$$
$$= p \sum_{j=1}^n \log j + n \log \alpha_p + \frac{\alpha_{p-1}}{\alpha_p} \log e \sum_{j=1}^n \frac{1}{j} + \beta_n,$$

where (β_n) is a convergent sequence. The Euler summation formula, for instance, shows that the sequences

$$\left(\sum_{j=1}^n \log j - \tfrac{1}{2} \log n - n \log n + n \log e\right) \quad \text{and} \quad \left(\log e \sum_{j=1}^n \frac{1}{j} - \log n\right)$$

both converge. It follows that the sequence

$$\left(\sum_{j=1}^n \log g(j) - pn \log n - n(\log \alpha_p - p \log e) - \left(\frac{p}{2} + \frac{\alpha_{p-1}}{\alpha_p}\right) \log n\right)$$

is convergent as well. With Proposition 4.6(iii), $\left(\prod_{j=1}^n a_j\right)$ is Benford if and only if $\left(h(n)\right)$ is u.d. mod 1, where $h(t) = pt \log t + t(\log \alpha_p - p \log e)$, and an application of [93, Exc. I.2.26] shows that $\left(h(n)\right)$ is u.d. mod 1. ∎

EXAMPLE 6.45. **(i)** Let $f_1(x) = x$, and for $n \geq 2$ consider the linear maps $f_n(x) = F_n/F_{n-1}x$. Thus $a_n = F_n/F_{n-1}$ and $\lim_{n\to\infty} a_n = \varphi = \frac{1}{2}(1 + \sqrt{5})$, and since $\log \varphi$ is irrational, $\big(f_n \circ \ldots \circ f_1(x_0)\big) = (F_n x_0)$ is Benford for all $x_0 \neq 0$. Specifically choosing $x_0 = 1$ provides yet another proof that (F_n) is Benford.

(ii) For $f_n(x) = nx$, Lemma 6.44 implies that $\big(f_n \circ \ldots \circ f_1(x_0)\big) = (n! x_0)$ is Benford for every $x_0 \neq 0$, as already suggested by Figure 6.1.   ✠

The remainder of this section focuses on nonautonomous analogues of the maps leading to super-exponential growth (as studied in Section 6.4). This also provides a natural opportunity to present proofs for (more general versions of) two results stated and used earlier (Propositions 6.20 and 6.31).

To formulate a nonautonomous analogue of Proposition 6.20, consider the maps $f_n : \mathbb{R}^+ \to \mathbb{R}^+$ with

$$f_n(x) = a_n x^{b_n}, \quad n \in \mathbb{N}, \tag{6.14}$$

where (a_n) and (b_n) are sequences of positive and non-zero real numbers, respectively. The following is a generalization of Proposition 6.20, to which it reduces if $a_n \equiv a$ and $b_n \equiv b$. Note that unlike for the latter, $b_n > 1$ is not required to hold for every n in (6.14).

THEOREM 6.46. *For every $n \in \mathbb{N}$, let $f_n(x) = a_n x^{b_n}$ with $a_n > 0$ and $b_n \neq 0$. If $\liminf_{n\to\infty} |b_n| > 1$ then $\big(f_n \circ \ldots \circ f_1(x_0)\big)$ is Benford for almost all $x_0 > 0$, but every non-empty open interval in $\mathbb{R}^+$ contains uncountably many x_0 for which $\big(f_n \circ \ldots \circ f_1(x_0)\big)$ is not Benford.*

PROOF. For every $n \in \mathbb{N}$, let $h_n(y) = \log f_n(10^y) = b_n y + \log a_n$. Note that if $x_0 > 0$ then $x_n = f_n \circ \ldots \circ f_1(x_0) > 0$ for all n, and

$$y_n = \log x_n = \log f_n \circ \ldots \circ f_1(x_0) = h_n \circ \ldots \circ h_1(\log x_0).$$

Clearly, the map $H_n = h_n \circ \ldots \circ h_1$ is linear for every n; in fact

$$H_n(y) = h_n \circ \ldots \circ h_1(y) = b_1 b_2 \cdots b_n \left(y + \sum_{j=1}^{n} \frac{\log a_j}{b_1 b_2 \cdots b_j} \right).$$

Since $\liminf_{n\to\infty} |b_n| > 1$ and H_n has non-zero slope, it can be assumed without loss of generality that $|b_n| \geq 10^\alpha > 1$ for some $\alpha > 0$ and all $n \in \mathbb{N}$. Also, $H_m' - H_n'$ is constant (hence monotone), and for all $m > n$,

$$|H_m' - H_n'| = |b_{n+1} \cdots b_m - 1| \cdot |b_1 \cdots b_n| \geq (10^\alpha - 1) 10^\alpha > 0.$$

By Proposition 4.14, $\big(H_n(y)\big)$ is u.d. mod 1 for almost all $y \in \mathbb{R}$. Thus (x_n) is Benford for almost all $x_0 > 0$.

It remains to establish the claims regarding the exceptional points. To this end, for every $0 < \delta < \frac{1}{4}$ denote by N_δ the smallest integer such that $\delta 10^{\alpha N_\delta} \geq 1$, i.e., $N_\delta = -\lfloor \alpha^{-1} \log \delta \rfloor$, and fix $\delta > 0$ so small that $5\delta N_\delta < 1$. Fix any $y \in \mathbb{R}$ and any $0 < \varepsilon < \frac{1}{4}$. Since $|b_1 b_2 \cdots b_{N_\varepsilon}| 2\varepsilon \geq 2 \cdot 10^{\alpha N_\varepsilon} \varepsilon \geq 2$, the interval

$H_{N_\varepsilon}([y-\varepsilon,y+\varepsilon])$ has length at least 2. Consequently, there exist an integer k_1 and a closed interval $J_1 \subset [y-\varepsilon,y+\varepsilon]$ such that $H_{N_\varepsilon}(J_1) = [k_1+\frac{1}{2}-\delta, k_1+\frac{1}{2}+\delta]$. But then $H_{N_\delta+N_\varepsilon}(J_1)$ again has length at least 2, and so there exist $k_2 \in \mathbb{Z}$ and a closed interval $J_2 \subset J_1$ such that

$$H_{N_\varepsilon}(J_2) \subset [k_1+\tfrac{1}{2}-\delta, k_1+\tfrac{1}{2}+\delta] \quad \text{and} \quad H_{N_\delta+N_\varepsilon}(J_2) = [k_2+\tfrac{1}{2}-\delta, k_2+\tfrac{1}{2}+\delta]\,.$$

Continuing in this manner, there exist integers $k_1, k_2, \ldots$ and closed intervals $J_1 \supset J_2 \supset \ldots$ such that, for every $n \in \mathbb{N}$,

$$H_{(j-1)N_\delta+N_\varepsilon}(J_n) \subset [k_j + \tfrac{1}{2} - \delta, k_j + \tfrac{1}{2} + \delta] \quad \text{for all } j = 1, 2, \ldots, n-1\,,$$

and

$$H_{(n-1)N_\delta+N_\varepsilon}(J_n) = [k_n + \tfrac{1}{2} - \delta, k_n + \tfrac{1}{2} + \delta]\,.$$

Since it is the intersection of a sequence of nested compact (and non-empty) intervals, the set $\bigcap_{n=1}^\infty J_n$ is not empty, and for every $y^* \in \bigcap_{n=1}^\infty J_n$,

$$H_{(n-1)N_\delta+N_\varepsilon}(y^*) \in [k_n + \tfrac{1}{2} - \delta, k_n + \tfrac{1}{2} + \delta] \quad \text{for all } n \in \mathbb{N}\,.$$

It follows that

$$\liminf_{N \to \infty} \frac{\#\{1 \le n \le N : \langle H_n(y^*)\rangle \in [\frac{1}{2} - \delta, \frac{1}{2} + \delta]\}}{N} \ge \frac{1}{N_\delta} > 5\delta\,, \quad (6.15)$$

and so $\big(H_n(y^*)\big)$ is not u.d. mod 1. For convenience, let

$$U = \big\{y \in \mathbb{R} : \big(H_n(y)\big) \text{ is u.d. mod } 1\big\}\,.$$

With this, the above argument shows that $\mathbb{R} \setminus U$ is dense. Moreover, consider the continuous, 1-periodic function $\psi : \mathbb{R} \to [0,1]$ given by

$$\psi(t) = \begin{cases} 0 & \text{if } |\langle t\rangle - \frac{1}{2}| \ge 2\delta\,, \\ 2 - |\frac{1}{2} - \langle t\rangle|/\delta & \text{if } \delta \le |\langle t\rangle - \frac{1}{2}| < 2\delta\,, \\ 1 & \text{if } |\langle t\rangle - \frac{1}{2}| < \delta\,. \end{cases}$$

Since $\int_0^1 \psi(t)\,\mathrm{d}t = 3\delta$, (6.15) implies $\liminf_{N \to \infty} \frac{1}{N}\sum_{n=1}^N \psi\big(H_n(y^*)\big) \ge 5\delta$, and therefore

$$\liminf_{N \to \infty} \left| \frac{1}{N} \sum_{n=1}^N \psi\big(H_n(y^*)\big) - \int_0^1 \psi(t)\,\mathrm{d}t \right| \ge 2\delta\,. \quad (6.16)$$

For every $m \in \mathbb{N}$, define the set $U_m \subset \mathbb{R}$ as

$$U_m = \left\{ y \in \mathbb{R} : \left| \frac{1}{N} \sum_{n=1}^N \psi\big(H_n(y)\big) - \int_0^1 \psi(t)\,\mathrm{d}t \right| \le \delta \text{ for all } N \ge m \right\}\,,$$

and note that U_m is closed because the functions ψ and $H_1, H_2, \ldots$ are continuous. Also, $U_1 \subset U_2 \subset \ldots$, and (6.16) shows that U_m does not contain any open intervals, that is, U_m has empty interior. Finally, note that if

$y \in U$, i.e., if $(H_n(y))$ is u.d. mod 1, then it follows from [93, Cor. I.1.2] that $\lim_{N \to \infty} \frac{1}{N} \sum_{n=1}^{N} \psi(H_n(y)) = \int_0^1 \psi(t) \, dt$, and hence $y \in U_m$ for all sufficiently large m. In other words, $U \subset \bigcup_{m=1}^{\infty} U_m$. Thus the set U is contained in the countable union of the nowhere dense sets $U_1, U_2, \ldots$. For every non-empty open interval $J \subset \mathbb{R}$, $J = \bigcup_{m \in \mathbb{N}} (J \cap U_m) \cup \bigcup_{x \in J \setminus U} \{x\}$, so since $J \cap U_m$ and $\{x\}$ are nowhere dense for all $m \in \mathbb{N}$ and $x \in \mathbb{R}$, the Baire Category Theorem implies that $J \setminus U$ is uncountable. It follows that the set $\{x_0 \in J : (x_n) \text{ is not Benford}\}$ is uncountable for every non-empty open interval $J \subset \mathbb{R}^+$. ∎

EXAMPLE 6.47. **(i)** Let $f_n(x) = 2^n x^2$ for every $n \in \mathbb{N}$. By Theorem 6.46, $(x_n) = (f_n \circ \ldots \circ f_1(x_0))$ is Benford for almost all, but not all, $x_0 \in \mathbb{R}$. Note that in order to draw this conclusion it is not necessary to know that $\lim_{n \to \infty} x_n = 0$ if and only if $|x_0| \le \frac{1}{4}$, and $\lim_{n \to \infty} x_n = +\infty$ otherwise.

(ii) Let $f_n(x) = 2^n / x^2$ for every $n \in \mathbb{N}$. Again, Theorem 6.46 applies and shows that $(x_n) = (f_n \circ \ldots \circ f_1(x_0))$ is Benford for almost all, but not all, $x_0 \in \mathbb{R}$. As in (i), this conclusion does not depend on the specific behavior of (x_n) which is now slightly more complicated: If $|x_0| > 2^{2/9}$ then $\lim_{n \to \infty} x_{2n} = +\infty$ and $\lim_{n \to \infty} x_{2n-1} = 0$, whereas for $|x_0| < 2^{2/9}$ the roles of (x_{2n}) and (x_{2n-1}) are reversed. Moreover, $x_n = 2^{(2+3n)/9} \to +\infty$ if $|x_0| = 2^{2/9}$. ✠

In Theorem 6.46, it is not enough to assume that $b_n > 1$ for all (sufficiently large) $n \in \mathbb{N}$. Under the latter, weaker assumption, $(f_n \circ \ldots \circ f_1(x_0))$ may be Benford for every $x_0 > 0$, or for none at all, as the following example shows.

EXAMPLE 6.48. **(i)** Consider the maps $f_n(x) = 10^n x^{3^{1/n}}$, i.e., $a_n = 10^n$ and $b_n = 3^{1/n} > 1 + n^{-1}$, for which

$$x_n = f_n \circ \ldots \circ f_1(x_0) = 10^{\sum_{j=1}^{n} j \prod_{i=j+1}^{n} 3^{1/i}} x_0^{3^{\sum_{j=1}^{n} 1/j}}, \quad n \in \mathbb{N}.$$

Recall the well-known fact that $\sum_{j=1}^{n} \frac{1}{j} - \ln n - \gamma - \frac{1}{2} n^{-1} + \frac{1}{12} n^{-2} = \mathcal{O}(n^{-4})$ as $n \to \infty$, where $\gamma = 0.5772$ is Euler's constant. With this, it is straightforward to show that

$$\log x_n = \frac{n^2}{2 - \ln 3} + \alpha_1 n^{\ln 3} + \alpha_2 n + \alpha_3 n^{\ln 3 - 1} + \beta_n \,,$$

with a convergent sequence (β_n) and the appropriate $\alpha_1, \alpha_2, \alpha_3 \in \mathbb{R}$, where the latter depend on $\log x_0$ but, more importantly, are independent of n. It now follows from [93, Thm. I.3.1] that $(\log x_n)$ is u.d. mod 1, i.e., (x_n) is Benford for every $x_0 > 0$.

(ii) On the other hand, if $a_n \equiv 1$ and $b_n = 3^{1/n^2} > 1 + n^{-2}$, and thus $f_n(x) = x^{3^{1/n^2}}$, then

$$x_n = f_n \circ \ldots \circ f_1(x_0) = x_0^{\prod_{j=1}^{n} 3^{1/j^2}} = x_0^{3^{\sum_{j=1}^{n} 1/j^2}}, \quad n \in \mathbb{N},$$

and so, for every $x_0 > 0$ the sequence (x_n) is not Benford since it converges to the finite positive limit $\lim_{n \to \infty} x_n = x_0^{\alpha}$ with $\alpha = 3^{\sum_{j=1}^{\infty} 1/j^2} = 3^{\pi^2/6} = 6.093$. ✠

In analogy to Section 6.4, maps more general than $f_n(x) = a_n x^{b_n}$ are now considered. In situations where most of the maps $f_n : \mathbb{R}^+ \to \mathbb{R}^+$ are strongly expanding, the following generalized version of Proposition 6.31 may be useful. Only an outline of the proof is given here, under an additional assumption, and the reader is referred to [13] for a complete proof.

THEOREM 6.49. *Let $c \geq 0$ and, for every $n \in \mathbb{N}$, let $f_n : \mathbb{R}^+ \to \mathbb{R}^+$ be a map such that both of the following conditions hold:*

(i) *The function $\log f_n(10^x)$ is convex on $(c, +\infty)$;*

(ii) $\dfrac{\log f_n(10^x) - \log f_n(10^c)}{x - c} \geq b_n > 0$ *for all $x > c$.*

If $\liminf_{n \to \infty} b_n > 1$ then $\big(f_n \circ \ldots \circ f_1(x_0)\big)$ is Benford for almost all sufficiently large x_0, but there also exist uncountably many $x_0 > c$ for which $\big(f_n \circ \ldots \circ f_1(x_0)\big)$ is not Benford.

OUTLINE OF PROOF. For every $n \in \mathbb{N}$, let $g_n(x) = \log f_n(10^x)$. Then, by assumption, every map g_n is convex on $(c, +\infty)$, and $g_n(x) - g_n(c) \geq b_n(x - c)$. The main idea of the proof is given here only for the special case where, in addition, $x^{-1} g_n(x)$ is non-decreasing, and $b_n \geq b > 1$, the complete details of which are in [15, Sec. 5]. First, fix $b > 1$. By Theorem 4.2, $\big(f_n \circ \ldots \circ f_1(x_0)\big)$ is Benford if and only if $\big(\log f_n \circ \ldots \circ f_1(x_0)\big)$ is u.d. mod 1. Since $\log$ maps sets of measure zero into sets of measure zero, setting

$$S_n(x) = \log f_n \circ \ldots \circ f_1(10^x) = g_n \circ \ldots \circ g_1(x),$$

it suffices to show that for all sufficiently large $j \in \mathbb{N}$,

$$\big(S_n(x)\big) \text{ is u.d. mod 1 for almost all } x \in [j-1, j]. \tag{6.17}$$

Fix $0 < s < 1$ and let $Y_n = \mathbb{1}_{[0,s)}(\langle S_n \rangle)$, i.e., $Y_n = 1$ if $\langle S_n \rangle < s$, and $Y_n = 0$ otherwise. Since a random variable X is uniformly distributed on $[j-1, j]$ if and only if $\mathbb{P}(X \leq j - 1 + s) = s$ for all rational $0 < s < 1$, and since countable unions of sets of measure zero have measure zero themselves, to establish (6.17), it suffices to show that

$$\frac{Y_1 + \cdots + Y_n}{n} \to s \quad \text{a.s. as } n \to \infty. \tag{6.18}$$

In general the (Y_n) are neither independent nor identically distributed, so the classical Strong Law of Large Numbers does not apply. However, using the fact that if $\log f$ and $\log g$ are convex, non-decreasing, and nonnegative, then so are f, g, and $\log(f \circ g)$, and the fact [15, Lem. 5.6] that if $f : [0, 1] \to \mathbb{R}$ is convex, non-decreasing, and nonnegative, then for all $0 < s < 1$,

$$s - \frac{1}{f'^+(0)} \leq \lambda\big(\{x \in [0, 1] : \langle f(x) \rangle \leq s\}\big) \leq s + \frac{2}{f'^+(0)},$$

(6.18) can be established [15, Thm. 5.5] using a strong law for averages of bounded random variables satisfying an $\mathcal{O}(N^{-3})$-growth constraint on their correlations [100, p. 154]. ∎

Remark. As in the autonomous context of Section 6.4, when applied to the maps $f_n(x) = a_n x^{b_n}$ with positive a_n, b_n, Theorem 6.49 only yields a weaker form of Theorem 6.46.

EXAMPLE 6.50. **(i)** Let f_n be given by

$$f_n(x) = \begin{cases} x^2 & \text{if } n \text{ is a prime number}, \\ 2^x & \text{if } n \text{ is not a prime number}. \end{cases}$$

By Theorem 6.49, $(x_n) = \bigl(f_n \circ \ldots \circ f_n(x_0)\bigr)$ is Benford for almost all, but not all, sufficiently large x_0, and in fact for almost all $x_0 \in \mathbb{R}$, since $x_4 > 1$ and in any case $\lim_{n\to\infty} x_n = +\infty$.

(ii) For $f_n(x) = x^{2n} + 1$, $n \in \mathbb{N}$, Theorem 6.49 applies with $c = 0$ and $b_n = n$. Hence $\bigl(f_n \circ \ldots \circ f_n(x_0)\bigr)$ is Benford for almost all, but not all, $x_0 \in \mathbb{R}$.

(iii) With the map $f_n : \mathbb{R}^+ \to \mathbb{R}^+$ given by

$$f_n(x) = \begin{cases} x & \text{if } n \text{ is a prime number}, \\ x^2 & \text{if } n \text{ is not a prime number}, \end{cases}$$

assumptions (i) and (ii) in Theorem 6.49 hold with $c = 0$ and $b_n = 1$ or $b_n = 2$, depending on whether n is prime or not. Consequently, $\liminf_{n\to\infty} b_n = 1$, and the theorem does not apply. However, it follows from [10, Thm. 3.1] that $\bigl(f_n \circ \ldots \circ f_n(x_0)\bigr)$ is Benford for almost all $x_0 \in \mathbb{R}^+$. ✠

In Theorem 6.49, the assumptions of convexity and $\liminf_{n\to\infty} b_n > 1$ can be relaxed somewhat [13]. However, as the following example shows, the conclusion of the theorem may fail if one of its hypotheses is violated for even a single n.

EXAMPLE 6.51. The functions f_n given by

$$f_n(x) = \begin{cases} x^2 & \text{if } n \neq 2015, \\ 10 & \text{if } n = 2015, \end{cases}$$

satisfy (i) and (ii) in Theorem 6.49 for every $n \neq 2015$, but do not satisfy (ii) for $n = 2015$. Clearly, (x_n) is not Benford for any $x_0 > 0$ because $D_1(x_n) = 1$ whenever $n \geq 2015$. ✠

6.7   CHAOTIC SYSTEMS: TWO EXAMPLES

The scenarios studied so far for their conformance to Benford's law have all been dynamically very simple indeed: In Theorems 6.13, 6.23, 6.40, and 6.49, $\lim_{n\to\infty} x_n = +\infty$ holds automatically for all relevant initial values x_0, whereas $\lim_{n\to\infty} x_n = 0$ in Corollaries 6.18 and 6.28. While this dynamical simplicity of (x_n) does not necessarily force the behavior of $\bigl(S(x_n)\bigr)$ to be equally simple (recall Example 6.21), it raises the question of what may happen under more general circumstances, that is, in situations where (some or most) orbits exhibit a less trivial long-term behavior. The present section presents two simple examples in this regard. Both systems are *chaotic* in the sense that nearby orbits diverge quickly, exhibiting what appears to be a completely erratic pattern of recurrence. While the latter is very intricate mathematically, as far as Benford's law is concerned, fortunately, the discussion of the two examples can be kept quite informal and non-technical. Nevertheless, the examples illustrate how Benford sequences, though not completely absent, may be less prevalent here than for the simple dynamical systems studied in earlier sections.

EXAMPLE 6.52. Let $f : \mathbb{R} \to \mathbb{R}$ be the classical tent map defined in Example 2.8(i), i.e., $f(x) = 1 - |2x - 1|$. Since $f(x) = 2x$ for all $x \leq \frac{1}{2}$, it is clear that $(f^n(x_0))$ is Benford whenever $x_0 \notin [0, 1]$. Also, $f(0) = f(1) = 0$. As far as Benford's law is concerned, therefore, it only remains to analyze $(f^n(x_0))$ for $0 < x_0 < 1$. To this end, similarly to Section 6.4, consider the set

$$B = \left\{ x \in [0, 1] : (f^n(x)) \text{ is Benford} \right\}.$$

From the graph of f and its iterates (see Figure 6.6), it is evident that f has *many* periodic points, and the latter are actually dense in $[0, 1]$. Concretely, it is not hard to see that x_0 is a p-periodic point $(p \in \mathbb{N})$ whenever

$$x_0 \in \left\{ \frac{2j}{2^p + 1} : j = 1, 2, \ldots, 2^{p-1} \right\}.$$

For example, $x_0 = \frac{2}{3}$ is a fixed point, and $x_0 = \frac{4}{9}$ is 3-periodic, its orbit being $\left(\frac{8}{9}, \frac{2}{9}, \frac{4}{9}, \frac{8}{9}, \frac{2}{9}, \ldots \right)$. Thus $[0, 1] \setminus B$ contains *many* points. Moreover, note that for every $0 < a < b < 1$,

$$\lambda_{0,1} \circ f^{-1}([a, b]) = \lambda_{0,1} \left(\left[\tfrac{1}{2}a, \tfrac{1}{2}b \right] \cup \left[1 - \tfrac{1}{2}b, 1 - \tfrac{1}{2}a \right] \right) = b - a = \lambda_{0,1}([a, b]),$$

which shows that $\lambda_{0,1} \circ f^{-1} = \lambda_{0,1}$, i.e., f is $\lambda_{0,1}$-preserving. In fact, it can be shown that f is even *ergodic* with respect to $\lambda_{0,1}$. By the Birkhoff Ergodic Theorem, $(x_n) = (f^n(x_0))$ is distributed according to $\lambda_{0,1}$ for (Lebesgue) almost all $x_0 \in [0, 1]$. Recall from Example 3.10 (i) that for every such x_0 the sequence $(S(x_n))$ is uniformly distributed in $[1, 10)$, and hence is *not* Benford. It follows that $\lambda_{0,1}(B) = 0$, and the reader may wonder whether $B \neq \varnothing$, i.e., whether B contains any points at all. As the following argument shows, it does indeed.

Let I_L and I_R denote, respectively, the left and right half of $[0, 1]$, that is, $I_L = [0, \frac{1}{2}]$ and $I_R = [\frac{1}{2}, 1]$, and note that $f(x) = 2x$ whenever $x \in I_L$. Thus $x_n = 2x_{n-1}$, and (x_n) would clearly be Benford if only $x_n = f^n(x_0) \in I_L$ held for all n. Unfortunately, though, the latter is impossible, since the only point x_0 with $f^n(x_0) \in I_L$ for all $n \in \mathbb{N}$ is the fixed point $x_0 = 0$, the orbit of which trivially is *not* Benford. Recall, however, that being Benford is an asymptotic property, and hence in order for (x_n) to be Benford it is enough for $x_n \in I_L$

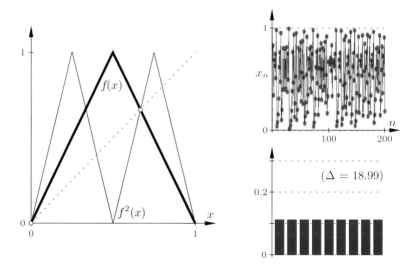

Figure 6.6: For the tent map $f(x) = 1 - |2x - 1|$, typical orbits $(x_n) = (f^n(x_0))$ with $0 < x_0 < 1$ are uniformly distributed in $[0, 1]$ (top right) and hence have a non-Benford asymptotic distribution of first significant digits (bottom right); see Example 6.52.

to hold for *most* n. To be concrete, consider any sequence (ω_n) in $\{L, R\}$, i.e., (ω_n) is a sequence made up of the two symbols L and R, of the form

$$(\omega_n) = \Big(\underbrace{L, L, \ldots, L}_{N_1 \text{ times}}, R, \underbrace{L, L, \ldots, L}_{N_2 \text{ times}}, R, \underbrace{L, L, \ldots, L}_{N_3 \text{ times}}, R, L \ldots \Big), \qquad (6.19)$$

where (N_n) is any sequence of positive integers. Since, for every $n \in \mathbb{N}$, the set

$$J_n = I_L \cap \bigcap_{j=1}^{n} (f^j)^{-1}(I_{\omega_j}) = \left\{ x \in [0, \tfrac{1}{2}] : f^j(x) \in I_{\omega_j} \text{ for all } j = 1, 2, \ldots, n \right\}$$

is a closed interval of length $2^{-(n+1)}$ with $J_1 \supset J_2 \supset \ldots$, it is clear that there exists a unique point $x_\omega \in [0, \tfrac{1}{2}]$ such that $f^n(x_\omega) \in I_{\omega_n}$ for every $n \in \mathbb{N}$. Specifically, if (N_n) is increasing, and hence $\lim_{n \to \infty} N_n = \infty$, it follows from [13, Lem. 2.7(i)], and can also be verified directly, that $(f^n(x_\omega))$ is Benford. For a concrete example, choose $(N_n) = (n)$, i.e., take

$$(\omega_n) = (L, R, L, L, R, L, L, L, R, L, L, L, L, R, L, L, L, L, L, R, L \ldots),$$

in which case

$$x_\omega = \sum_{n=1}^{\infty} 2^{-n(n+3)/2}(-1)^{n+1} = 2^{-2} - 2^{-5} + 2^{-9} - 2^{-14} + 2^{-20} - \ldots = 0.2206,$$

and $\left(f^n(x_\omega)\right)$ is Benford. Notice that if $f^p(x_0) = x_\omega$ for some $p \in \mathbb{N}$, then $\left(f^n(x_0)\right)$ is Benford as well. It follows that B is dense in $[0, 1]$. Finally, observe that (6.19) with arbitrary increasing (N_n) yields *uncountably many different* points x_ω. Thus B is uncountable. Despite being a nullset, therefore, the set B is actually quite large in that it is uncountable and dense in $[0, 1]$. ✠

Remark. Conclusions similar to those in Example 6.52 also hold for the popular (*full*) *logistic map* $f(x) = 4x(1 - x)$; see [13]. For an alternative approach to Benford's law for tent and logistic maps, the reader is referred to [151].

EXAMPLE 6.53. Dynamically, the smooth map $f : \mathbb{R} \to \mathbb{R}$ given by

$$f(x) = \begin{cases} 1 - e^{8x(x-1)/(2x-1)^2} & \text{if } x \neq \frac{1}{2}, \\ 1 & \text{if } x = \frac{1}{2}, \end{cases}$$

has much in common with the tent map in Example 6.52: If $x_0 \notin [0, 1]$ then $(x_n) = \left(f^n(x_0)\right)$ is very regular in that $\lim_{n\to\infty} x_n = x^*$, with $x^* = -6.903$ denoting the unique attracting fixed point of f. In this case, (x_n) clearly is not Benford. (Using Lemma 6.36, it is not hard to see that $(x_n - x^*)$ and $(x_{n+1} - x_n)$ are Benford.) Also, $f(0) = f(1) = 0$, and once again it only remains to consider the case $0 < x_0 < 1$, for which the orbit $\left(f^n(x_0)\right)$ is typically quite chaotic. However, as Figure 6.7 (top right) suggests, this chaotic behavior is significantly different from the one observed in Example 6.52 in that typical orbits now seem to be close to $x = 0$ unproportionally often, that is, $x_n \approx 0$ for many n. More formally, it can be proved that for every $\varepsilon > 0$ and almost all $0 < x_0 < 1$,

$$\lim_{N\to\infty} \frac{\#\{1 \leq n \leq N : f^n(x_0) < \varepsilon\}}{N} = 1. \tag{6.20}$$

Note next that $f(x) \approx f'(0)x = 8x$ for $x \approx 0$, and (6.20), together with the fact that $\log 8$ is irrational, strongly suggests that (x_n) is Benford whenever (6.20) holds, namely, for almost all x_0. A rigorous argument [13] shows that indeed

$$\lambda_{0,1}\left(\left\{x_0 \in [0, 1] : (x_n) \text{ is Benford}\right\}\right) = 1.$$

Unlike in the previous example, Benford orbits of f now constitute a set of full measure. The deeper reason for this, and indeed for (6.20), can be seen in the fact that, unlike the tent map, the map f does not preserve any absolutely continuous probability measure, but does preserve an absolutely continuous *infinite* measure with well-understood properties [161]. ✠

6.8 DIFFERENTIAL EQUATIONS

By presenting a few results on, and examples of, differential equations, i.e., continuous-time deterministic processes, this section aims at convincing the reader that the emergence of Benford's law is not at all restricted to discrete-time

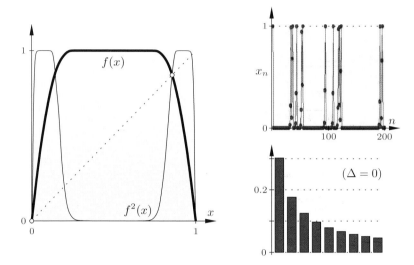

Figure 6.7: The map f in Example 6.53 has a flat critical point, and $\left(f^n(x_0)\right)$ is Benford for almost all $x_0 \in [0, 1]$.

dynamics. Rather, solutions of ordinary or partial differential equations often are Benford as well. Recall that a (Borel measurable) function $f : [0, +\infty) \to \mathbb{R}$ is *Benford* if and only if $\log |f|$ is u.d. mod 1.

Let $F : \mathbb{R} \to \mathbb{R}$ be a continuously differentiable function, and, given any $x_0 \in \mathbb{R}$, consider the *initial value problem* (IVP)

$$\dot{x} = F(x), \quad x(0) = x_0 ; \tag{6.21}$$

here, as usual, $\dot{x}$ denotes the first derivative of $x = x(t)$ with respect to t, that is, $\dot{x} = \frac{\mathrm{d}}{\mathrm{d}t} x$. To keep the presentation simple, it will be assumed throughout that

For every $x_0 \in \mathbb{R}$, the IVP (6.21) has a unique solution $x = x(t)$ for all $t \geq 0$. $$\tag{6.22}$$

The following simple, explicitly solvable examples illustrate how solutions of (6.21) may or may not be Benford.

EXAMPLE 6.54. **(i)** If $F(x) \equiv 1$ then the solution of (6.21) is $x(t) = x_0 + t$, and hence not Benford, by Example 4.9(ii).

(ii) For $F(x) = ax$ with $a \in \mathbb{R}$, one finds $x(t) = x_0 e^{at}$, and so x is Benford unless $ax_0 = 0$, by Example 4.9(iii).

(iii) If $F(x) = -x^3$ then $x(t) = x_0/\sqrt{1 + 2tx_0^2}$, and Proposition 4.8(ii) shows that no solution of (6.21) is Benford. ✠

Note that if $F(x_0) = 0$ then $x(t) \equiv x_0$ is the unique solution of (6.21), referred to as an *equilibrium* (or *stationary*) solution. Clearly, no equilibrium solution is Benford. On the other hand, if $F(x_0) > 0$, say, then by (6.22) either $\lim_{t \to +\infty} x(t) = +\infty$ or $\lim_{t \to +\infty} x(t) = x^*$ for some $x^* \in \mathbb{R}$ with $x^* > x_0$. In the first case, $F(x) > 0$ for all $x > x_0$, and it can be assumed without loss of generality that $x_0 = 0$. (Otherwise replace x by $x + x_0$.) In the second case, $F(x) > 0$ for all $x \in (x_0, x^*)$, but $F(x^*) = 0$, and $x^* - x$ solves (6.21) with $F(x)$ and x_0 replaced by $\widetilde{F}(x) = -F(x^* - x)$ and $x^* - x_0$, respectively. Note that $\widetilde{F}(0) = 0$ and $\widetilde{F}(x) < 0$ for all $x \in (0, x^* - x_0)$. The case of $F(x_0) < 0$ can be dealt with similarly. As far as Benford's law is concerned, therefore, only the following two special cases of (6.21) have to be considered with $x_0 > 0$: Either $F(x) > 0$ for all $x > 0$ (and hence $\lim_{t \to +\infty} x(t) = +\infty$), or else $F(0) = 0$ and $F(x) < 0$ for all $x > 0$ (and hence $\lim_{t \to +\infty} x(t) = 0$). Simple results pertaining to these two cases wil now be discussed separately.

Assume first that $F(x) > 0$ for all $x > 0$. As suggested by Example 6.54(i), if $F(x)$ only grows slowly as $x \to +\infty$, or does not grow at all, then the solutions of (6.21) are not Benford. More formally, the following continuous-time analogue of Theorem 6.4 holds.

THEOREM 6.55. *Let $F : \mathbb{R} \to \mathbb{R}$ be C^1, and assume that $F(x) > 0$ for all $x > 0$. If $F = o(x^{1-\varepsilon})$ as $x \to +\infty$ with some $\varepsilon > 0$, then, for every $x_0 > 0$, the solution of (6.21) is not Benford.*

PROOF. Since $\lim_{t \to +\infty} x(t) = +\infty$, there exists $t_0 \geq 0$ with $\dot{x}(t) \leq x(t)^{1-\varepsilon}$ for all $t \geq t_0$. But then, for every $t \geq t_0$,

$$x(t)^\varepsilon = x(t_0)^\varepsilon + \int_{t_0}^t \frac{\mathrm{d}}{\mathrm{d}\tau} x^\varepsilon \, \mathrm{d}\tau = x(t_0)^\varepsilon + \int_{t_0}^t \varepsilon x(\tau)^{\varepsilon-1} \dot{x}(\tau) \, \mathrm{d}\tau \leq x(t_0)^\varepsilon + \varepsilon(t - t_0),$$

which in turn shows that $\limsup_{t \to +\infty} \log x(t)/\log t \leq \varepsilon^{-1} < +\infty$, and hence, by Proposition 4.8(ii), $\log x$ is not u.d. mod 1, i.e., x is not Benford. ∎

EXAMPLE 6.56. If $F(x) = x^3/(1+x^4)$ then $F = \mathcal{O}(x^{-1})$ as $x \to +\infty$, and so Theorem 6.55 applies, showing that no solution of (6.21) with $x_0 > 0$ is Benford. Since $\frac{\mathrm{d}}{\mathrm{d}t}(-x) = -\dot{x} = -F(x) = F(-x)$, the same is true for $x_0 < 0$. Finally, $x_0 = 0$ is an equilibrium, so no solution of (6.21) is Benford. Note that rather than by invoking Theorem 6.55, this conclusion could also have been reached via an explicit calculation: For every $x_0 \neq 0$,

$$x(t)^2 = \frac{x_0^4 + 2tx_0^2 - 1 + \sqrt{(x_0^4 + 2tx_0^2 - 1)^2 + 4x_0^4}}{2x_0^2}, \quad t \geq 0,$$

which again shows that $\limsup_{t \to +\infty} \log |x(t)|/\log t < +\infty$ and, as in the proof of Theorem 6.55, implies that x is not Benford. ✠

In complete analogy to the discrete-time case (Example 6.7(ii)), the assumption $F = o(x^{1-\varepsilon})$ as $x \to +\infty$ in Theorem 6.55 cannot be weakened to $F = o(x)$, as the following simple example shows.

EXAMPLE 6.57. Let $F : \mathbb{R} \to \mathbb{R}^+$ be any C^1-function F with $F(x) = x/\log x$ for all $x \geq 2$. If $x_0 \geq 2$ then

$$\big(\log x(t)\big)^2 = (\log x_0)^2 + \int_0^t 2\log e \log x(\tau) \frac{\dot{x}(\tau)}{x(\tau)} \, \mathrm{d}\tau = (\log x_0)^2 + 2t\log e \,, \quad t \geq 0 \,.$$

Since $\log x(t) = \sqrt{(\log x_0)^2 + 2t\log e}$ is u.d. mod 1 by Example 4.7(iii) and Proposition 4.8(i), every solution of (6.21) is Benford. Thus the conclusion of Theorem 6.55 may fail if $F = o(x)$ as $x \to +\infty$. ✠

As the example $F(x) = ax$ with $a > 0$ illustrates, if $F(x)$ grows faster than $o(x^{1-\varepsilon})$ as $x \to +\infty$, then all solutions of (6.21) may be Benford. This is true in greater generality, as the following theorem, a continuous-time analogue of Theorem 6.13, shows. To fully appreciate the result, notice that the rate of growth of $F(x)$ as $x \to +\infty$ is limited by (6.22) in that the latter implies that $\liminf_{x \to +\infty} F(x)/x^{1+\varepsilon} = 0$ for every $\varepsilon > 0$.

THEOREM 6.58. *Let $F : \mathbb{R} \to \mathbb{R}$ be C^1, and assume that $F(x) > 0$ for all $x > 0$. If $F(x)/x$ converges to a finite positive limit as $x \to +\infty$, or if $F(x)/x$ is non-decreasing on $[c, +\infty)$ for some $c > 0$, then the solution of (6.21) is Benford for every $x_0 > 0$.*

PROOF. Assume first that $\lim_{x \to +\infty} F(x)/x = a \in \mathbb{R}^+$. With $x(t) \to +\infty$ as $t \to +\infty$, and letting $y = \log x$, it follows that

$$\dot{y}(t) = \log e \frac{\dot{x}(t)}{x(t)} = \log e \frac{F\big(x(t)\big)}{x(t)} \xrightarrow{t \to +\infty} a \log e \,. \tag{6.23}$$

Fix any $\delta \in (0, 1)$ and let $y_n = y(n\delta)$ for all $n \in \mathbb{N}$. With this,

$$y_{n+1} - y_n = \int_{n\delta}^{(n+1)\delta} \dot{y}(t) \, \mathrm{d}t = \int_0^\delta \dot{y}(n\delta + t) \, \mathrm{d}t \xrightarrow{n \to \infty} a\delta \log e \,.$$

For all but countably many δ, therefore, the sequence (y_n) is u.d. mod 1, and Proposition 4.8(i) implies that y is u.d. mod 1 as well. In other words, x is Benford.

To complete the proof, assume that $F(x)/x$ is non-decreasing on $[c, +\infty)$. For all sufficiently large t, the function $y = y(t)$ has a positive non-decreasing derivative, by (6.23). Hence y is u.d. mod 1 by [93, Exc. I.9.13], and again x is Benford. ∎

EXAMPLE 6.59. Let $F(x) = \sqrt{x^2 + 1}$. For every $x > 0$,

$$0 < \frac{F(x)}{x} - 1 = \frac{\sqrt{x^2 + 1}}{x} - 1 = \frac{1}{x^2 + x\sqrt{x^2 + 1}} \,,$$

hence $\lim_{x \to +\infty} F(x)/x = 1$. (Note that $F(x)/x$ is actually *decreasing* on $\mathbb{R}^+$.) By Theorem 6.58, every solution of (6.21) is Benford. Again, this can be confirmed by an explicit calculation, since

$$x(t) = x_0 \cosh t + \sqrt{x_0^2 + 1} \sinh t = \cosh t \left(x_0 + \sqrt{x_0^2 + 1} \tanh t \right), \quad t \geq 0,$$

is Benford for every $x_0 \in \mathbb{R}$; see Example 4.13(ii). ✠

Remark. If, in the setting of Theorem 6.58, $\lim_{x \to +\infty} F(x)/x$ equals 0 or $+\infty$, then the solutions of (6.21) may or may not be Benford. For the case $\lim_{x \to +\infty} F(x)/x = 0$, for instance, this can be seen from Examples 6.57 (where every solution is Benford) and 6.56 (where none is).

Now consider the second relevant scenario for the IVP (6.21) above. That is, assume $F(0) = 0$ and $F(x) < 0$ for all $x > 0$. As might be expected, with regard to Benford's law this case is, in a sense, reciprocal to the scenario studied earlier. For instance, if $|F(x)|$ does not become small too fast as $x \to 0$, then all solutions of (6.21) with x_0 close to 0 are Benford. More precisely, the following continuous-time analogue of Corollary 6.18 holds. Note that the result is a simple corollary of Theorem 6.58, via taking reciprocals.

THEOREM 6.60. *Let* $F : \mathbb{R} \to \mathbb{R}$ *be* C^1 *with* $F(0) = 0$. *If* $F'(0) < 0$ *then, for every* $x_0 \neq 0$ *sufficiently close to 0, the solution of (6.21) is Benford.*

PROOF. Let $x_0 \neq 0$ be so close to zero that $xF(x) < 0$ for all $0 < |x| \leq |x_0|$. Assume without loss of generality that $x_0 > 0$. (The case $x_0 < 0$ is analogous.) Hence $F(x) < 0$ for every $x \in (0, x_0)$. Since

$$\frac{\mathrm{d}}{\mathrm{d}t}(x^{-1}) = -\frac{\dot{x}}{x^2} = -\frac{F(x)}{x^2},$$

the function x^{-1} is a solution of (6.21) with $F(x)$ and x_0 replaced by, respectively, $\widetilde{F}(x) = -x^2 F(x^{-1})$ and x_0^{-1}. Note that $\widetilde{F}$ is C^1, with $\widetilde{F}(x) > 0$ for all sufficiently large x, and

$$\lim_{x \to +\infty} \frac{\widetilde{F}(x)}{x} = -\lim_{x \to +\infty} xF(x^{-1}) = -\lim_{x \to 0} \frac{F(x)}{x} = -F'(0) > 0.$$

Hence by Theorem 6.58, x^{-1} is Benford, and so is x, by Theorem 4.4. ∎

EXAMPLE 6.61. **(i)** Let $F(x) = -x/(x^2 + 1)$. By Theorem 6.60, every solution of (6.21) with $x_0 \neq 0$ is Benford.

(ii) Consider the smooth function $F(x) = -\sin(\pi x)$. Clearly $x(t) \equiv k$ is an equilibrium for every integer k. Since $F'(k) = (-1)^{k+1}\pi$, this equilibrium is attracting if k is even, and repelling (i.e., attracting for $t \to -\infty$) if k is odd. Note that if x is a solution of $\dot{x} = F(x)$ then so is $x - 2k$, since $F(x+2) = F(x)$ for all x. It follows from Theorem 6.60 that $x - 2k$ is Benford whenever x is a solution of (6.21) with $2k - 1 < x_0 < 2k + 1$. ✠

If $F(0) = 0$ and $F(x) < 0$ for all $x > 0$, yet $|F(x)|$ decreases rapidly as $x \to 0$, then this may prevent the solutions of (6.21) from being Benford. A case in point is $F(x) = -x^3$, where no solution is Benford, as seen in Example 6.54(iii). The general observation is as follows. (Note that the smoothness assumption on F is stronger here than anywhere else in this section.)

THEOREM 6.62. *Let* $F : \mathbb{R} \to \mathbb{R}$ *be* C^2 *with* $xF(x) \leq 0$ *for all* x *in a neighborhood of* 0. *If* $F'(0) = 0$ *then, for every* x_0 *sufficiently close to* 0, *the solution of* (6.21) *is not Benford.*

PROOF. Let $x_0 \neq 0$ be so close to 0 that $xF(x) \leq 0$ for all $0 \leq |x| \leq |x_0|$. Again, assume without loss of generality that $x_0 > 0$. If $F(x^*) = 0$ for some $0 < x^* \leq x_0$ then $\lim_{t \to +\infty} x(t) \geq x^* > 0$, and clearly x is not Benford. Thus it only remains to consider the case where $F(x) < 0$ for all $0 < x \leq x_0$. By assumption, there exists a continuous function $G : \mathbb{R} \to \mathbb{R}$ with $G(x) > 0$ for all $0 < x \leq x_0$ such that $F(x) = -x^2 G(x)$. Let $G_0 = \max_{0 \leq x \leq x_0} G(x) > 0$, and consider the function $y = -\log x$. Then $\lim_{t \to +\infty} y(t) = +\infty$, and

$$\dot{y} = -\log e \frac{\dot{x}}{x} = \log e\, xG(x) = \log e\, 10^{-y} G(10^{-y}) \leq G_0 \log e\, 10^{-y}\,,$$

which in turn yields, for every $t \geq 0$,

$$10^{y(t)} - 10^{y(0)} = \int_0^t \frac{10^{y(\tau)}}{\log e} \dot{y}(\tau)\, \mathrm{d}\tau \leq G_0 t\,,$$

and consequently $\limsup_{t \to +\infty} y(t)/\log t \leq 1$. Thus by Proposition 4.8(ii), y is not u.d. mod 1, i.e., x is not Benford. ∎

EXAMPLE 6.63. For the smooth function $F(x) = -\pi x + \sin(\pi x)$, clearly $F(0) = F'(0) = 0$, and $xF(x) < 0$ for all $x \neq 0$. By Theorem 6.62, no solution of (6.21) is Benford. ✠

In Theorem 6.62, the requirement that F be C^2 can be weakened somewhat [15, Thm. 6.7]. Very similarly to its discrete-time counterpart (Corollary 6.9; cf. also Example 6.11), however, the conclusion may fail if F is merely C^1.

EXAMPLE 6.64. For the C^1-function F with $F(0) = 0$ and

$$F(x) = -\frac{x}{\sqrt{(\log|x|)^2 + 1}}\,, \quad x \neq 0\,,$$

$xF(x) < 0$ for all $x \neq 0$, and $F'(0) = 0$. Moreover, fix any $x_0 \neq 0$ and let $y = -\log|x|$. Then

$$\dot{y} = \log e \frac{\dot{x}}{x} = \frac{\log e}{\sqrt{y^2 + 1}}\,,$$

from which it is straightforward to deduce that the function $\eta(t) = y(t+1) - y(t)$ is decreasing with $\lim_{t \to +\infty} \eta(t) = 0$ and $\lim_{t \to +\infty} t\eta(t) = +\infty$. Hence y is u.d. mod 1 by [93, Thm. I.9.4], and x is Benford. Thus the conclusion of Theorem 6.62 may fail if F is only C^1. ✠

Finally, it should be mentioned that at present little seems to be known about the Benford property for solutions of *partial differential equations* or more general functional equations such as delay or integro-differential equations. Quite likely, it will be hard to decide in any generality whether many, or even most, solutions of such systems exhibit the Benford property in one form or another.

Only one simple but fundamental example of a partial differential equation is discussed briefly here, namely, the so-called one-dimensional *heat* (or *diffusion*) *equation*

$$\frac{\partial u}{\partial t} = \frac{\partial^2 u}{\partial x^2},\tag{6.24}$$

a linear second-order equation for $u = u(t, x)$. Physically, (6.24) describes, for example, the diffusion over time of heat in a homogeneous one-dimensional medium. Without further conditions, (6.24) has *many* solutions, of which

$$u(t, x) = cx^2 + 2ct,$$

with any constant $c \neq 0$, is Benford neither in t ("time") nor in x ("space"), by Example 4.9(ii), whereas by Example 4.9(i) and Theorem 4.10, the solution

$$u(t, x) = e^{-c^2 t} \sin(cx)$$

is Benford (or identically zero) in t but not in x, and

$$u(t, x) = e^{c^2 t + cx}$$

is Benford in both t and x. Usually, to specify a unique solution an equation like (6.24) has to be supplemented with initial and/or boundary conditions.

EXAMPLE 6.65. **(i)** A prototypical example of an *Initial-boundary Value Problem* (IBVP) consists of (6.24) together with

$$\begin{aligned} u(0, x) &= u_0(x) && \text{for all } 0 < x < 1, \\ u(t, 0) &= u(t, 1) = 0 && \text{for all } t > 0. \end{aligned}\tag{6.25}$$

Physically, the conditions (6.25) may be interpreted as both ends of the medium, at $x = 0$ and $x = 1$, being kept at a reference temperature $u = 0$ while the initial distribution of heat is given by the function $u_0 : [0, 1] \to \mathbb{R}$. It turns out that, under very mild assumptions on u_0, the IBVP consisting of (6.24) and (6.25) has a unique solution which, for any $t > 0$, can be written as a Fourier series,

$$u(t, x) = \sum_{n=1}^{\infty} u_n e^{-\pi^2 n^2 t} \sin(\pi n x),$$

where $u_n = 2 \int_0^1 u_0(s) \sin(\pi n s) \, ds$. From this it is easy to see that, for every fixed $0 \leq x \leq 1$, the function $u = u(t, x)$ either vanishes identically or else is Benford (in time), by Theorem 4.12.

(ii) Another possible set of initial and boundary data is

$$u(0, x) = u_0(x) \qquad \text{for all } x > 0 \,,$$
$$u(t, 0) = 0 \qquad \qquad \text{for all } t > 0 \,,$$
$$(6.26)$$

corresponding to a semi-infinite one-dimensional medium kept at zero temperature at its left end $x = 0$, with an initial heat distribution given by the (integrable) function $u_0 : [0, +\infty) \to \mathbb{R}$. Again, (6.24) together with (6.26) has a unique solution, given by

$$u(t, x) = \frac{1}{2\sqrt{\pi t}} \int_0^{+\infty} u_0(y) \left(e^{-(x-y)^2/(4t)} - e^{-(x+y)^2/(4t)} \right) \mathrm{d}y \quad \text{for all } t > 0 \,.$$

Assuming $\int_0^{+\infty} y |u_0(y)| \, \mathrm{d}y < +\infty$, it is not hard to see that, for every $x \geq 0$,

$$\lim_{t \to +\infty} t^{3/2} u(t, x) = \frac{x}{2\sqrt{\pi}} \int_0^{+\infty} y u_0(y) \, \mathrm{d}y \,,$$

and hence, for any fixed $x \geq 0$, the function u is not Benford in time, except possibly in the case of $\int_0^{+\infty} y u_0(y) \, \mathrm{d}y = 0$. On the other hand, if, for example, $u_0(x) = x e^{-x}$, then a short calculation confirms that, for every $t > 0$,

$$\lim_{x \to +\infty} \frac{e^x u(t, x)}{x} = e^t \,,$$

showing that u is Benford in space. Similarly, if $u_0(x) = \mathbb{1}_{[0,1)}(x)$ then

$$\lim_{x \to +\infty} x e^{(x-1)^2/(4t)} u(t, x) = \sqrt{\frac{t}{\pi}}$$

for every $t > 0$, and again u is Benford in space.   ✠

Chapter Seven

Multi-dimensional Linear Processes

For many applications, models based solely on the one-dimensional processes studied in the previous chapter are often too simple, and have to be replaced with or complemented by more sophisticated multi-dimensional models. The purpose of this chapter is to study Benford's law in the simplest deterministic multi-dimensional processes, namely, *linear* processes in discrete and continuous time. Despite their simplicity, these systems provide important models for many areas of science. Through far-reaching generalizations of results from earlier chapters, they will be shown to very often conform to Benford's law in that their dynamics is an abundant source of Benford sequences and functions. As in the previous chapter, the properties of continuous-time systems (i.e., differential equations) are analogous to those of discrete-time systems, and the chapter focuses on the latter in every but its last section. Again, recall throughout that by Theorem 4.2 a sequence or function is Benford if and only if its (decimal) logarithm is uniformly distributed modulo one.

7.1 LINEAR PROCESSES, OBSERVABLES, AND DIFFERENCE EQUATIONS

Recall the perhaps simplest example of a Benford sequence, first encountered in Chapter 4, namely, the sequence

$$(a^n) = (a, a^2, a^3, \ldots),\qquad(7.1)$$

where a is any real number with $\log|a|$ irrational. It is natural to ask what happens if the number a in (7.1) is replaced by a real $d \times d$-matrix A with $d \geq 2$. For instance, is it possible for the entries in, say, the upper-left corner (or at any other fixed position) in the sequence of matrices

$$(A^n) = (A, A^2, A^3, \ldots)$$

to be Benford? (Here and throughout, the powers of A are denoted by A, A^2, A^3, etc., and A^0 is understood to equal I_d, the $d \times d$ identity matrix.) For a concrete example, consider the 2×2-matrix

$$A = \begin{bmatrix} 1 & 1 \\ 1 & 0 \end{bmatrix},\qquad(7.2)$$

for which it is easy to verify that

$$A^n = \begin{bmatrix} F_{n+1} & F_n \\ F_n & F_{n-1} \end{bmatrix}, \quad n \geq 2, \tag{7.3}$$

where (F_n) is again the sequence of Fibonacci numbers. With $[A]_{jk}$ denoting the entry of A at position (j,k), i.e., in the jth row and k^{th} column, all four sequences $([A^n]_{jk})$ with $j,k \in \{1,2\}$ are Benford. In fact, by means of the explicit formula for F_n from Example 4.18, it is not hard to see that every positive linear combination of the entries of (A^n) is also Benford. More formally, given any *positive* numbers $h_{11}, h_{12}, h_{21}, h_{22}$, the sequence (x_n) with

$$x_n = h_{11}[A^n]_{11} + h_{12}[A^n]_{12} + h_{21}[A^n]_{21} + h_{22}[A^n]_{22}, \quad n \in \mathbb{N}, \tag{7.4}$$

is Benford. Still more generally, except for the trivial case $x_n \equiv 0$, which results, for example, from the choice $h_{11} = -2$, $h_{12} = h_{21} = 1$, $h_{22} = 2$, the sequence (x_n) turns out to be Benford even if the coefficients h_{jk} in (7.4) are *arbitrary* real numbers.

Next consider the 2×2-matrix

$$B = \begin{bmatrix} -4 & -3 \\ 6 & 5 \end{bmatrix}. \tag{7.5}$$

Note that, unlike the matrix A above, B has negative as well as positive entries. Again, it is easy to check that

$$B^n = \begin{bmatrix} -2^n + 2(-1)^n & -2^n + (-1)^n \\ 2^{n+1} - 2(-1)^n & 2^{n+1} - (-1)^n \end{bmatrix}, \quad n \in \mathbb{N},$$

and as before, all sequences $([B^n]_{jk})$ with $j,k \in \{1,2\}$ are Benford; see Theorem 4.16. However, the example

$$2[B^n]_{11} + [B^n]_{22} = 3(-1)^n, \quad n \in \mathbb{N},$$

shows that not every analogue of (7.4) is Benford or trivial (identically zero).

Linear observables

The above observations can be formalized in a simple and effective way. To see how to do this, recall that the sequence $(x_n) = (a^n)$ is uniquely defined via the recursion $x_n = ax_{n-1}$ for $n \geq 2$, together with $x_1 = a$. Similarly, (A^n) is the unique solution of the (matrix) recursion $X_n = AX_{n-1}$ with $X_1 = A$. Thus one may think of (A^n) as a dynamical process in the (d^2-dimensional) *phase space* $\mathbb{R}^{d \times d}$, perhaps providing a simple model for some real-world process. From a physicist's or engineer's point of view it may not be desirable or even possible to observe or record the entire sequence of *matrices* (A^n), especially if d is very large. Rather, what matters is the behavior of certain sequences of *numbers* distilled from (A^n). To formalize this, call any function $h : \mathbb{R}^{d \times d} \to \mathbb{R}$ an *observable* (on $\mathbb{R}^{d \times d}$). This notion is extremely flexible. For example,

$$h(A) = \sum_{j,k=1}^{d} [A]_{jk}^2 \quad \text{and} \quad h(A) = \max\{|\lambda| : \lambda \text{ is an eigenvalue of } A\} \quad (7.6)$$

are two basic examples of (continuous) observables. With this terminology (which is motivated by similar usage in quantum mechanics and ergodic theory [34, 69]), what really matters from an applied scientist's point of view is the behavior of $(h(A^n))$ for specific observables h that are relevant to the system or process being described by (A^n).

In the context of linear processes, a special role is naturally played by *linear observables*, i.e., by observables h on $\mathbb{R}^{d \times d}$ satisfying $h(A + B) = h(A) + h(B)$ and $h(aA) = ah(A)$ for all $A, B \in \mathbb{R}^{d \times d}$ and all $a \in \mathbb{R}$. Neither of the two observables in (7.6) is linear. On the other hand, the observable $h(A) = [A]_{jk}$ is linear for all $j, k \in \{1, 2, \ldots, d\}$. In fact, given any linear observable h on $\mathbb{R}^{d \times d}$, there exists a unique matrix $[h_{jk}] \in \mathbb{R}^{d \times d}$ such that $h(A) = \sum_{j,k=1}^{d} h_{jk}[A]_{jk}$ for all $A \in \mathbb{R}^{d \times d}$. Henceforth, for convenience, denote by $\mathcal{L}_d$ the set of all linear observables on $\mathbb{R}^{d \times d}$. With this, the d^2 linear observables $[\cdot]_{jk}$ form a basis of the linear space $\mathcal{L}_d$. Thus (7.4) represents every possible sequence of the form $(h(A^n))$ with $h \in \mathcal{L}_2$. With A from (7.2), the sequence $(h(A^n))$ is Benford for every *nonnegative* observable $h \neq 0$ on $\mathbb{R}^{2 \times 2}$ (see Section 7.2 for the formal definition of nonnegative observables), and in fact for *every* $h \in \mathcal{L}_2$ unless $h(I_2) = h(A) = 0$, in which case $h(A^n) \equiv 0$. On the other hand, with B from (7.5) and the linear observable $h = 2[\cdot]_{11} + [\cdot]_{22}$, the sequence $(h(B^n))$ is neither Benford nor identically zero.

Two of the main theorems of this chapter allow the reader to easily draw similar conclusions for arbitrary nonnegative and for general real $d \times d$-matrices A (Theorems 7.3 and 7.21, respectively). They provide necessary and sufficient conditions for $(h(A^n))$ to be, respectively, Benford for every nonnegative linear observable h on $\mathbb{R}^{d \times d}$, and Benford or trivial (that is, zero for all $n \geq d$) for every $h \in \mathcal{L}_d$. These conditions generalize, and in the case $d = 1$ reduce to, the fact that for (7.1) to be Benford (or identically zero) it is necessary and sufficient that $\log |a|$ be irrational (or $a = 0$).

Linear difference equations

In applied sciences as well as in mathematics, linear processes often present themselves in the form of (autonomous) linear *difference* or *differential equations*. These important models are directly amenable to the results for sequences $(h(A^n))$ outlined above, or to their continuous-time analogues; the reader is referred to Sections 7.5 and 7.6 for all pertinent details. A simple but also quite prominent example is the second-order linear difference equation (or two-step recursion)

$$x_n = x_{n-1} + x_{n-2}, \quad n \geq 3. \quad (7.7)$$

This difference equation can be rewritten in matrix-vector form as

$$\begin{bmatrix} x_n \\ x_{n-1} \end{bmatrix} = \begin{bmatrix} 1 & 1 \\ 1 & 0 \end{bmatrix} \begin{bmatrix} x_{n-1} \\ x_{n-2} \end{bmatrix}, \quad n \ge 3\,.$$

With the (invertible) matrix A from (7.2) this leads to

$$\begin{bmatrix} x_{n+1} \\ x_n \end{bmatrix} = A \begin{bmatrix} x_n \\ x_{n-1} \end{bmatrix} = \ldots = A^{n-1} \begin{bmatrix} x_2 \\ x_1 \end{bmatrix} = A^n \begin{bmatrix} x_1 \\ x_2 - x_1 \end{bmatrix}, \quad n \ge 1\,,$$

which in turn shows that every sequence (x_n) satisfying (7.7) is of the form $\big(h(A^n)\big)$, with the observable $h \in \mathcal{L}_2$ determined by the initial values x_1, x_2, namely, $h = x_1[\,\cdot\,]_{21} + (x_2 - x_1)[\,\cdot\,]_{22}$. Thus any result concerning the Benford property of sequences $\big(h(A^n)\big)$ in general leads directly to a corresponding result for difference equations; see, e.g., Theorems 7.39 and 7.41. In particular, it turns out that *every* solution (x_n) of (7.7) is Benford — except, of course, for the trivial case $x_1 = x_2 = 0$; see Example 7.42 below.

For another simple example, recall from Example 6.2(i) that every solution (x_n) of

$$x_n = -2x_{n-1}\,, \quad n \ge 2\,, \tag{7.8}$$

is Benford unless $x_1 = 0$. It is natural to ask whether this clear-cut situation persists if (7.8) is modified to, for instance,

$$x_n = -2x_{n-1} - 3x_{n-2}\,, \quad n \ge 3\,. \tag{7.9}$$

Theorem 7.41 below, one of the chapter's main results on difference equations, reduces this question to a familiar open problem in number theory; see Example 7.44(iii). Again, this result relies heavily on a careful analysis of sequences $\big(h(A^n)\big)$ for any $A \in \mathbb{R}^{d \times d}$ and any $h \in \mathcal{L}_d$. Thus the latter analysis, to be carried out in Sections 7.2 and 7.3, constitutes the core of this chapter.

In the study of linear observables and powers of matrices in the subsequent sections, standard linear algebra notions regarding complex numbers, matrices, and vectors are employed. Specifically, for every $z \in \mathbb{C}$, the numbers $\overline{z}$, $\Re z$, $\Im z$, and $|z|$ are the complex conjugate, real part, imaginary part, and absolute value (modulus) of z, respectively. Let $\mathbb{S}$ be the unit circle (circle of radius 1) in $\mathbb{C}$, i.e., $\mathbb{S} = \{z \in \mathbb{C} : |z| = 1\}$. The argument $\arg z$ of $z \ne 0$ is the unique number in $(-\pi, \pi]$ for which $z = |z|e^{\imath \arg z}$. For any set $Z \subset \mathbb{C}$ and number $w \in \mathbb{C}$, define $w + Z = \{w + z : z \in Z\}$ and $wZ = \{wz : z \in Z\}$. Thus, for instance, $w + \mathbb{S} = \{z \in \mathbb{C} : |z - w| = 1\}$ and $w\mathbb{S} = \{z \in \mathbb{C} : |z| = |w|\}$ for every $w \in \mathbb{C}$. Given any set $Z \subset \mathbb{C}$, denote by $\mathrm{span}_{\mathbb{Q}} Z$ the smallest subspace of $\mathbb{C}$ (over $\mathbb{Q}$) containing Z; equivalently, if $Z \ne \varnothing$ then $\mathrm{span}_{\mathbb{Q}} Z$ is the set of all *rational* linear combinations of elements of Z, i.e.,

$$\mathrm{span}_{\mathbb{Q}} Z = \big\{\rho_1 z_1 + \rho_2 z_2 + \ldots + \rho_n z_n : n \in \mathbb{N}, \rho_1, \rho_2, \ldots, \rho_n \in \mathbb{Q}, z_1, z_2, \ldots, z_n \in Z\big\};$$

note that $\mathrm{span}_{\mathbb{Q}} \varnothing = \{0\}$. With this terminology, recall that the numbers $z_1, z_2, \ldots, z_n \in \mathbb{C}$ are $\mathbb{Q}$-*independent* (or *rationally independent*) if the space

$\mathrm{span}_{\mathbb{Q}}\{z_1, z_2, \ldots, z_n\}$ is n-dimensional, or, equivalently, if $\sum_{k=1}^{n} p_k z_k = 0$ with integers $p_1, p_2, \ldots, p_n$ implies that $p_1 = p_2 = \ldots = p_n = 0$.

Throughout, d is a fixed but usually unspecified positive integer. For every $u \in \mathbb{R}^d$, the number $|u| \geq 0$ is the *Euclidean norm* of u, i.e., $|u| = \sqrt{\sum_{j=1}^{d} u_j^2}$. A vector $u \in \mathbb{R}^d$ is a *unit* vector if $|u| = 1$. For every $A \in \mathbb{R}^{d \times d}$, the transpose of A is $A^\top$, thus $[A^\top]_{jk} = [A]_{kj}$ for all $j, k \in \{1, 2, \ldots, d\}$. The *spectrum* of A, i.e., the set of its eigenvalues, is denoted by $\sigma(A)$. Hence $\sigma(A) \subset \mathbb{C}$ is non-empty, contains at most d numbers, and is symmetric with respect to the real axis, i.e., all non-real elements of $\sigma(A)$ occur in complex-conjugate pairs. The number $\rho(A) = \max\{|\lambda| : \lambda \in \sigma(A)\} \geq 0$ is referred to as the *spectral radius* of A. Note that $\rho(A) > 0$ unless A is *nilpotent*, i.e., unless $A^N = 0$ (zero matrix) for some $N \in \mathbb{N}$; in the latter case $A^d = 0$ as well. For every $A \in \mathbb{R}^{d \times d}$, the number $|A|$ is the *(spectral) norm* of A induced by $|\cdot|$, i.e., $|A| = \max\{|Au| : u \in \mathbb{R}^d, |u| = 1\}$. It is well known that $|A| = \sqrt{\rho(A^\top A)} \geq \rho(A) = \lim_{n \to \infty} |A^n|^{1/n}$; see [79, Sec. 5.6].

7.2   NONNEGATIVE MATRICES

A real $d \times d$-matrix A is *nonnegative* (or *positive*), in symbols $A \geq 0$ (or $A > 0$), if $[A]_{jk} \geq 0$ (or $[A]_{jk} > 0$) for all $j, k \in \{1, 2, \ldots, d\}$. Nonnegative matrices play an important role in many areas of mathematics, including game theory, combinatorics, optimization, operations research, and economics. The Benford properties of sequences (A^n) with $A \in \mathbb{R}^{d \times d}$ are much simpler for $A \geq 0$ than for the general case, that is, for A having both positive and negative entries. As the reader will learn through the present section, with regard to Benford's law, powers of nonnegative matrices behave much like one-dimensional sequences, and hence provide a natural bridge between the latter and the general multi-dimensional linear processes to be studied in Section 7.3.

The main results of this section, Theorems 7.3 and 7.11 below, make use of well-known facts regarding nonnegative matrices. These classical facts, due to O. Perron and F. G. Frobenius, are stated here for the reader's convenience in a form targeted at their subsequent use. (For proofs as well as comprehensive overall accounts of the theory and applications of nonnegative matrices, the reader may wish to consult [6], [25], or [79, Ch. 8].) Recall that a nonnegative matrix A is *irreducible* if for every $j, k \in \{1, 2, \ldots, d\}$ there exists $N \in \mathbb{N}$ with $[A^N]_{jk} > 0$, and A is *primitive* if $A^N > 0$ for some $N \in \mathbb{N}$. Clearly, $A \geq 0$ is primitive whenever positive, and irreducible whenever primitive; as simple examples show, neither of the converses holds in general if $d \geq 2$.

PROPOSITION 7.1. *Let $A \in \mathbb{R}^{d \times d}$ be nonnegative. Then:*

(i) *The spectral radius $\rho(A) \geq 0$ is an eigenvalue of A;*

(ii) *If A is irreducible then $\rho(A) > 0$, and the eigenvalue $\rho(A)$ is (algebraically) simple, i.e., a simple root of the characteristic polynomial of A.*

PROPOSITION 7.2. *Let $A \in \mathbb{R}^{d \times d}$ be nonnegative. Then the following are equivalent:*

(i) *A is primitive;*

(ii) *$A^{d^2 - 2d + 2} > 0$;*

(iii) *A^n is irreducible for every $n \in \mathbb{N}$;*

(iv) *A is irreducible, and $|\lambda| < \rho(A)$ for every eigenvalue $\lambda \neq \rho(A)$ of A;*

(v) *A is irreducible, and the limit $\lim_{n \to \infty} A^n / \rho(A)^n$ exists and is positive.*

Recall from Theorem 4.16 that (a^n) with $a \geq 0$ is Benford if and only if $\log a$ is irrational. The following theorem, the first main result of this section, generalizes this simple fact to arbitrary finite dimension. For a concise statement, call $h \in \mathcal{L}_d$ nonnegative if $h(A) \geq 0$ for every nonnegative $A \in \mathbb{R}^{d \times d}$. Equivalently, h is nonnegative precisely if $[h_{jk}] \geq 0$, where $h = \sum_{j,k=1}^{d} h_{jk}[\,\cdot\,]_{jk}$. For example, every observable $[\,\cdot\,]_{jk}$ is nonnegative.

THEOREM 7.3. *Let $A \in \mathbb{R}^{d \times d}$ be nonnegative. Then the following are equivalent:*

(i) *$([A^n]_{jk})$ is Benford for all $j, k \in \{1, 2, \ldots, d\}$;*

(ii) *$\big(h(A^n)\big)$ is Benford for every nonnegative linear observable $h \neq 0$ on $\mathbb{R}^{d \times d}$;*

(iii) *A is primitive, and $\log \rho(A)$ is irrational;*

(iv) *$A^{d^2 - 2d + 2} > 0$, and $\log \rho(A)$ is irrational.*

PROOF. Assume first that (i) holds. If, for some $N \in \mathbb{N}$, the nonnegative matrix A^N were reducible (i.e., not irreducible), then there would exist numbers $j, k \in \{1, 2, \ldots, d\}$ with $[A^{Nn}]_{jk} = 0$ for all $n \in \mathbb{N}$, so $([A^n]_{jk})$ would not be Benford. As this contradicts (i), A^n is irreducible for every $n \in \mathbb{N}$, and hence is primitive by Proposition 7.2(iii), and $\rho(A) > 0$. Proposition 7.2(v) shows that $Q := \lim_{n \to \infty} A^n / \rho(A)^n$ exists and is positive. It follows easily that $Q^2 = Q$ and $QA = AQ = \rho(A)Q$, so

$$A^n = \rho(A)^n Q + B^n, \quad n \in \mathbb{N}, \tag{7.10}$$

where $B = A - \rho(A)Q$. Note that $\lim_{n \to \infty} B^n / \rho(A)^n = 0$, and that consequently $\lim_{n \to \infty} h(B^n / \rho(A)^n) = 0$ for every $h \in \mathcal{L}_d$. (All linear observables on $\mathbb{R}^{d \times d}$ are continuous.) Moreover, if $h \neq 0$ is nonnegative then $h(Q) > 0$. For every nonnegative linear observable $h \neq 0$ on $\mathbb{R}^{d \times d}$, therefore,

$$\frac{h(A^n)}{\rho(A)^n} = h(Q) + h\left(\frac{B^n}{\rho(A)^n}\right) \stackrel{n \to \infty}{\longrightarrow} h(Q) > 0. \tag{7.11}$$

By Theorem 4.16, $\big(h(A^n)\big)$ is Benford only if $\log \rho(A)$ is irrational. Hence (iii) follows from (i).

Next assume that (iii) holds. As seen above, this implies (7.10) and (7.11) for every nonnegative linear observable $h \neq 0$ on $\mathbb{R}^{d \times d}$. Consequently, $(h(A^n))$ is Benford, again by Theorem 4.16, and (ii) holds. Clearly (ii) implies (i); simply let $h = [\,\cdot\,]_{jk}$ for any $j, k \in \{1, 2, \ldots, d\}$. The statements (i), (ii), (iii), therefore, all are equivalent. Since (iii)$\Leftrightarrow$(iv) also, by Proposition 7.2, the proof is complete. ∎

EXAMPLE 7.4. **(i)** For the nonnegative matrix A from (7.2), $A^2 = \begin{bmatrix} 2 & 1 \\ 1 & 1 \end{bmatrix}$ is positive, and $\rho(A) = \varphi$, where $\varphi = \frac{1}{2}(1 + \sqrt{5})$ as usual. Since $\log \varphi$ is irrational, $(h(A^n))$ is Benford for every nonnegative linear observable $h \neq 0$ on $\mathbb{R}^{2 \times 2}$. In particular, choosing $h = [\,\cdot\,]_{12}$ shows once again that $([A^n]_{12}) = (F_n)$ is Benford. On the other hand, note that the linear observable $h = [\,\cdot\,]_{11} - [\,\cdot\,]_{12}$ is not nonnegative, and hence Theorem 7.3 cannot be used to decide whether $(h(A^n)) = (0, 1, 1, 2, 3, \ldots)$ is Benford; Theorem 7.21 below shows that indeed it is.

(ii) The matrix $B = \begin{bmatrix} 2 & 0 \\ 0 & 1 \end{bmatrix}$ is nonnegative but not irreducible, let alone primitive. Even though $\log \rho(B) = \log 2$ is irrational, by Theorem 7.3 there exist $j, k \in \{1, 2\}$ for which $([B^n]_{jk})$ is not Benford. In fact, except for $j = k = 1$, each sequence $([B^n]_{jk})$ is constant and hence not Benford. ✠

EXAMPLE 7.5. **(i)** For the (symmetric) nonnegative matrix

$$A = \begin{bmatrix} 1 & 1 & 0 \\ 1 & 0 & 1 \\ 0 & 1 & 1 \end{bmatrix} \in \mathbb{R}^{3 \times 3},$$

$A^2 > 0$ and $\rho(A) = 2$. Hence Theorem 7.3 shows that $(h(A^n))$ is Benford for every nonnegative linear observable $h \neq 0$ on $\mathbb{R}^{3 \times 3}$.

(ii) Consider the positive 3×3-matrix

$$B = \tfrac{1}{10} \begin{bmatrix} 6 & 3 & 1 \\ 3 & 4 & 3 \\ 1 & 3 & 6 \end{bmatrix}.$$

Notice that the entries in each row of B sum to 1, so B is a so-called (row-) *stochastic* matrix. (Due to its being symmetric, B is also column-stochastic.) It is not hard to see that $\rho(B) = 1$ for every (row- or column-) stochastic matrix B. By Theorem 7.3, $(h(B^n))$ is not Benford for *some* nonnegative linear observable $h \neq 0$ on $\mathbb{R}^{3 \times 3}$. In fact, the explicit formula

$$B^n = \tfrac{1}{3} \begin{bmatrix} 1 & 1 & 1 \\ 1 & 1 & 1 \\ 1 & 1 & 1 \end{bmatrix} + 2^{-(n+1)} \begin{bmatrix} 1 & 0 & -1 \\ 0 & 0 & 0 \\ -1 & 0 & 1 \end{bmatrix} + \tfrac{1}{6} \cdot 10^{-n} \begin{bmatrix} 1 & -2 & 1 \\ -2 & 4 & -2 \\ 1 & -2 & 1 \end{bmatrix}$$

shows that $\lim_{n \to \infty} h(B^n)$ exists and is positive for *every* such observable, so clearly $(h(B^n))$ is never Benford. ✠

From (7.11) it is evident that if A is primitive and $\log \rho(A)$ is rational, then $\left(h(A^n)\right)$ is not Benford for *any* nonnegative $h \in \mathcal{L}_d$, as seen in Example 7.5(ii) above. It can be shown that primitivity of A is actually not needed for this conclusion: If $A \geq 0$ is *irreducible* and $\log \rho(A)$ is rational, then $\left(h(A^n)\right)$ is not Benford for any nonnegative $h \in \mathcal{L}_d$. (Still, $\left(h(A^n)\right)$ may be Benford for *some* $h \in \mathcal{L}_d$ but no such h can be nonnegative; see Example 7.7 below.) On the other hand, if A is reducible, or if A is irreducible but not primitive, and $\log \rho(A)$ is irrational, then the behavior of $\left(h(A^n)\right)$ may be less uniform.

EXAMPLE 7.6. **(i)** For every $a > 0$, the nonnegative matrix $A = \begin{bmatrix} a & 0 \\ 0 & 10 \end{bmatrix}$ is reducible. Suppose $\log a$ is irrational. Depending on the choice of $h \in \mathcal{L}_2$, the sequence $\left(h(A^n)\right)$ may be Benford (e.g., for $h = [\,\cdot\,]_{11}$), non-Benford but not trivial (e.g., for $h = [\,\cdot\,]_{22}$), or trivial (e.g., for $h = [\,\cdot\,]_{21}$). This non-uniformity of behavior exists regardless of whether $\log \rho(A)$ is irrational (if $a > 10$) or is rational (if $a \leq 10$).

(ii) The nonnegative matrix $B = \begin{bmatrix} 0 & 2 \\ 2 & 0 \end{bmatrix}$, with $\log \rho(B) = \log 2$ irrational, is irreducible. However, since B is not primitive, any sequence $\left(h(B^n)\right)$ may or may not be Benford. For instance, $([B^n]_{11} + [B^n]_{12}) = (2^n)$ is Benford whereas $(2[B^n]_{11}) = (2^n + (-2)^n)$ is not. ✠

Let A be a primitive matrix. When $\log \rho(A)$ is irrational, the sequence $\left(h(A^n)\right)$ may nevertheless fail to be Benford for some nontrivial $h \in \mathcal{L}_d$. On the other hand, even when $\log \rho(A)$ is rational, $\left(h(A^n)\right)$ may be Benford. The following example illustrates both situations. Note that, as a consequence of Theorem 7.3, h cannot be nonnegative in either case.

EXAMPLE 7.7. Let $h = [\,\cdot\,]_{11} - [\,\cdot\,]_{12} \in \mathcal{L}_2$. Clearly, h is not nonnegative. The matrix $A = \begin{bmatrix} 5 & 15 \\ 15 & 5 \end{bmatrix}$ is positive, $\log \rho(A) = 1 + \log 2$ is irrational, and yet $\left(h(A^n)\right) = \left((-10)^n\right)$ is not Benford. On the other hand, $B = \begin{bmatrix} 6 & 4 \\ 4 & 6 \end{bmatrix}$ is also positive, with $\log \rho(B) = \log 10 = 1$ rational. Nevertheless, $\left(h(B^n)\right) = (2^n)$ is Benford. ✠

As demonstrated by the next example, the conclusion of Theorem 7.3 may fail if even a single entry of A is negative.

EXAMPLE 7.8. For $A = \begin{bmatrix} 1 & 1 \\ -1 & 3 \end{bmatrix}$ it is clear from Theorem 4.16 and

$$A^n = \begin{bmatrix} 2^n - n2^{n-1} & n2^{n-1} \\ -n2^{n-1} & 2^n + n2^{n-1} \end{bmatrix}, \quad n \in \mathbb{N},$$

that $([A^n]_{jk})$ is Benford for every $j, k \in \{1, 2\}$, whereas $[A^n]_{12} + [A^n]_{21} \equiv 0$, and A^n is not nonnegative for any $n \in \mathbb{N}$. Thus the implications (i)⇒(ii), (i)⇒(iii) and (i)⇒(iv) of Theorem 7.3 all fail in this case. ✠

In addition to linear observables, the Benford property of $\left(h(A^n)\right)$ may also be of interest for certain nonlinear observables h on $\mathbb{R}^{d \times d}$, notably for $h(A) = |A|$ and $h(A) = |Au|$, where $u \in \mathbb{R}^d \setminus \{0\}$ is a fixed vector. To conveniently formulate a result regarding these observables, call $u \in \mathbb{R}^d$ *semi-definite* if $u_j u_k \geq 0$ for all $j, k \in \{1, 2, \ldots, d\}$.

THEOREM 7.9. *Let $A \in \mathbb{R}^{d \times d}$ be nonnegative. If A satisfies* (i)–(iv) *in Theorem 7.3 then:*

(i) $(|A^n|)$ *is Benford;*

(ii) $(|A^n u|)$ *is Benford for every semi-definite $u \in \mathbb{R}^d \setminus \{0\}$.*

PROOF. Both assertions follow immediately from (7.10): Since $|Q| > 0$,

$$\frac{|A^n|}{\rho(A)^n} = \left| Q + \frac{B^n}{\rho(A)^n} \right| \overset{n \to \infty}{\longrightarrow} |Q| > 0 \,, \tag{7.12}$$

and since $\log \rho(A)$ is irrational, $(|A^n|)$ is Benford. Similarly, note that $Qu \neq 0$ whenever $u \neq 0$ is semi-definite. Hence

$$\frac{|A^n u|}{\rho(A)^n} = \left| Qu + \frac{B^n u}{\rho(A)^n} \right| \overset{n \to \infty}{\longrightarrow} |Qu| > 0 \,, \tag{7.13}$$

and $(|A^n u|)$ is Benford as well. ∎

Except for the trivial case $d = 1$, the converse of Theorem 7.9 is not true in general: Even if $(|A^n|)$ and $(|A^n u|)$ for every semi-definite $u \in \mathbb{R}^d \setminus \{0\}$ are Benford, A may not satisfy any of the conditions (i)–(iv) in Theorem 7.3, even if A is irreducible.

EXAMPLE 7.10. For the irreducible matrix $A = \begin{bmatrix} 0 & 2 \\ 2 & 0 \end{bmatrix}$, the sequences $(|A^n|) = (2^n)$ and $(|A^n u|) = (2^n |u|)$ are Benford, yet none of the conditions (i)–(iv) in Theorem 7.3 holds for A; see Example 7.6(ii). ✠

As evidenced by Example 7.10, properties (i) and (ii) in Theorem 7.9 are generally weaker than properties (i)–(iv) in Theorem 7.3, even if A is irreducible. If, however, A is assumed to be *primitive*, then all those properties turn out to be equivalent. Informally put, the following theorem shows that as far as Benford's law is concerned, for a primitive matrix A, the sequence (A^n) behaves just like the one-dimensional sequence $(\rho(A)^n)$.

THEOREM 7.11. *Let $A \in \mathbb{R}^{d \times d}$ be nonnegative. If A is primitive then the following are equivalent:*

(i) $\log \rho(A)$ *is irrational;*

(ii) $([A^n]_{jk})$ *is Benford for all $j, k \in \{1, 2, \ldots, d\}$;*

(iii) $\left(h(A^n)\right)$ *is Benford for every nonnegative linear observable $h \neq 0$ on $\mathbb{R}^{d \times d}$;*

(iv) $(|A^n|)$ *is Benford;*

(v) $(|A^n u|)$ *is Benford for every semi-definite $u \in \mathbb{R}^d \setminus \{0\}$.*

PROOF. Using Theorem 4.16, equivalences (i)⇔(iii), (i)⇔(iv), and (i)⇔(v) follow immediately from (7.11), (7.12), and (7.13), respectively. Moreover, (7.11) with $h = [\,\cdot\,]_{jk}$ for any $j, k \in \{1, 2, \ldots, d\}$ shows that (ii)⇒(i), and since clearly (iii)⇒(ii), the proof is complete. ∎

Remark. Similarly to Theorem 7.3, if A is primitive and $\log \rho(A)$ is rational, then $(h(A^n))$ and $(|A^n u|)$ are not Benford for *any* nonnegative linear observable h on $\mathbb{R}^{d \times d}$ and *any* semi-definite $u \in \mathbb{R}^d$, respectively.

Applying Theorems 7.3 and 7.11 is especially easy when $A \in \mathbb{R}^{d \times d}$ is a nonnegative integer (or rational) matrix, i.e., $[A]_{jk}$ is a nonnegative integer (or rational number) for all $j, k \in \{1, 2, \ldots, d\}$. Using Proposition 7.2(ii), it is straightforward to check whether A is primitive. If it is, deciding whether $\log \rho(A)$ is irrational does not require explicitly determining the number $\rho(A)$, let alone the entire set $\sigma(A)$. The next example illustrates this; for more details, the interested reader is referred to [16, Ex. 3.14].

EXAMPLE 7.12. The nonnegative integer matrix

$$A = \begin{bmatrix} 0 & 1 & 0 \\ 0 & 0 & 1 \\ 6 & 1 & 0 \end{bmatrix}$$

is not positive, and neither are A^2, A^3, A^4. However,

$$A^5 = \begin{bmatrix} 6 & 1 & 6 \\ 36 & 12 & 1 \\ 6 & 37 & 12 \end{bmatrix} > 0 \, ;$$

hence A is primitive. (In this example, the exponent $d^2 - 2d + 2$ of Proposition 7.2(ii) is smallest possible, but often smaller exponents suffice; see [79, Sec. 8.5].) Moreover, $\det A = 6$ and therefore $\rho(A) \geq \sqrt[3]{6} > 1$. To decide whether $\log \rho(A)$ is rational, suppose $\rho(A) = 10^{p/q}$ for some relatively prime $p \in \mathbb{Z}$ and $q \in \mathbb{N}$. Then $p \geq 1$ since $\rho(A) > 1$, and 10^p is an eigenvalue of A^q. Hence 10^p divides $\det A^q = 6^q$. This, however, is impossible for $p \geq 1$. It follows that $\log \rho(A)$ is irrational (in fact, transcendental), and by Theorem 7.11, $(h(A^n))$ is Benford for every nonnegative linear observable $h \neq 0$ on $\mathbb{R}^{3 \times 3}$. Note that in order for this argument to work it is not at all necessary to know that $\rho(A) = 2$. ✠

Remarks. (i) Under the assumption that $A \in \mathbb{R}^{d \times d}$ satisfies $A^N > 0$ for some $N \in \mathbb{N}$, properties (i)–(v) in Theorem 7.11 remain equivalent even if A is not nonnegative; see [16, Thm. 3.2].

(ii) Theorem 7.11(iii) can be replaced by the stronger condition that $(h(A^n))$ is Benford for every nonnegative *polynomial* observable h on $\mathbb{R}^{d \times d}$ that satisfies $h(0) = 0$ and is not constant.

(iii) Since (7.12) and (7.13) also follow from (7.10) if $|\cdot|$ is replaced by any other norm on $\mathbb{R}^{d \times d}$ and $\mathbb{R}^d$, respectively, Theorems 7.9(i, ii) and 7.11(iv, v) remain valid for arbitrary norms as well.

7.3   GENERAL MATRICES

This section studies the Benford properties of sequences $(h(A^n))$, where A is an arbitrary real $d \times d$-matrix and h is any linear observable on $\mathbb{R}^{d \times d}$. To compare this completely general situation to the more special one considered in the previous section (where A was assumed to be nonnegative), observe that while the assumption of nonnegativity yields elegant results with short proofs (via classical facts about nonnegative matrices), it also naturally limits their scope. For instance, Theorems 7.3 and 7.11 are inconclusive whenever $h \in \mathcal{L}_d$ fails to be nonnegative. To illustrate how a more general setting may be relevant even if one is primarily interested in nonnegative matrices A and observables h, recall from Proposition 7.2 that if A is primitive then the matrix

$$Q = \lim_{n \to \infty} \frac{A^n}{\rho(A)^n} \tag{7.14}$$

exists and is positive. Often one is interested in the (Benford) properties of the sequences $(A^{n+1} - \rho(A)A^n)$ and $(A^n - \rho(A)^n Q)$, both of which in some sense measure the speed of convergence in (7.14). With a view towards Theorems 7.3 and 7.11, note that, given any $h \in \mathcal{L}_d$, the sequence $(h(A^{n+1} - \rho(A)A^n))$ is actually of the form $(g(A^n))$ with $g \in \mathcal{L}_d$ defined as

$$g(C) = h(C(A - \rho(A)I_d)), \quad C \in \mathbb{R}^{d \times d}. \tag{7.15}$$

Similarly, $(h(A^n - \rho(A)^n Q))$ equals $(h(B^n))$ with $B = A - \rho(A)Q$. In general, however, neither the linear observable g in (7.15) nor the matrix B will be nonnegative, even if h and A are.

EXAMPLE 7.13. The nonnegative matrix $A = \frac{1}{2}\begin{bmatrix} 1 & 1 \\ 2 & 0 \end{bmatrix}$ has spectral radius $\rho(A) = 1$ and is primitive since $A^2 > 0$; it is easy to see that

$$Q = \lim_{n \to \infty} A^n = \frac{1}{3}\begin{bmatrix} 2 & 1 \\ 2 & 1 \end{bmatrix} \quad \text{and} \quad B = A - Q = \frac{1}{6}\begin{bmatrix} -1 & 1 \\ 2 & -2 \end{bmatrix}.$$

In accordance with Theorem 7.11, the sequence $(h(A^n))$ is not Benford for any nonnegative linear observable h on $\mathbb{R}^{2 \times 2}$. (In fact, (7.10) with $B^n = (-\frac{1}{2})^{n-1}B$ shows that $(h(A^n))$, for any $h \in \mathcal{L}_2$, is not Benford unless $h(B) \neq 0$ and $h(Q) = 0$; the latter is impossible if h is nonnegative.) While A is nonnegative (in fact, row-stochastic) and Q is positive, clearly the matrix B is neither. Also, with the nonnegative $h = [\cdot]_{11}$, for instance, the linear observable g in (7.15) is $g = -\frac{1}{2}[\,\cdot\,]_{11} + [\,\cdot\,]_{12}$ and hence fails to be nonnegative. Nevertheless, a short calculation yields

$$A^n - Q = \left(-\tfrac{1}{2}\right)^{n-1} B \quad \text{and} \quad A^{n+1} - A^n = 3\left(-\tfrac{1}{2}\right)^n B\,, \quad n \in \mathbb{N}\,.$$

From this it is clear that, given any $h \in \mathcal{L}_2$, both sequences $\big(h(A^n - Q)\big)$ and $\big(h(A^{n+1} - A^n)\big)$ are either Benford or identically zero, depending on whether $h(B) \neq 0$ or $h(B) = 0$. Theorem 7.32 below explains this phenomenon in general. ✠

Recall from Theorem 7.11 that, given any primitive nonnegative $d \times d$-matrix A, the sequence $\big(h(A^n)\big)$ is Benford for all nonnegative linear observables $h \neq 0$ on $\mathbb{R}^{d \times d}$ (or for none) precisely if $\log \rho(A)$ is irrational (or rational). Such a clear-cut all-or-nothing situation cannot be expected when arbitrary $h \in \mathcal{L}_d$ are considered, simply because $\big(h(A^n)\big)$ may be trivial (identically zero) irrespective of any specific properties of A. The most that can be expected in general, and hence the natural analogue of the setting of Theorem 7.11, is that every *nontrivial* sequence $\big(h(A^n)\big)$ is Benford. By illustrating this point, the next two examples also motivate the phrasing of the main results, Theorem 7.21 and Proposition 7.31 below.

EXAMPLE 7.14. For the positive matrix A from (7.2) and any $h \in \mathcal{L}_2$, it follows from (7.3) that

$$h(A^n) = F_n h(A) + F_{n-1} h(I_2)\,, \quad n \geq 2\,,$$

which in turn shows, via Theorem 4.16 and Example 4.18, that $\big(h(A^n)\big)$ is Benford unless $h(A) = h(I_2) = 0$; in the latter case clearly $h(A^n) \equiv 0$. Thus, for instance, $h_0(A^n) \equiv 0$ for $h_0 = [\,\cdot\,]_{12} - [\,\cdot\,]_{21}$.

Observe that, similarly, $h_0(B^n) \equiv 0$ for the matrix B in Example 7.7. Unlike in the case for A, however, for arbitrary $h \in \mathcal{L}_2$ not every sequence $\big(h(B^n)\big)$ is Benford since, for example, $(2[B^n]_{11}) = (10^n + 2^n)$, which is not Benford by Theorem 4.16; on the other hand, $([B^n]_{11} - [B^n]_{12}) = (2^n)$ is Benford. In summary, while the sequence $\big(h(A^n)\big)$ is Benford or trivial for every $h \in \mathcal{L}_2$, this is clearly not the case for $\big(h(B^n)\big)$. ✠

EXAMPLE 7.15. Consider the nonnegative 4×4-matrix

$$A = \begin{bmatrix} 1 & 1 & 1 & 1 \\ 1 & 1 & 1 & 0 \\ 1 & 1 & 0 & 0 \\ 2 & 2 & 4 & 2 \end{bmatrix}\,,$$

for which it is readily confirmed that $A^2 > 0$ and $\rho(A) = 4$. By Theorem 7.11, $\big(h(A^n)\big)$ is Benford for every nonnegative linear observable $h \neq 0$ on $\mathbb{R}^{4 \times 4}$. As in the previous example, this property largely persists for arbitrary $h \in \mathcal{L}_4$: The explicit formula

$$A^n = 4^{n-3} \begin{bmatrix} 22 & 22 & 22 & 11 \\ 10 & 10 & 10 & 5 \\ 8 & 8 & 8 & 4 \\ 48 & 48 & 48 & 24 \end{bmatrix} = 4^{n-3} A^3, \quad n \geq 3, \qquad (7.16)$$

shows that, given any $h \in \mathcal{L}_4$, the sequence $(h(A^n))$ is Benford precisely if $h(A^3) \neq 0$. Note that, unlike in the previous example, $(h(A^n))$ may fail to be Benford and yet may not vanish identically. For instance, this happens for $h = 5[\,\cdot\,]_{31} - 4[\,\cdot\,]_{21}$, where $(h(A^n)) = (1, -2, 0, 0, 0, \ldots)$. However, (7.16) guarantees that in this case $h(A^n) = 0$ for all $n \geq 3$.

Next consider the nonnegative 4×4-matrix

$$B = \begin{bmatrix} 1 & 1 & 1 & 1 \\ 1 & 1 & 1 & 0 \\ 1 & 1 & 0 & 0 \\ 14 & 14 & 16 & 8 \end{bmatrix}.$$

Again, $B^2 > 0$, and now $\rho(B) = 10$. In fact, as before, $B^n = 10^{n-3}B^3$ for every $n \geq 3$, and hence $(h(B^n))$ is not Benford for any $h \in \mathcal{L}_4$. ✠

The main goal of this section is to characterize, for arbitrary $A \in \mathbb{R}^{d \times d}$, the situation encountered for the matrices A in the two examples above: For every $h \in \mathcal{L}_d$, the sequence $(h(A^n))$ is either Benford or else vanishes identically from some n onward. For convenience, from now on any sequence $(h(A^n))$ with $h(A^n) = 0$ for all $n \geq d$ is referred to as *terminating*. To provide the reader with some intuition as to which features of A may affect the Benford property of $(h(A^n))$, first a few simple examples are discussed.

EXAMPLE 7.16. **(i)** Consider the matrix $A = \begin{bmatrix} 1 & -1 \\ 1 & 0 \end{bmatrix}$. From the explicit formula

$$A^n = \cos(\tfrac{1}{3}\pi n)I_2 + \frac{\sin(\tfrac{1}{3}\pi n)}{\sqrt{3}}(2A - I_2), \quad n \in \mathbb{N},$$

it is clear that, given any $h \in \mathcal{L}_2$, the sequence $(h(A^n))$ is 6-periodic, i.e., $h(A^{n+6}) = h(A^n)$ for all $n \in \mathbb{N}$. For no choice of $h \in \mathcal{L}_2$, therefore, is $(h(A^n))$ Benford. The oscillatory behavior of $(h(A^n))$ corresponds to the fact that the eigenvalues of A are $\lambda = e^{\pm \pi i/3}$ and hence lie on the unit circle $\mathbb{S}$.

(ii) For the matrix B in Example 7.6(ii),

$$B^n = 2^{n-2}(B + 2I_2) - (-2)^{n-2}(B - 2I_2), \quad n \in \mathbb{N},$$

and so for any $h \in \mathcal{L}_2$ the sequence $(h(B^n))$ is unbounded, provided that

$$h(I_2) \neq 0 \quad \text{or} \quad h(B) \neq 0. \qquad (7.17)$$

Even if h satisfies (7.17), however, $\big(h(B^n)\big)$ may not be Benford, as the examples $h = [\,\cdot\,]_{11}$ and $h = [\,\cdot\,]_{12}$ show, for which $h(B^n) = 0$ for all odd and all even $n \in \mathbb{N}$, respectively. This failure of $\big(h(B^n)\big)$, for every $h \in \mathcal{L}_2$, to be either Benford or trivial is caused by B's having two eigenvalues with the same modulus but opposite signs, namely, $\lambda = \pm 2$.

(iii) Let $\gamma = \cos(\pi \log 2) = 0.5851$ and consider the matrix $C = \begin{bmatrix} 4\gamma & -4 \\ 1 & 0 \end{bmatrix}$. As in (i) and (ii), an explicit formula for C^n is easily derived, namely,

$$C^n = 2^n \cos(\pi n \log 2) I_2 + 2^n \frac{\sin(\pi n \log 2)}{2\sqrt{1-\gamma^2}}(C - 2\gamma I_2)\,, \quad n \in \mathbb{N}\,.$$

Although somewhat oscillatory, the sequence $\big(h(C^n)\big)$ is unbounded for most $h \in \mathcal{L}_2$. As will be shown now, however, it is not Benford. While the argument is essentially the same for every $h \in \mathcal{L}_2$, for convenience assume specifically that $h(I_2) = 0$ and $h(C) = 2\sqrt{1 - \gamma^2}$, which in turn yields

$$\log|h(C^n)| = \log\big(2^n|\sin(\pi n \log 2)|\big) = n\log 2 + \log|\sin(\pi n \log 2)|\,, \quad n \in \mathbb{N}\,.$$

Therefore, with the (Borel measurable) map $f : [0, 1) \to [0, 1)$ defined as

$$f(s) = \langle s + \log|\sin(\pi s)|\rangle\,, \quad 0 \le s < 1\,,$$

$\langle \log|h(C^n)|\rangle = f(\langle n \log 2\rangle)$ for all $n \in \mathbb{N}$. Recall from Example 4.7(i) that $(n \log 2)$ is u.d. mod 1, and hence the sequence $\big(\langle \log|h(C^n)|\rangle\big)$ is distributed according to the probability measure $\lambda_{0,1} \circ f^{-1}$. Consequently, $\big(h(C^n)\big)$ is Benford if and only if $\lambda_{0,1} \circ f^{-1}$ is equal to $\lambda_{0,1}$. The latter, however, is not the case. While this is clear intuitively, an easy way to see it formally is to observe that f is piecewise smooth and has a unique local maximum $s_0 \in (0, 1)$. (Specifically, $s_0 = 1 - \frac{1}{\pi}\arctan(\pi \log e) = 0.7013$.) Thus if $\lambda_{0,1} \circ f^{-1} = \lambda_{0,1}$ then for all sufficiently small $\varepsilon > 0$,

$$\frac{f(s_0) - f(s_0 - \varepsilon)}{\varepsilon} = \frac{\lambda_{0,1}\big(\big[f(s_0 - \varepsilon), f(s_0)\big)\big)}{\varepsilon} = \frac{\lambda_{0,1} \circ f^{-1}\big(\big[f(s_0 - \varepsilon), f(s_0)\big)\big)}{\varepsilon}$$

$$\ge \frac{\lambda_{0,1}\big(\big[s_0 - \varepsilon, s_0\big)\big)}{\varepsilon} = 1\,,$$

which is impossible because $f'(s_0) = 0$; see Figure 7.1. Hence $\big(h(C^n)\big)$ is not Benford. The reason for this can be seen in the fact that while the number $\log|\lambda| = \log 2$ is irrational for the eigenvalues $\lambda = 2e^{\pm \pi i \log 2}$ of C, there clearly is a rational dependence between the two real numbers $\log|\lambda|$ and $\frac{1}{2\pi}\arg\lambda$, namely, $\log|\lambda| \mp 2\big(\frac{1}{2\pi}\arg\lambda\big) = 0$.

Notice also that if γ were chosen as, say, $\gamma = \cos(\pi \log 3) = 0.07181$ or $\gamma = \cos(\pi \log 201) = 0.5796$, then no such rational dependence would exist, and Theorem 7.21 below shows that in both these cases $\big(h(C^n)\big)$ would indeed be Benford for all $h \in \mathcal{L}_2$ unless $h(C^n) \equiv 0$; see Figure 7.1. ✠

$$x_n = 2^n \sin(\pi n \delta) \qquad\qquad f(s) = \langle s + \log|\sin(\pi s)|\rangle$$

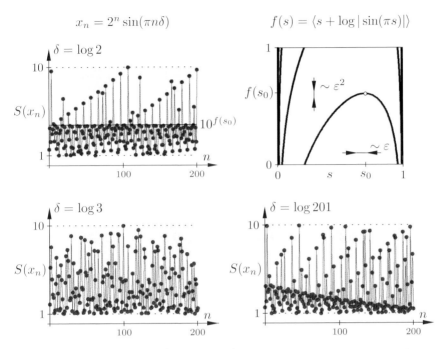

Figure 7.1: The sequence $\big(2^n \sin(\pi n \delta)\big)$ is not Benford for $\delta = \log 2$ (top left), but is Benford for $\delta = \log 3$ and $\delta = \log 201$ (bottom). The former corresponds to the fact that the map f does not preserve $\lambda_{0,1}$; see Example 7.16(iii).

The above examples indicate that, from the perspective of Benford's law, the main difficulty when dealing with multi-dimensional systems is their potential for cyclic behavior, either of the orbits themselves or of their significands. (In the case of primitive nonnegative matrices, as seen in the previous section, cyclicality does not occur or, more correctly, remains hidden.) To clarify this difficulty precisely, the following terminology is useful. Recall that, given any set $Z \subset \mathbb{C}$, $\mathrm{span}_{\mathbb{Q}} Z$ denotes the smallest linear subspace of $\mathbb{C}$ (over $\mathbb{Q}$) containing Z.

DEFINITION 7.17. A non-empty set $Z \subset \mathbb{C}$ with $|z| = r$ for some $r > 0$ and all $z \in Z$, i.e., $Z \subset r\mathbb{S}$, is *nonresonant* if the associated set

$$\Delta_Z := \left\{ 1 + \frac{\arg z - \arg w}{2\pi} : z, w \in Z \right\} \subset \mathbb{R}$$

satisfies the following two conditions:

(i) $\Delta_Z \cap \mathbb{Q} = \{1\}$;

(ii) $\log r \notin \mathrm{span}_{\mathbb{Q}} \Delta_Z$.

An arbitrary set $Z \subset \mathbb{C}$ is *nonresonant* if, for every $r > 0$, the set $Z \cap r\mathbb{S}$ is either nonresonant or empty; otherwise Z is *resonant*.

Note that the set Δ_Z automatically satisfies $1 \in \Delta_Z \subset (0,2)$ and is symmetric with respect to the point 1. The empty set $\varnothing$ and the singleton $\{0\}$ are nonresonant. Also, if Z is nonresonant then so is every $W \subset Z$. On the other hand, Z is certainly resonant if either $\#(Z \cap r\mathbb{S} \cap \mathbb{R}) = 2$ for some $r > 0$, in which case (i) is violated, or $Z \cap \mathbb{S} \neq \varnothing$, which causes (ii) to fail. Finally, it is easily checked that if $Z \subset \mathbb{C}$ is nonresonant then so is $Z^n := \{z^n : z \in Z\}$ for every $n \in \mathbb{N}$. The converse is not true in general, as can be seen from the resonant set $Z = \{-2, 2\}$, for which $Z^2 = \{4\}$ is nonresonant.

EXAMPLE 7.18. The singleton $\{z\}$ is nonresonant if and only if either $z = 0$ or $\log |z| \notin \mathbb{Q}$. Similarly, the set $\{z, \overline{z}\}$ with $z \in \mathbb{C} \setminus \mathbb{R}$ is nonresonant if and only if 1, $\log |z|$, and $\frac{1}{2\pi} \arg z$ are $\mathbb{Q}$-independent.   ✠

As one learns in linear algebra, the asymptotic behavior of (A^n) is completely determined by the eigenvalues of A, together with the corresponding (generalized) eigenvectors. As far as Benford's law is concerned, the key question turns out to be whether or not the set $\sigma(A)$ is nonresonant. Note that $\log \rho(A)$ is irrational whenever $\sigma(A)$ is nonresonant (and A is not nilpotent), but the converse is not true in general.

EXAMPLE 7.19. For the matrix A from (7.2), $\sigma(A) = \{-\varphi^{-1}, \varphi\}$ is nonresonant. On the other hand, the spectrum of the matrix B from (7.5) is $\sigma(B) = \{-1, 2\}$, and hence resonant.   ✠

EXAMPLE 7.20. The 2×2-matrices considered in Example 7.16,

$$A = \begin{bmatrix} 1 & -1 \\ 1 & 0 \end{bmatrix}, \quad B = \begin{bmatrix} 0 & 2 \\ 2 & 0 \end{bmatrix}, \quad C = \begin{bmatrix} 4\gamma & -4 \\ 1 & 0 \end{bmatrix},$$

all have resonant spectrum. Indeed $\sigma(A) = \{e^{\pm \pi \imath/3}\}$, and $\Delta_{\sigma(A)} = \{\frac{2}{3}, 1, \frac{4}{3}\}$ contains rational points other than 1, which violates (i) in Definition 7.17. In addition, $\log |e^{\pm \pi \imath/3}| = 0$, and so (ii) is also violated. Similarly, $\sigma(B) = \{\pm 2\}$; hence $\Delta_{\sigma(B)} = \{\frac{1}{2}, 1, \frac{3}{2}\}$, and (i) fails whereas $\log 2 \notin \mathbb{Q} = \mathrm{span}_\mathbb{Q} \Delta_{\sigma(B)}$, i.e., (ii) holds. (Note, however, that $\sigma(B^2) = \{4\}$ is nonresonant, and so is $\sigma(B^{2n})$ for all $n \in \mathbb{N}$.) Finally, with $\sigma(C) = \{2e^{\pm \pi \imath \log 2}\}$, the set $\Delta_{\sigma(C)} = \{1, 1 \pm \log 2\}$ satisfies (i), but $\log 2 \in \mathrm{span}_\mathbb{Q}\{1, \log 2\} = \mathrm{span}_\mathbb{Q} \Delta_{\sigma(C)}$, violating (ii).   ✠

The following theorem is the main result in this section. Like Theorems 7.3 and 7.11 above, but without any further assumptions on A, it extends to arbitrary dimension the simple fact that for the sequence (a^n) with $a \in \mathbb{R}$ to be either Benford or trivial, it is necessary and sufficient that $\log |a|$ be irrational or $a = 0$. By Example 7.18, the latter is equivalent to the singleton $\{a\}$ being nonresonant.

THEOREM 7.21 ([17]). *For every $A \in \mathbb{R}^{d \times d}$ the following are equivalent:*

(i) *The set $\sigma(A)$ is nonresonant;*

(ii) *For every $h \in \mathcal{L}_d$ the sequence $\big(h(A^n)\big)$ is Benford or terminating.*

A proof of Theorem 7.21 (and also of Theorem 7.28 below) will be given here only under an additional assumption on the matrix A. While this assumption, (7.18) below, holds for *most* matrices, it clearly fails for *some*, and the interested reader may want to consult [17] for a complete proof not making use of (7.18). The argument presented here relies on the Benford properties of sequences of a particular form, stated below for ease of reference; with Theorem 4.2 and Proposition 4.6, assertion (i) is an immediate consequence of [11, Cor. 2.6], and (ii) is analogous to [16, Lem. 2.6]. Note that (i) contains the "if" part of Theorem 4.16 as a special case.

PROPOSITION 7.22. *Let $a > 0$, $b \in \mathbb{R}$, and (ε_n) be a sequence in $\mathbb{R}$ with $\lim_{n \to \infty} \varepsilon_n = 0$. Suppose $\theta \in \mathbb{R}$ is irrational, and suppose the 1-periodic function $f : [0, +\infty) \to \mathbb{R}$ is continuous, with $f(t) \neq 0$ for almost all $t \geq 0$. Then, for the sequence (x_n) defined by*

$$x_n = a^n n^b (f(n\theta) + \varepsilon_n), \quad n \in \mathbb{N},$$

the following hold:

(i) *If $\log a \notin \operatorname{span}_{\mathbb{Q}}\{1, \theta\}$ then (x_n) is Benford;*

(ii) *If $\log a \in \operatorname{span}_{\mathbb{Q}}\{1, \theta\}$ and f is differentiable with $f(t) = 0$ for some $t \geq 0$ then (x_n) is not Benford.*

As indicated above, a proof of Theorem 7.21 is given here only under an additional assumption. Concretely, it will be assumed that

$$\#(\sigma(A) \cap r\mathbb{S}) \leq 2 \quad \text{for all } r > 0, \tag{7.18}$$

i.e., for every $r > 0$ the matrix A has at most two eigenvalues of modulus r, which may take the form of the real pair $-r, r$, or a non-real pair $\lambda, \overline{\lambda}$, with $|\lambda| = r$. Note that (7.18) holds for *most* $A \in \mathbb{R}^{d \times d}$ in that the matrices A *not* satisfying (7.18) form a nowhere dense nullset in $\mathbb{R}^{d \times d}$. (If $d < 3$ then clearly (7.18) holds for *all* $A \in \mathbb{R}^{d \times d}$.) For convenience, let

$$\sigma^+(A) = \{\lambda \in \sigma(A) : \Im \lambda \geq 0\} \setminus \{0\},$$

and recall from the Jordan Normal Form Theorem (e.g., [79, Sec. 3.2]) that A^n can be written in the form

$$A^n = \Re\left(\sum\nolimits_{\lambda \in \sigma^+(A)} P_\lambda(n) \lambda^n\right), \quad n \geq d, \tag{7.19}$$

where for every $\lambda \in \sigma^+(A)$, P_λ is a (possibly non-real) matrix-valued polynomial of degree at most $d-1$, i.e., for all $j, k \in \{1, 2, \ldots, d\}$ the entry $[P_\lambda]_{jk}$ is a complex polynomial in n of degree at most $d - 1$. Moreover, P_λ is real whenever $\lambda \in \mathbb{R}$. The representation (7.19) will be used repeatedly in what follows.

PROOF OF THEOREM 7.21. If $\sigma^+(A) = \varnothing$ then A is nilpotent, $\sigma(A) = \{0\}$ is nonresonant, and every sequence $(h(A^n))$ is identically zero for $n \geq d$. Thus (i) is equivalent to (ii) trivially in this case. From now on, assume that $\sigma^+(A)$ is not empty. Given any $h \in \mathcal{L}_d$, it follows from (7.19) that

$$h(A^n) = \Re\left(\sum\nolimits_{\lambda \in \sigma^+(A)} h\big(P_\lambda(n)\big)\lambda^n\right) = \Re\left(\sum\nolimits_{\lambda \in \sigma^+(A)} p_\lambda(n)\lambda^n\right), \quad n \geq d,$$
(7.20)

where $p_\lambda = h(P_\lambda) := h(\Re P_\lambda) + ih(\Im P_\lambda)$ is a (possibly non-real) polynomial in n of degree at most $d-1$.

To establish the asserted equivalence, assume first that $\sigma(A)$ is nonresonant and, given any $h \in \mathcal{L}_d$, that $p_\lambda \neq 0$ for some $\lambda \in \sigma^+(A)$. (Otherwise $h(A^n) = 0$ for all $n \geq d$.) Let

$$r = \max\{|\lambda| : \lambda \in \sigma^+(A), p_\lambda \neq 0\} > 0.$$

Recall from (7.18) that $\sigma(A) \cap r\mathbb{S}$ contains at most two elements. Note also that r and $-r$ cannot both be eigenvalues of A, since otherwise $\sigma(A)$ would be resonant. Hence either exactly one of the two numbers $r, -r$ is an eigenvalue of A, and $\log r$ is irrational, or else $\sigma(A) \cap r\mathbb{S} = \{re^{\pm\pi\imath\vartheta}\}$ with the appropriate irrational $0 < \vartheta < 1$, and $\log r \notin \mathrm{span}_{\mathbb{Q}}\{1, 1 \pm \vartheta\} = \mathrm{span}_{\mathbb{Q}}\{1, \vartheta\}$.

In the former case, assume without loss of generality that r is an eigenvalue. (The case of $-r$ being an eigenvalue is completely analogous.) Recall that $|\lambda| < r$ for every other eigenvalue λ of A with $p_\lambda \neq 0$. Denote by $k \in \{0, 1, \ldots, d-1\}$ the degree of the polynomial p_r, and let $c = \lim_{n\to\infty} p_r(n)/n^k$. Note that c is non-zero and real. From (7.20), it follows that

$$h(A^n) = p_r(n)r^n + \Re\left(\sum\nolimits_{\lambda \in \sigma^+(A):|\lambda|<r} p_\lambda(n)\lambda^n\right) = r^n n^k(c + \varepsilon_n), \quad n \geq d,$$

with the (real) sequence (ε_n) given by

$$\varepsilon_n = \frac{p_r(n)}{n^k} - c + \frac{1}{r^n n^k}\Re\left(\sum\nolimits_{\lambda \in \sigma^+(A):|\lambda|<r} p_\lambda(n)\lambda^n\right), \quad n \in \mathbb{N}.$$

Clearly, $\lim_{n\to\infty} \varepsilon_n = 0$. Since $\log r$ is irrational, Proposition 7.22(i) with $a = r$, $b = k$, $f(t) \equiv c$, and any $\theta \notin \mathrm{span}_{\mathbb{Q}}\{1, \log r\}$ shows that $(h(A^n))$ is Benford.

In the latter case, the matrix A has $\lambda_0 = re^{\pi\imath\vartheta}$ and its complex conjugate $\overline{\lambda_0} = re^{-\pi\imath\vartheta}$ as eigenvalues, and $|\lambda| < r$ for every other eigenvalue λ of A with $p_\lambda \neq 0$. With k denoting the degree of p_{λ_0}, again let $c = \lim_{n\to\infty} p_{\lambda_0}(n)/n^k$, and note that c may now be non-real, but is non-zero as before. Deduce from (7.20) that

$$h(A^n) = \Re\left(p_{\lambda_0}(n)r^n e^{\pi\imath n\vartheta} + \sum\nolimits_{\lambda \in \sigma^+(A):|\lambda|<r} p_\lambda(n)\lambda^n\right) = r^n n^k\big(\Re(ce^{\pi\imath n\vartheta}) + \varepsilon_n\big),$$

with the (real) sequence (ε_n), given by

$$\varepsilon_n = \Re\left(\left(\frac{p_{\lambda_0}(n)}{n^k} - c\right)e^{\pi\imath n\vartheta} + \frac{1}{r^n n^k}\sum\nolimits_{\lambda \in \sigma^+(A):|\lambda|<r} p_\lambda(n)\lambda^n\right), \quad n \in \mathbb{N},$$

again satisfying $\lim_{n\to\infty} \varepsilon_n = 0$. Recall that $\vartheta \in \mathbb{R} \setminus \mathbb{Q}$, and $\log r \notin \operatorname{span}\{1, \vartheta\}$. Hence Proposition 7.22(i) with $a = r$, $b = k$, $\theta = \frac{1}{2}\vartheta$, and $f(t) = \Re(ce^{2\pi i t})$ shows that $(h(A^n))$ is Benford. This completes the proof of (i)$\Rightarrow$(ii).

To establish the reverse implication, assume that $\sigma(A)$ is resonant. Then, for some $r_0 > 0$ and with $Z = \sigma(A) \cap r_0 \mathbb{S}$, the set Δ_Z contains rational numbers other than 1, or else $\log r_0 \in \operatorname{span}_{\mathbb{Q}} \Delta_Z$, or both. Assume first that $1 + \rho \in \Delta_Z$ for some rational number $\rho > 0$. With (7.18) this implies that Z contains exactly two elements, either r_0 and $-r_0$, or $r_0 e^{\pm \pi i \rho}$. In the former case, let $u, v \in \mathbb{R}^d$ be unit eigenvectors of A corresponding to the eigenvalues r_0 and $-r_0$, respectively, and let $h_0 = \sum_{j,k=1}^d (u_j - v_j)(u_k + v_k)[\,\cdot\,]_{jk} \in \mathcal{L}_d$. Then

$$h_0(A^n) = (u - v)^\top A^n (u + v) = (1 - u^\top v)\big(r_0^n - (-r_0)^n\big), \quad n \in \mathbb{N}.$$

By the Cauchy–Schwarz Inequality, $1 - u^\top v > 0$. Hence $h_0(A^n) = 0$ for all even n and $h_0(A^n) > 0$ for all odd n, so $(h_0(A^n))$ is neither Benford nor terminating. In the latter case of non-real eigenvalues, there exist linearly independent unit vectors $u, v \in \mathbb{R}^d$ such that, for all $n \in \mathbb{N}$,

$$A^n u = r_0^n \cos(\pi n\rho)u - r_0^n \sin(\pi n\rho)v, \quad A^n v = r_0^n \sin(\pi n\rho)u + r_0^n \cos(\pi n\rho)v. \tag{7.21}$$

Hence with $h_0 = \sum_{j,k=1}^d (u_j - v_j)(u_k + v_k)[\,\cdot\,]_{jk} \in \mathcal{L}_d$ as above,

$$\begin{aligned} h_0(A^n) &= r_0^n (u - v)^\top \big((\cos(\pi n\rho) + \sin(\pi n\rho))u + (\cos(\pi n\rho) - \sin(\pi n\rho))v\big) \\ &= 2(1 - u^\top v)r_0^n \sin(\pi n\rho), \end{aligned}$$

and again $h_0(A^n) = 0$ periodically but not identically since $0 < \rho < 1$ is rational. Thus $(h_0(A^n))$ is neither Benford nor terminating.

It remains to consider the case where $\Delta_Z \cap \mathbb{Q} = \{1\}$ for every $Z = \sigma(A) \cap r\mathbb{S}$ and $r > 0$, but $\log r_0 \in \operatorname{span}_{\mathbb{Q}} \Delta_Z$ for some $r_0 > 0$. Again, it is helpful to distinguish two cases: Either $\sigma(A) \cap r_0 \mathbb{S} \subset \mathbb{R}$ or $\sigma(A) \cap r_0 \mathbb{S} \subset \mathbb{C} \setminus \mathbb{R}$. In the former case, exactly one of the two numbers r_0 and $-r_0$ is an eigenvalue of A. The argument for $-r_0$ being analogous, assume without loss of generality that $\sigma(A) \cap r_0 \mathbb{S} = \{r_0\}$. Then $\Delta_Z = \{1\}$ and hence $\log r_0$ is rational. In particular, taking $h_0 = \sum_{j,k=1}^d u_j u_k [\,\cdot\,]_{jk} \in \mathcal{L}_d$, where $u \in \mathbb{R}^d$ is any unit eigenvector of A corresponding to the eigenvalue r_0, yields $h_0(A^n) = r_0^n$, thus $(h_0(A^n))$ is neither Benford nor terminating. In the latter case, i.e., for $Z = \sigma(A) \cap r_0 \mathbb{S} = \{r_0 e^{\pm \pi i \vartheta}\}$ with some irrational $0 < \vartheta < 1$, again pick linearly independent unit vectors $u, v \in \mathbb{R}^d$ such that (7.21) holds for all $n \in \mathbb{N}$ with ρ replaced by ϑ. With $h_0 = \sum_{j,k=1}^d (u_j + v_j)(u_k + v_k)[\,\cdot\,]_{jk}$, it follows that

$$h_0(A^n) = 2r_0^n(1 + u^\top v)\cos(\pi n\vartheta), \quad n \in \mathbb{N}.$$

Recall that $\log r_0 \in \operatorname{span}_{\mathbb{Q}}\{1, \vartheta\}$, by assumption. Hence Proposition 7.22(ii) with $a = r_0$, $b = 0$, $\theta = \frac{1}{2}\vartheta$, $f(t) = 2(1 + u^\top v)\cos(2\pi t)$, and $\varepsilon_n \equiv 0$ shows that the sequence $(h_0(A^n))$ is not Benford. Clearly, it is not terminating either. In summary, (ii) fails whenever $\sigma(A)$ is resonant. Thus (ii)$\Rightarrow$(i), and the proof is complete.   ∎

Remark. Weaker forms of the implication (i)$\Rightarrow$(ii) in Theorem 7.21 or special cases thereof can be traced back at least to [138, 142] and may also be found in [11, 20, 87, 110]. Especially in the early literature, the corresponding results often apply only to the special situation of linear difference equations; cf. Section 7.5. The reverse implication (ii)$\Rightarrow$(i) seems to have been addressed previously only for $d < 4$; see [20, Thm. 5.37].

The following examples review matrices encountered earlier in the light of Theorem 7.21. Note that if $A \in \mathbb{R}^{d \times d}$ is invertible then (7.19) holds for all $n \in \mathbb{N}$, and hence "terminating" in Theorem 7.21(ii) can be replaced by "identically zero" in this case; see Proposition 7.31 below.

EXAMPLE 7.23. **(i)** As seen in Example 7.19, for the matrix A from (7.2) the set $\sigma(A)$ is nonresonant; note also that A is invertible. Thus for every $h \in \mathcal{L}_2$ the sequence $\big(h(A^n)\big)$ is Benford unless $h(I_2) = h(A) = 0$; in the latter case $h(A^n) \equiv 0$. Choosing specifically $h = [\,\cdot\,]_{21}$ yields $h(A) = 1 \neq 0$ and hence shows again that $([A^n]_{21}) = (F_n)$ is Benford. Similarly, $h = -[\,\cdot\,]_{11} + 3[\,\cdot\,]_{12}$ yields $h(A) = 2 \neq 0$, and so $\big(h(A^n)\big) = (2, 1, 3, 4, 7, \ldots)$, traditionally referred to as the sequence of *Lucas numbers* [117, A000032], is Benford as well. Note that the latter conclusion could not have been reached using Theorem 7.3 because h is not nonnegative.
(ii) For the (invertible) matrix B from (7.5), the set $\sigma(B)$ is resonant. By Theorem 7.21, there exists $h \in \mathcal{L}_2$ for which $\big(h(B^n)\big)$ is neither Benford nor identically zero. One example of such an h has already been seen in Section 7.1, namely, $h = 2[\,\cdot\,]_{11} + [\,\cdot\,]_{22}$, for which $h(B^n) = 3(-1)^n$ for all $n \in \mathbb{N}$.   ✠

EXAMPLE 7.24. **(i)** The spectrum of the (invertible) matrix A in Example 7.5(i), $\sigma(A) = \{-1, 1, 2\}$, is resonant. For some $h \in \mathcal{L}_3$, therefore, $\big(h(A^n)\big)$ is neither Benford nor identically zero. A simple example is $h = [\,\cdot\,]_{11} - [\,\cdot\,]_{13}$, for which $h(A^n) \equiv 1$. Recall from Example 7.5(i), however, that $\big(h(A^n)\big)$ is Benford whenever $h \neq 0$ is nonnegative.
(ii) Since the (invertible) matrix B in Example 7.5(ii) is stochastic, $\sigma(B)$ is resonant. This in turn corroborates the observation made there that $\big(h(B^n)\big)$ is neither Benford nor identically zero whenever $h \in \mathcal{L}_3 \setminus \{0\}$ is nonnegative. Recall from Example 7.5(ii) that

$$\lim_{n \to \infty} B^n = \tfrac{1}{3} \begin{bmatrix} 1 & 1 & 1 \\ 1 & 1 & 1 \\ 1 & 1 & 1 \end{bmatrix} =: Q\,.$$

The question whether sequences $\big(h(B^n - Q)\big)$ and $\big(h(B^{n+1} - B^n)\big)$ can be Benford is addressed by Theorem 7.32 below.   ✠

EXAMPLE 7.25. The spectra of the (invertible) matrices A and B in Example 7.7, $\sigma(A) = \{-10, 20\}$ and $\sigma(B) = \{2, 10\}$, are both resonant. However, with

$$A^n = \tfrac{2}{3} \cdot 20^{n-1}(A + 10I_2) + \tfrac{1}{3} \cdot (-10)^{n-1}(A - 20I_2), \quad n \in \mathbb{N},$$

it is clear from Theorem 4.16 that $\big(h(A^n)\big)$ is Benford whenever $h(A+10I_2) \neq 0$. For *most* $h \in \mathcal{L}_2$, therefore, $\big(h(A^n)\big)$ is Benford. On the other hand,

$$B^n = \tfrac{5}{4} \cdot 10^{n-1}(B - 2I_2) - 2^{n-3}(B - 10I_2), \quad n \in \mathbb{N}.$$

Thus the sequence $\big(h(B^n)\big)$ can be Benford *only* if $h(B - 2I_2) = 0$, and hence, for most $h \in \mathcal{L}_2$, is neither Benford nor identically zero. ✠

EXAMPLE 7.26. Theorem 7.3 does not apply to the (invertible) matrix A in Example 7.8. Since $\sigma(A) = \{2\}$ is nonresonant, by Theorem 7.21 the sequence $\big(h(A^n)\big)$ is Benford for every $h \in \mathcal{L}_2$ unless $h(I_2) = h(A) = 0$; in the latter case it is identically zero. ✠

While Theorem 7.21 neatly characterizes the Benford property of $\big(h(A^n)\big)$ in terms of $\sigma(A)$, it should be noted that deciding whether $\sigma(A)$ is resonant or not can be a challenge. The next example illustrates the basic difficulty.

EXAMPLE 7.27. The (invertible, nonnegative, integer) matrix A in Example 7.12 has spectrum $\sigma(A) = \{-1 \pm \imath\sqrt{2}, 2\}$. Since $\log 2$ is irrational, $\sigma(A)$ is nonresonant precisely if $\{-1 \pm \imath\sqrt{2}\} = \sigma(A) \cap \sqrt{3}\mathbb{S}$ is. Note that

$$\mathrm{span}_{\mathbb{Q}} \Delta_{\sigma(A) \cap \sqrt{3}\mathbb{S}} = \mathrm{span}_{\mathbb{Q}} \left\{ 1, \frac{1}{\pi} \arctan \sqrt{2} \right\},$$

so $\sigma(A)$ is nonresonant if and only if $\log 3 \notin \mathrm{span}_{\mathbb{Q}} \left\{ 1, \frac{1}{\pi} \arctan \sqrt{2} \right\}$. Using standard number theory tools, it is not hard to check that $\log 3$ and $\frac{1}{\pi} \arctan \sqrt{2}$ are both irrational (in fact, transcendental). However, it seems to be unknown whether the numbers 1, $\log 3$, and $\frac{1}{\pi} \arctan \sqrt{2}$ are $\mathbb{Q}$-independent. (The $\mathbb{Q}$-independence of these three numbers would follow easily from a prominent but as yet unresolved conjecture of S. Schanuel [158].) In other words, it is unknown whether $\sigma(A)$ is resonant or not. The reader may want to compare this delicate situation to the ease with which it was ascertained, in Example 7.12, that $\big(h(A^n)\big)$ is Benford for every *nonnegative* observable $h \neq 0$ on $\mathbb{R}^{3 \times 3}$. Thus, for instance,

$$([A^n]_{12} + [A^n]_{13}) = (1, 1, 1, 7, 7, 13, 49, \ldots)$$

is Benford. On the other hand,

$$Q = \lim_{n \to \infty} \frac{A^n}{2^n} = \frac{1}{11} \begin{bmatrix} 3 & 2 & 1 \\ 6 & 4 & 2 \\ 12 & 8 & 4 \end{bmatrix},$$

and hence, as just explained, it is not known whether the integer sequence

$$(-[A^n]_{11} + [A^n]_{12} + [A^n]_{13}) = (1, 1, -5, 7, 1, -23, 43, \ldots)$$

is Benford; cf. [117, A087455]. Experimental evidence suggests that it is; see Figure 7.2. Also note that even if $\sigma(A)$ is resonant, the sequence $\big(h(A^n)\big)$ is nevertheless Benford whenever $h(Q) \neq 0$. For *most* $h \in \mathcal{L}_3$, therefore, $\big(h(A^n)\big)$ is Benford regardless of whether or not $\sigma(A)$ turns out to be resonant.   ✠

		1	2	3	4	5	6	7	8	9	Δ
$[\cdot]_{12} + [\cdot]_{13}$	$N = 10$	50.00	0.00	10.00	10.00	10.00	0.00	20.00	0.00	0.00	19.89
	$N = 10^2$	32.00	16.00	13.00	10.00	7.00	6.00	7.00	4.00	5.00	1.89
	$N = 10^3$	30.30	17.50	12.50	9.70	7.80	6.50	6.10	4.90	4.70	0.30
	$N = 10^4$	30.12	17.59	12.50	9.69	7.91	6.69	5.82	5.09	4.59	0.02
$-[\cdot]_{11} + [\cdot]_{12} + [\cdot]_{13}$	$N = 10$	40.00	20.00	0.00	10.00	10.00	0.00	10.00	0.00	10.00	12.49
	$N = 10^2$	25.00	21.00	12.00	6.00	7.00	9.00	7.00	5.00	8.00	5.10
	$N = 10^3$	29.40	17.80	12.40	9.40	7.70	7.10	6.40	4.60	5.20	0.70
	$N = 10^4$	30.17	17.55	12.64	9.40	8.24	6.64	5.61	5.23	4.52	0.32

Figure 7.2: Relative frequencies of the first significant digit for the first N terms of $\big(h(A^n)\big)$, with A as in Examples 7.12 and 7.27: For $h = [\cdot]_{12} + [\cdot]_{13}$, the sequence is Benford; the empirical data suggest that for $h = -[\cdot]_{11} + [\cdot]_{12} + [\cdot]_{13}$ this is the case also.

As with nonnegative matrices in the previous section, the Benford property of $\big(h(A^n)\big)$ may also be of interest for the nonlinear observables $h(A) = |A|$ and $h(A) = |Au|$ on $\mathbb{R}^{d \times d}$.

THEOREM 7.28 ([17]). *Let $A \in \mathbb{R}^{d \times d}$. If $\sigma(A)$ is nonresonant then:*

(i) *The sequence $(|A^n|)$ is Benford, or $A^d = 0$;*

(ii) *For every $u \in \mathbb{R}^d$ the sequence $(|A^n u|)$ is Benford or terminating.*

PROOF. As in the case of Theorem 7.21 above, the argument presented here makes use of the additional assumption (7.18); see [17] for a complete proof without this assumption.

To prove (i), deduce from (7.19) that either $A^d = 0$ or else, with the appropriate $r > 0$, $Z \subset \sigma^+(A) \cap r\mathbb{S}$, and $k \in \{0, 1, \ldots, d-1\}$,

$$A^n = r^n n^k \left(\Re \left(\sum\nolimits_{\lambda \in Z} C_\lambda e^{\imath n \arg \lambda} \right) + B_n \right), \quad n \geq d,$$

where each $C_\lambda \neq 0$ is a (possibly non-real) $d \times d$-matrix, and (B_n) is a sequence in $\mathbb{R}^{d \times d}$ for which $(n|B_n|)$ is bounded. By (7.18), the set $\sigma(A) \cap r\mathbb{S}$ contains at most two elements; it is also nonresonant, by assumption. Hence either the set $\sigma(A) \cap r\mathbb{S}$ contains exactly one of the two numbers $\pm r$ and $\log r$ is irrational, or else $\sigma(A) \cap r\mathbb{S}$ equals $\{re^{\pm\pi\iota\vartheta}\}$ with some irrational $0 < \vartheta < 1$ for which $\log r \notin \mathrm{span}_{\mathbb{Q}}\{1, \vartheta\}$.

Assume first that $\sigma(A) \cap r\mathbb{S}$ equals $\{r\}$ or $\{-r\}$. Then, with $\lambda = r$ or $\lambda = -r$, the matrix C_λ is real, i.e., $C_\lambda \in \mathbb{R}^{d \times d}$, and

$$\log|A^n| = n \log r + k \log n + \log \left| C_\lambda + B_n(-1)^{n \arg \lambda / \pi} \right|, \quad n \in \mathbb{N}.$$

Together with Propositions 4.3(i) and 4.6(ii, iii), this shows that $(\log|A^n|)$ is u.d. mod 1 since $\log r$ is irrational. Hence $(|A^n|)$ is Benford.

Assume now that $\sigma(A) \cap r\mathbb{S} = \{re^{\pm\pi\iota\vartheta}\}$. Then, with $\lambda = re^{\pi\iota\vartheta}$,

$$|A^n| = r^n n^k \left| \Re\left(C_\lambda e^{\pi\iota n\vartheta}\right) + B_n \right| = r^n n^k \left(f(\tfrac{1}{2}n\vartheta) + \varepsilon_n \right), \quad n \in \mathbb{N},$$

with the 1-periodic continuous function $f(t) = \left| \Re\left(C_\lambda e^{2\pi\iota t}\right) \right|$ and the (real) sequence (ε_n) given by

$$\varepsilon_n = \left| \Re\left(C_\lambda e^{\pi\iota n\vartheta}\right) + B_n \right| - \left| \Re\left(C_\lambda e^{\pi\iota n\vartheta}\right) \right|, \quad n \in \mathbb{N}.$$

Note that $f(t) > 0$ for all but (at most) countably many $t \geq 0$. Also $|\varepsilon_n| \leq |B_n|$, so $\lim_{n \to \infty} \varepsilon_n = 0$. Consequently, Proposition 7.22(i) with $a = r$, $b = k$, and $\theta = \frac{1}{2}\vartheta$ shows that $(|A^n|)$ is Benford.

The completely analogous proof of (ii) is left to the reader. ∎

Remark. As is the case with Theorems 7.9 and 7.11, Theorem 7.28 (i) and (ii) both hold with $|\cdot|$ replaced by any other norm on $\mathbb{R}^{d \times d}$ and $\mathbb{R}^d$, respectively. Moreover, the conclusions remain valid under the weaker assumption that $\sigma(A^N)$ is nonresonant for some $N \in \mathbb{N}$; see [17].

EXAMPLE 7.29. **(i)** The matrix A from (7.2) has nonresonant spectrum $\sigma(A)$. By Theorem 7.28, $(|A^n|)$ is Benford. This is also clear since $|A^n| \equiv \varphi^n$.

(ii) For the matrix B from (7.5), $\sigma(B)$ is resonant, so Theorem 7.28 does not apply. Still, it is readily checked that $\lim_{n \to \infty} |B^n|/2^n = \sqrt{10} \neq 0$, so $(|B^n|)$ is Benford nevertheless, by Theorem 4.16. ✠

Example 7.29(ii) above indicates that, in contrast to Theorem 7.11, the converse of Theorem 7.28 is not true in general: Even if $\sigma(A)$ is resonant, $(|A^n|)$ may nevertheless be Benford or terminating, and so may $(|A^n u|)$ for every $u \in \mathbb{R}^d$. In fact, as the following example illustrates, it is impossible to characterize properties (i) and (ii) in Theorem 7.28 solely in terms of $\sigma(A)$ — except, of course, in the trivial case $d = 1$.

EXAMPLE 7.30. Consider the (invertible) 2×2-matrix

$$A = 10^\pi \begin{bmatrix} \cos(\pi^2) & -\sin(\pi^2) \\ \sin(\pi^2) & \cos(\pi^2) \end{bmatrix},$$

for which $\sigma(A) = \left\{ 10^\pi e^{\pm \iota \pi^2} \right\}$ is resonant because

$$\pi = \log 10^\pi \in \mathrm{span}_{\mathbb{Q}} \Delta_{\sigma(A)} = \mathrm{span}_{\mathbb{Q}}\{1, \pi\}\,.$$

Nevertheless, $(|A^n|) = (10^{\pi n})$ is Benford, and so is $(|A^n u|) = (10^{\pi n}|u|)$ for every $u \in \mathbb{R}^2 \setminus \{0\}$. This again shows that the nonresonance assumption in Theorem 7.28 is not necessary for the conclusion.

Next consider the (invertible) matrix

$$B = \frac{10^\pi}{\sqrt{3}} \left[\begin{array}{cc} \sqrt{3}\cos(\pi^2) & -3\sin(\pi^2) \\ \sin(\pi^2) & \sqrt{3}\cos(\pi^2) \end{array} \right],$$

so $\sigma(B) = \left\{ 10^\pi e^{\pm \iota \pi^2} \right\} = \sigma(A)$. Thus, as far as their spectra are concerned, the matrices A and B are indistinguishable. (In fact, A and B are similar.) In particular, $\sigma(B)$ is resonant as well. From

$$B^n = \frac{10^{\pi n}}{\sqrt{3}} \left[\begin{array}{cc} \sqrt{3}\cos(\pi^2 n) & -3\sin(\pi^2 n) \\ \sin(\pi^2 n) & \sqrt{3}\cos(\pi^2 n) \end{array} \right], \quad n \in \mathbb{N},$$

a straightforward calculation yields

$$|B^n| = \frac{10^{\pi n}}{\sqrt{3}} \sqrt{4 - \cos(2\pi^2 n) + |\sin(\pi^2 n)|\sqrt{14 - 2\cos(2\pi^2 n)}}\,, \quad n \in \mathbb{N},$$

which in turn shows that $\langle \log |B^n| \rangle = f(\langle n\pi \rangle)$ for all $n \in \mathbb{N}$, with the smooth function $f : [0, 1) \to [0, 1)$ given by

$$f(s) = \left\langle s - \tfrac{1}{2}\log 3 + \tfrac{1}{2}\log\left(4 - \cos(2\pi s) + \sin(\pi s)\sqrt{14 - 2\cos(2\pi s)}\right)\right\rangle.$$

Recall from Example 4.7(i) that $(n\pi)$ is u.d. mod 1. Since f is a smooth bijection of $[0, 1)$ with non-constant derivative, it follows that $\big(f(\langle n\pi \rangle)\big)$ is not u.d. mod 1, simply because $\lambda_{0,1} \circ f^{-1} \neq \lambda_{0,1}$. Thus $(|B^n|)$ is neither Benford nor terminating. In a similar manner, it can be shown that $(|B^n u|)$ is neither Benford nor terminating for any $u \in \mathbb{R}^2 \setminus \{0\}$. ✠

As noted earlier, if $A \in \mathbb{R}^{d \times d}$ is invertible then (7.19) holds for all $n \in \mathbb{N}$; Theorems 7.21 and 7.28 therefore have the following corollary.

PROPOSITION 7.31. *Let* $A \in \mathbb{R}^{d \times d}$ *be invertible. If* $\sigma(A)$ *is nonresonant then:*

(i) *For every* $h \in \mathcal{L}_d$ *the sequence* $\big(h(A^n)\big)$ *is Benford or identically zero;*

(ii) *The sequence* $(|A^n|)$ *is Benford;*

(iii) *For every* $u \in \mathbb{R}^d \setminus \{0\}$ *the sequence* $(|A^n u|)$ *is Benford.*

Remark. By Proposition 7.31, for any invertible $A \in \mathbb{R}^{d \times d}$, nonresonance of $\sigma(A)$ guarantees an abundance of Benford sequences of the form $(h(A^n))$. Most $d \times d$-matrices are invertible with nonresonant spectrum, from a topological as well as a measure-theoretic perspective. To put this more formally, let

$$\mathcal{G}_d = \left\{ A \in \mathbb{R}^{d \times d} : A \text{ is invertible and } \sigma(A) \text{ is nonresonant} \right\}.$$

Thus, for example, $\mathcal{G}_1 = \{[a] : a \in \mathbb{R} \setminus \{0\}, \log|a| \notin \mathbb{Q}\}$. While the complement of $\mathcal{G}_d$ is dense in $\mathbb{R}^{d \times d}$, it is a topologically small set: $\mathbb{R}^{d \times d} \backslash \mathcal{G}_d$ is *of first category*, i.e., a countable union of nowhere dense sets. A (topologically) typical ("generic") $d \times d$-matrix therefore belongs to $\mathcal{G}_d$. Also, $\mathbb{R}^{d \times d} \setminus \mathcal{G}_d$ is a (Lebesgue) nullset. Thus if A is an $\mathbb{R}^{d \times d}$-valued random variable, that is, a random matrix, whose distribution is a.c. with respect to the d^2-dimensional Lebesgue measure on $\mathbb{R}^{d \times d}$, then $\mathbb{P}(A \in \mathcal{G}_d) = 1$, i.e., with probability one A is invertible and $\sigma(A)$ is nonresonant.

Cancellations of resonance

If A is a real $d \times d$-matrix and $\log \rho(A)$ is rational then, for most linear observables h on $\mathbb{R}^{d \times d}$, the sequence $(h(A^n))$ is *not* Benford. Even in this situation, however, it is quite possible for the sequence $(h(A^{n+1} - \rho(A)A^n))$ to be Benford. This phenomenon has already been observed in Example 7.13 for the (row-stochastic) matrix $A = \frac{1}{2} \begin{bmatrix} 1 & 1 \\ 2 & 0 \end{bmatrix}$, where $\log \rho(A) = \log 1 = 0$ is rational, and yet

$$A^{n+1} - A^n = \left(-\tfrac{1}{2}\right)^n (A - I_2), \quad n \in \mathbb{N}.$$

Hence every sequence $(h(A^{n+1} - A^n))$ is Benford or identically zero. The remainder of the present section studies this "cancellation of resonance" scenario and demonstrates how it can be easily understood by utilizing the results from above. The scenario was first described in, and is of particular interest for, the case of *stochastic* matrices [22]; this important special case is the subject of the next section. However, as explained here, "cancellation of resonance" may occur much more generally, namely, whenever A has a simple dominant eigenvalue.

Assume that the real $d \times d$-matrix A has a dominant eigenvalue λ_0 that is (algebraically) simple, i.e., $|\lambda| < |\lambda_0|$ for every $\lambda \in \sigma(A) \setminus \{\lambda_0\}$, and λ_0 is a simple root of the characteristic polynomial of A. Note that λ_0 is necessarily a real number, and $\rho(A) = |\lambda_0|$. For $\lambda_0 \neq 0$, it is not hard to see that the limit

$$Q_A = \lim_{n \to \infty} \frac{A^n}{\lambda_0^n} \tag{7.22}$$

exists. Moreover, $Q_A A = A Q_A = \lambda_0 Q_A$ and $Q_A^2 = Q_A$. In fact, Q_A is simply the spectral projection associated with λ_0 and can also be represented in the form

$$Q_A = \frac{uv^\top}{u^\top v}, \tag{7.23}$$

where u, v are eigenvectors of A and $A^\top$, respectively, corresponding to the eigenvalue λ_0. A simple dominant eigenvalue is often observed in practice and occurs, for instance, whenever A is nonnegative and primitive; see Proposition 7.2. (In this case, Q_A is in fact positive.) But it also occurs for matrices such as $\begin{bmatrix} 1 & -1 \\ -1 & 1 \end{bmatrix}$ and $\begin{bmatrix} 2 & 1 \\ 0 & 1 \end{bmatrix}$, which are not nonnegative and nonnegative but not primitive, respectively.

Now consider the sequences $(A^{n+1} - \lambda_0 A^n)$ and $(A^n - \lambda_0^n Q_A)$, both of which in some sense measure the speed of convergence in (7.22), and therefore are often of interest in their own right. With the help of Theorem 7.21, the Benford properties of these sequences are easy to understand.

THEOREM 7.32. *Let $A \in \mathbb{R}^{d \times d}$ have a dominant eigenvalue λ_0 that is algebraically simple, and let Q_A be the associated projection* (7.23). *Then the following are equivalent:*

(i) *The set $\sigma(A) \setminus \{\lambda_0\}$ is nonresonant;*

(ii) *For every $h \in \mathcal{L}_d$ the sequence $\big(h(A^{n+1} - \lambda_0 A^n)\big)$ is Benford or terminating;*

(iii) *For every $h \in \mathcal{L}_d$ the sequence $\big(h(A^n - \lambda_0^n Q_A)\big)$ is Benford or terminating.*

PROOF. As the theorem clearly holds for $d = 1$, henceforth assume $d \geq 2$ and thus $\lambda_0 \neq 0$. As in the proof of Theorem 7.3, let $B = A - \lambda_0 Q_A$ and observe that $AB = BA$ and $Q_A B = 0 = B Q_A$, so

$$A^n = \lambda_0^n Q_A + B^n, \quad n \in \mathbb{N}. \tag{7.24}$$

It will first be shown that $\sigma(B) = (\sigma(A) \setminus \{\lambda_0\}) \cup \{0\}$. To this end, note that for every $\lambda \in \sigma(A) \setminus \{\lambda_0\}$ and $w \in \mathbb{R}^d \setminus \{0\}$ with $(A^2 - 2\Re\lambda A + |\lambda|^2 I_d)^d w = 0$ (i.e., w is a generalized eigenvector of A corresponding to the eigenvalue $\lambda \neq \lambda_0$ or $\overline{\lambda} \neq \lambda_0$),

$$0 = v^\top (A^2 - 2\Re\lambda A + |\lambda|^2 I_d)^d w = \big(\big((A^\top)^2 - 2\Re\lambda A^\top + |\lambda|^2 I_d\big)^d v\big)^\top w$$
$$= (\lambda_0 - \lambda)^d (\lambda_0 - \overline{\lambda})^d v^\top w.$$

Thus $v^\top w = 0$, which in turn implies $Q_A w = 0$, and hence $A^n w = B^n w$ for all $n \in \mathbb{N}$, by (7.24). Consequently, $\lambda \in \sigma(B)$. Also, $Au = \lambda_0 u = \lambda_0 Q_A u$ and therefore $Bu = 0$. Thus $0 \in \sigma(B)$, and hence $(\sigma(A) \setminus \{\lambda_0\}) \cup \{0\} \subset \sigma(B)$. Moreover, if $Bw = \lambda_0 w$ for some $w \in \mathbb{R}^d$, then (7.24) yields

$$Q_A w = \lim_{n \to \infty} \frac{A^n w}{\lambda_0^n} = Q_A w + w;$$

hence $w = 0$, which shows that $\lambda_0 \notin \sigma(B)$. Conversely, if $\lambda \in \sigma(B) \setminus \{0\}$ then $(B^2 - 2\Re\lambda B + |\lambda|^2 I_d)w = 0$ for some $w \in \mathbb{R}^d \setminus \{0\}$, which in turn implies

$0 = Q_A(B^2 - 2\Re\lambda B + |\lambda|^2 I_d)w = |\lambda|^2 Q_A w$ and hence $Q_A w = 0$. It follows that $(A^2 - 2\Re\lambda A + |\lambda|^2 I_d)w = 0$ as well, i.e., $\lambda \in \sigma(A)$. In summary, therefore, $\sigma(B) = (\sigma(A) \setminus \{\lambda_0\}) \cup \{0\}$, as claimed, and $\sigma(B)$ is nonresonant if and only if $\sigma(A) \setminus \{\lambda_0\}$ is nonresonant. Deduce from (7.24) that

$$A^{n+1} - \lambda_0 A^n = B^n(B - \lambda_0 I_d), \quad A^n - \lambda_0^n Q_A = B^n, \quad n \in \mathbb{N}.$$

Since $B - \lambda_0 I_d$ is invertible, the equivalence of (i), (ii), and (iii) now follows immediately from Theorem 7.21. ∎

EXAMPLE 7.33. **(i)** The (positive) matrix $A = \begin{bmatrix} 6 & 4 \\ 4 & 6 \end{bmatrix}$, first encountered in Example 7.7, has the simple dominant eigenvalue $\lambda_0 = 10$. Thus Theorem 7.32 applies with

$$Q_A = \lim_{n \to \infty} \frac{A^n}{\lambda_0^n} = \frac{1}{2}\begin{bmatrix} 1 & 1 \\ 1 & 1 \end{bmatrix}.$$

Since the set $\sigma(A) \setminus \{10\} = \{2\}$ is nonresonant, every sequence $(h(A^{n+1} - 10A^n))$ and $(h(A^n - 10^n Q_A))$ with $h \in \mathcal{L}_2$ is Benford or terminating (in fact, identically zero). This can also be seen directly from

$$A^{n+1} - 10A^n = 2^n(A - 10I_2), \quad A^n - 10^n Q_A = -2^{n-3}(A - 10I_2), \quad n \in \mathbb{N}.$$

(ii) For the (nonnegative) matrix $B = \begin{bmatrix} 19 & 20 \\ 1 & 0 \end{bmatrix}$, $\lambda_0 = 20$ is a simple dominant eigenvalue, and $Q_B = \frac{1}{21}\begin{bmatrix} 20 & 20 \\ 1 & 1 \end{bmatrix}$. However, $\sigma(B) \setminus \{20\} = \{-1\}$ is resonant, and hence some (in fact, most) sequences $(h(B^{n+1} - 20B^n))$ and $(h(B^n - 20^n Q_B))$ are neither Benford nor terminating. Again, this can be confirmed by an explicit calculation, which yields

$$B^{n+1} - 20B^n = (-1)^n(B - 20I_2), \quad B^n - 20^n Q_B = (-1)^n \frac{1}{21}(20I_2 - B), \quad n \in \mathbb{N}.$$

(iii) The matrix $C = \begin{bmatrix} 6 & -8 \\ 4 & -6 \end{bmatrix}$ does not have a dominant eigenvalue, since $\sigma(C) = \{\pm 2\}$. Hence Theorem 7.32 does not apply, and the limit $\lim_{n \to \infty} C^n/2^n$ does not exist. Note that even in this case, however, for every $h \in \mathcal{L}_2$, the sequence $(h(C^{n+1} - 2C^n))$ is Benford or identically zero, because

$$C^{n+1} - 2C^n = (-2)^n(2I_2 - C), \quad n \in \mathbb{N}.$$

(iv) Consider the (nonnegative) 3×3-matrix

$$D = \begin{bmatrix} 3 & 1 & 0 \\ 0 & 3 & 0 \\ 0 & 0 & 2 \end{bmatrix}.$$

Clearly, $\lambda_0 = 3$ is a dominant eigenvalue, and $\sigma(D) \setminus \{3\} = \{2\}$ is nonresonant. However,

$$D^n = \begin{bmatrix} 3^n & n3^{n-1} & 0 \\ 0 & 3^n & 0 \\ 0 & 0 & 2^n \end{bmatrix}, \quad n \in \mathbb{N},$$

so $\lim_{n \to \infty} D^n/3^n$ does not exist. The reason for this can be seen in the fact that the eigenvalue λ_0, although dominant, is not simple. Thus Theorem 7.32 does not apply. Nevertheless, $\big(h(D^{n+1} - 3D^n)\big)$ is Benford or identically zero for every $h \in \mathcal{L}_3$.   ✠

Remark. Close inspection of the proof of Theorem 7.32 shows that the assumption of algebraic simplicity for λ_0 can be relaxed somewhat: The equivalences in Theorem 7.32 remain unchanged if the dominant eigenvalue λ_0 is merely assumed to be *semi-simple*, meaning that its algebraic and geometric multiplicities coincide or, equivalently, that $A - \lambda_0 I_d$ and $(A - \lambda_0 I_d)^2$ have the same rank. For instance, the eigenvalue $\lambda_0 = 3$ of D in Example 7.33(iv) is not semi-simple.

7.4   AN APPLICATION TO MARKOV CHAINS

This brief section demonstrates how the results from earlier in the chapter can fruitfully be applied to the matrices of transition probabilities associated with finite-state Markov chains. Markov chains constitute the simplest, most fundamental class of stochastic processes with widespread applications throughout science [116]. Recall that a (time-homogeneous) d-state *Markov chain* is a discrete-time Markov process (X_n) on d states $s_1, s_2, \ldots, s_d$ such that $\mathbb{P}(X_{n+1} = s_k | X_n = s_j)$ is independent of n, for all $j, k \in \{1, 2, \ldots, d\}$. Thus

$$p_{jk} = \mathbb{P}(X_{n+1} = s_k | X_n = s_j), \quad j, k \in \{1, 2, \ldots, d\}, \tag{7.25}$$

defines a matrix $P = [p_{jk}] \in \mathbb{R}^{d \times d}$, the matrix of one-step transition probabilities of (X_n). Clearly, P is nonnegative and row-stochastic. Moreover, (7.25) implies that the N-step transition probabilities of (X_n) are simply given by the entries of P^N, that is, for all $n \in \mathbb{N}$,

$$\mathbb{P}(X_{n+N} = s_k | X_n = s_j) = [P^N]_{jk}, \quad j, k \in \{1, 2, \ldots, d\}.$$

Thus the long-term behavior of the stochastic process (X_n) is governed by the sequence of stochastic matrices (P^n).

A d-state Markov chain (X_n) is *irreducible* if, for every $j, k \in \{1, 2, \ldots, d\}$, there exists $N \in \mathbb{N}$ such that $\mathbb{P}(X_N = s_k | X_1 = s_j) > 0$. Note that this is equivalent to the associated matrix P being irreducible. Also (X_n) is *aperiodic* if, for every $j \in \{1, 2, \ldots, d\}$, the greatest common divisor of the numbers in the set $\big\{n \geq 2 : \mathbb{P}(X_n = s_j | X_1 = s_j) > 0\big\}$ equals 1; in this case, for convenience, call P aperiodic also. It is well known (and easy to check) that (X_n) is irreducible and aperiodic if and only if the associated matrix P of one-step transition probabilities is primitive.

Consider now an irreducible and aperiodic d-step Markov chain (X_n) with associated matrix P. Recall from Proposition 7.2 that $P^* = \lim_{n\to\infty} P^n$ exists and is positive. In fact, since P^* has rank one and is row-stochastic, all rows of P^* are identical. The following observation is an immediate consequence of Theorem 7.11.

PROPOSITION 7.34. *Let $P \in \mathbb{R}^{d\times d}$ be the matrix of one-step transition probabilities of an irreducible and aperiodic d-state Markov chain (X_n). Then, for each $j, k \in \{1, 2, \ldots, d\}$, the sequences $([P^n]_{jk})$ and $\big(\mathbb{P}(X_n = s_j)\big)$ are not Benford.*

A common problem in many Markov chain models is to estimate the matrix $P^* = \lim_{n\to\infty} P^n$ through numerical simulations. In this context, the sequences $(P^{n+1} - P^n)$ and $(P^n - P^*)$ are of special interest as they both in some sense measure the speed of convergence $P_n \to P^*$. As the following corollary of Theorem 7.32 shows, often these sequences are Benford. (Recall that every irreducible and aperiodic stochastic matrix has $\lambda_0 = 1$ as a simple dominant eigenvalue.)

PROPOSITION 7.35. [22, Thm. 12] *Let $P \in \mathbb{R}^{d\times d}$ be the matrix of one-step transition probabilities of an irreducible and aperiodic d-state Markov chain, and $P^* = \lim_{n\to\infty} P^n$. If the set $\sigma(P) \setminus \{1\}$ is nonresonant then, for each $j, k \in \{1, 2, \ldots, d\}$, the sequences $([P^{n+1} - P^n]_{jk})$ and $([P^n - P^*]_{jk})$ are Benford or terminating.*

EXAMPLE 7.36. **(i)** The stochastic matrix $P = \frac{1}{2} \begin{bmatrix} 1 & 1 \\ 2 & 0 \end{bmatrix}$, first encountered in Example 7.13, is irreducible and aperiodic, and since $\sigma(P) \setminus \{1\} = \{-\frac{1}{2}\}$ is nonresonant, Proposition 7.35 applies. In Example 7.13, it was seen that $\big(h(P^{n+1} - P^n)\big)$ and $\big(h(P^n - P^*)\big)$, with $P^* = \frac{1}{3} \begin{bmatrix} 2 & 1 \\ 2 & 1 \end{bmatrix}$, are Benford or identically zero for every $h \in \mathcal{L}_2$, depending on whether $h(I_2 - P) \neq 0$ or $h(I_2 - P) = 0$.

(ii) For the (irreducible and aperiodic) stochastic matrix

$$P = \frac{1}{5} \begin{bmatrix} 3 & 2 & 0 \\ 4 & 0 & 1 \\ 0 & 3 & 2 \end{bmatrix},$$

$\sigma(P) = \{\pm\frac{1}{\sqrt{5}}, 1\}$. Since $\sigma(P) \setminus \{1\} = \{\pm\frac{1}{\sqrt{5}}\}$ is resonant, Proposition 7.35 does not apply. A straightforward calculation confirms that

$$P^* = \frac{1}{10} \begin{bmatrix} 6 & 3 & 1 \\ 6 & 3 & 1 \\ 6 & 3 & 1 \end{bmatrix},$$

and, for instance, that $[P^n - P^*]_{11} = 5^{-(n+2)/2}\big(1 + (-1)^n\big)$ for all $n \in \mathbb{N}$; hence $([P^n - P^*]_{11})$ is neither Benford nor terminating. ✠

It is not hard to see that for $d = 2$ the converse of Proposition 7.35 is also true: If the sequences $([P^{n+1} - P^n]_{jk})$ and $([P^n - P^*]_{jk})$ are Benford or terminating for all $j, k \in \{1, 2\}$, then $\sigma(P) \setminus \{1\}$ is nonresonant. For $d \geq 3$ the analogous statement is no longer true. In fact, as the next example shows, for $d \geq 3$ the property that the sequences $([P^{n+1} - P^n]_{jk})$ and $([P^n - P^*]_{jk})$ are Benford or terminating for every $j, k \in \{1, 2, \ldots, d\}$ cannot be characterized in terms of $\sigma(P)$ alone.

EXAMPLE 7.37. Consider the (irreducible and aperiodic) stochastic matrix

$$P = \tfrac{1}{30} \begin{bmatrix} 14 & 11 & 5 \\ 11 & 14 & 5 \\ 5 & 5 & 20 \end{bmatrix},$$

for which $\sigma(P) \setminus \{1\} = \left\{ \tfrac{1}{10}, \tfrac{1}{2} \right\}$ is resonant. Hence Corollary 7.35 does not apply. However, from

$$P^n = \tfrac{1}{3} \begin{bmatrix} 1 & 1 & 1 \\ 1 & 1 & 1 \\ 1 & 1 & 1 \end{bmatrix} + \tfrac{1}{3} \cdot 2^{-(n+1)} \begin{bmatrix} 1 & 1 & -2 \\ 1 & 1 & -2 \\ -2 & -2 & 4 \end{bmatrix} + \tfrac{1}{2} \cdot 10^{-n} \begin{bmatrix} 1 & -1 & 0 \\ -1 & 1 & 0 \\ 0 & 0 & 0 \end{bmatrix},$$

it is straightforward to deduce that $([P^{n+1} - P^n]_{jk})$ and $([P^n - P^*]_{jk})$ are Benford for all $j, k \in \{1, 2, 3\}$. This shows that the converse of Proposition 7.35 does not hold in general if $d \geq 3$.

Next consider the (irreducible and aperiodic) stochastic matrix

$$Q = \tfrac{1}{10} \begin{bmatrix} 6 & 3 & 1 \\ 3 & 4 & 3 \\ 1 & 3 & 6 \end{bmatrix}.$$

Since $\sigma(Q) = \left\{ \tfrac{1}{10}, \tfrac{1}{2}, 1 \right\} = \sigma(P)$, the two matrices P and Q are indistinguishable with regard to their spectrum. (In fact, P and Q are similar.) Moreover, as P and Q are both symmetric, $Q^* = P^*$. As was essentially seen in Example 7.5(ii), for instance, the sequences $([Q^{n+1} - Q^n]_{22})$ and $([Q^n - Q^*]_{22})$ are neither Benford nor terminating. ✠

The hypotheses in Proposition 7.35 are satisfied by most stochastic matrices. To put this more formally, denote by $\mathcal{P}_d$ the family of all (row-) stochastic $d \times d$-matrices, that is,

$$\mathcal{P}_d = \left\{ P \in \mathbb{R}^{d \times d} : P \geq 0, \sum\nolimits_{k=1}^d [P]_{jk} = 1 \text{ for all } j = 1, 2, \ldots, d \right\}.$$

The set $\mathcal{P}_d$ is a compact and convex subset of $\mathbb{R}^{d \times d}$; for example,

$$\mathcal{P}_1 = \{[1]\} \quad \text{and} \quad \mathcal{P}_2 = \left\{ \begin{bmatrix} s & 1-s \\ 1-t & t \end{bmatrix} : 0 \leq s, t \leq 1 \right\}.$$

Note that $\mathcal{P}_d$ can be identified with a d-fold copy of the standard $(d-1)$-simplex, that is, $\mathcal{P}_d \simeq \left\{ u \in \mathbb{R}^d : u \geq 0, \sum_{j=1}^{d} u_j = 1 \right\}^d$, and hence $\mathcal{P}_d$ carries the (normalized) $d(d-1)$-dimensional Lebesgue measure Leb. Now let

$$\mathcal{H}_d = \left\{ P \in \mathcal{P}_d : P \text{ is irreducible and aperiodic, } \sigma(P) \setminus \{1\} \text{ is nonresonant} \right\},$$

which is exactly the family of stochastic matrices covered by Proposition 7.35. For instance, $\mathcal{H}_1 = \{[1]\} = \mathcal{P}_1$ and

$$\mathcal{H}_2 = \left\{ \begin{bmatrix} s & 1-s \\ 1-t & t \end{bmatrix} : 0 \leq s, t < 1, \; s+t = 1 \text{ or } \log|s+t-1| \notin \mathbb{Q} \right\},$$

and in both cases $\mathcal{H}_d$ makes up *most* of $\mathcal{P}_d$. More formally, it can be shown that, for every $d \in \mathbb{N}$, the complement of $\mathcal{H}_d$ in $\mathcal{P}_d$ is a set of first category and has Lebesgue measure zero. Thus if P is a $\mathcal{P}_d$-valued random variable, i.e., a random stochastic matrix, whose distribution is absolutely continuous (with respect to Leb, which means that $\mathbb{P}(P \in C) = 0$ whenever $C \subset \mathcal{P}_d$ and $\mathrm{Leb}(C) = 0$), then $\mathbb{P}(P \in \mathcal{H}_d) = 1$. Together with Proposition 7.35, this implies the following.

PROPOSITION 7.38. [22, Thm. 17] *If the random stochastic matrix P has an absolutely continuous distribution then, with probability one, P is irreducible and aperiodic, and the sequences $([P^{n+1} - P^n]_{jk})$ and $([P^n - P^*]_{jk})$ are Benford or terminating for each $j, k \in \{1, 2, \ldots, d\}$.*

Note, for instance, that a random stochastic matrix has an absolutely continuous distribution whenever its d rows are chosen independently according to the same density on the standard $(d-1)$-simplex. While the above generic properties are very similar to the corresponding results for arbitrary matrices (see the *Remark* on p. 159), they do not follow directly from the latter. In fact, they are somewhat harder to prove because they assert (topological as well as measure-theoretic) prevalence of $\mathcal{H}_d$ within the space $\mathcal{P}_d$, which, as a subset of $\mathbb{R}^{d \times d}$, is itself a nowhere dense nullset. The interested reader may want to consult [22] for details.

7.5   LINEAR DIFFERENCE EQUATIONS

In applied science as well as in mathematics, linear processes often emerge in the form of *linear difference equations*. Despite their simplicity, these equations form a sufficiently wide class of systems to provide fundamental models for disciplines as diverse as mechanical engineering, population dynamics, and signal processing [41, 55]. This section studies the Benford property for solutions of linear difference equations. The main results (Theorems 7.39 and 7.41) are direct applications of results presented earlier in this chapter.

Consider the (autonomous) linear difference equation (or recursion)

$$x_n = a_1 x_{n-1} + a_2 x_{n-2} + \ldots + a_d x_{n-d}, \quad n \geq d+1, \tag{7.26}$$

where, here as throughout, d is a fixed positive integer, referred to as the *order* of (7.26), and $a_1, a_2, \ldots, a_d$ are real numbers, with $a_d \neq 0$. Thus, (7.8) is a first-order equation with $a_1 = -2$, whereas (7.7) and (7.9) are second-order equations with $a_1 = a_2 = 1$ and $a_1 = -2$, $a_2 = -3$, respectively.

Once the initial values $x_1, x_2, \ldots, x_d$ are specified, (7.26) uniquely defines a sequence of real numbers (x_n), referred to as a *solution* of (7.26). Under what conditions is (x_n) Benford? In order to make this question amenable to results obtained earlier in this chapter, rewrite (7.26) by inserting $d-1$ redundant rows,

$$
\begin{bmatrix} x_n \\ x_{n-1} \\ x_{n-2} \\ \vdots \\ x_{n-d+1} \end{bmatrix} = \begin{bmatrix} a_1 & a_2 & \cdots & a_{d-1} & a_d \\ 1 & 0 & \cdots & 0 & 0 \\ 0 & 1 & 0 & \cdots & 0 \\ \vdots & \ddots & \ddots & \ddots & \vdots \\ 0 & \cdots & 0 & 1 & 0 \end{bmatrix} \begin{bmatrix} x_{n-1} \\ x_{n-2} \\ x_{n-3} \\ \vdots \\ x_{n-d} \end{bmatrix}, \quad n \geq d+1 .
$$

Hence with the $d \times d$-matrix

$$
A = \begin{bmatrix} a_1 & a_2 & \cdots & a_{d-1} & a_d \\ 1 & 0 & \cdots & 0 & 0 \\ 0 & 1 & 0 & \cdots & 0 \\ \vdots & \ddots & \ddots & \ddots & \vdots \\ 0 & \cdots & 0 & 1 & 0 \end{bmatrix}, \tag{7.27}
$$

for every $n \in \mathbb{N}$, the number x_n is simply the last (bottom) entry of

$$
\begin{bmatrix} x_{n+d-1} \\ \vdots \\ x_{n+1} \\ x_n \end{bmatrix} = A \begin{bmatrix} x_{n+d-2} \\ \vdots \\ x_n \\ x_{n-1} \end{bmatrix} = \ldots = A^{n-1} \begin{bmatrix} x_d \\ \vdots \\ x_2 \\ x_1 \end{bmatrix} . \tag{7.28}
$$

Recall that $a_d \neq 0$ and hence that A is invertible; specifically,

$$
A^{-1} \begin{bmatrix} x_d \\ \vdots \\ x_2 \\ x_1 \end{bmatrix} = \begin{bmatrix} x_{d-1} \\ \vdots \\ x_1 \\ (x_d - a_1 x_{d-1} - \ldots - a_{d-1} x_1)/a_d \end{bmatrix} .
$$

With the linear observable

$$
h = x_{d-1}[\cdot]_{d1} + \ldots + x_1[\cdot]_{dd-1} + \frac{x_d - a_1 x_{d-1} - \ldots - a_{d-1} x_1}{a_d}[\cdot]_{dd} \tag{7.29}
$$

on $\mathbb{R}^{d \times d}$, therefore, $(x_n) = \big(h(A^n)\big)$, and the Benford property of any solution (x_n) of (7.26) can indeed be studied by applying the results from earlier sections. To do this effectively, the following simple observations are useful.

First note that the matrix A in (7.27) is nonnegative if and only if $a_j \geq 0$ for all $j \in \{1, 2, \ldots, d\}$. Since $a_d \neq 0$, A is irreducible in this case. Next, recall from Sections 7.2 and 7.3 that the Benford behavior of $\big(h(A^n)\big)$ for any $h \in \mathcal{L}_d$ is determined by properties of $\sigma(A)$. With A given by (7.27),

$$\det(A - zI_d) = (-1)^d (z^d - a_1 z^{d-1} - \ldots - a_{d-1} z - a_d), \qquad (7.30)$$

which shows that $\sigma(A) = \{z \in \mathbb{C} : z^d = a_1 z^{d-1} + \ldots + a_{d-1} z + a_d\}$. Finally, it is important to note that not only can every solution (x_n) of (7.26) be written in the form $\big(h(A^n)\big)$ with the appropriate $h \in \mathcal{L}_d$, but also, conversely, the sequence $\big(h(A^n)\big)$ with A from (7.27) solves (7.26) for every $h \in \mathcal{L}_d$. Indeed, by (7.30) and the Cayley–Hamilton Theorem,

$$A^d = a_1 A^{d-1} + \ldots + a_{d-1} A + a_d I_d,$$

and hence, for every $h \in \mathcal{L}_d$,

$$\begin{aligned} h(A^n) = h(A^{n-d} A^d) &= h\big(A^{n-d}(a_1 A^{d-1} + \ldots + a_{d-1} A + a_d I_d)\big) \\ &= a_1 h(A^{n-1}) + \ldots + a_{d-1} h(A^{n-d+1}) + a_d h(A^{n-d}), \quad n \geq d+1. \end{aligned}$$

With these observations, the Benford property for solutions (x_n) of (7.26) is easily analyzed. First consider the case where the coefficients $a_1, a_2, \ldots, a_d$ are all *positive*. In this case, Theorem 7.11 has the following simple corollary.

THEOREM 7.39. *Let (x_n) be a solution of (7.26) with $a_1, a_2, \ldots, a_d > 0$. Assume that the initial values $x_1, x_2, \ldots, x_d$ are all nonnegative and at least one of them is positive. Then (x_n) is Benford if and only if $\log \zeta$ is irrational, where $z = \zeta$ is the root of $z^d = a_1 z^{d-1} + \ldots + a_{d-1} z + a_d$ with the largest real part.*

PROOF. Since $a_1, a_2, \ldots, a_d > 0$, the matrix A associated with (7.26) via (7.27) is primitive, and $\rho(A) = \zeta > 0$. Also, deduce from (7.28) that

$$x_{n+1} = x_d [A^n]_{d1} + \ldots + x_2 [A^n]_{dd-1} + x_1 [A^n]_{dd} =: g(A^n), \quad n \in \mathbb{N},$$

and note that, unlike h in (7.29), the linear observable g is nonnegative if $x_1, x_2, \ldots, x_d \geq 0$. Clearly, (x_{n+1}) is Benford if and only if (x_n) is.

To prove the theorem, assume first that $\log \zeta$ is irrational. Then the sequence $(x_{n+1}) = \big(g(A^n)\big)$ is Benford by Theorem 7.11. Conversely, if $\log \zeta$ is rational then (7.11) and Theorem 4.16 show that (x_{n+1}) is not Benford. ∎

EXAMPLE 7.40. **(i)** For the Fibonacci recursion (7.7), the root of $z^2 = z + 1$ with the largest real part is $\zeta = \varphi = \frac{1}{2}(1 + \sqrt{5})$. Since $\log \varphi$ is irrational, by Theorem 7.39 every solution of (7.7) with $x_1 x_2 \geq 0$ is Benford, except for the trivial case $x_n \equiv 0$. (For $x_1 \leq 0$ simply note that $(-x_n)$ is a solution of (7.7) as well.) Theorem 7.39 does not apply if $x_1 x_2 < 0$. As was essentially seen in Example 7.23, (x_n) is Benford in this case too. This also follows easily from Theorem 7.41 below.

(ii) Consider the third-order difference equation

$$x_n = x_{n-2} + x_{n-3}, \quad n \geq 4, \tag{7.31}$$

which resembles the Fibonacci recursion. Since $a_1 = 0$, Theorem 7.39 does not apply. However, close inspection of the proof of that theorem shows that its conclusion still holds because the matrix associated with (7.31),

$$A = \begin{bmatrix} 0 & 1 & 1 \\ 1 & 0 & 0 \\ 0 & 1 & 0 \end{bmatrix},$$

is primitive since $A^5 > 0$. Moreover $\det A = 1$, and essentially the same argument as that in Example 7.12 shows that $\log \zeta = \log \rho(A)$ is irrational. Hence every solution (x_n) of (7.31) with $x_1, x_2, x_3 \geq 0$ is Benford, unless $x_1 = x_2 = x_3 = 0$, in which case $x_n \equiv 0$. ✠

Remark. In Theorem 7.39 it is enough to assume that $a_1, a_2, \ldots, a_d \geq 0$ and $a_j a_d > 0$ for some $j \in \{1, 2, \ldots, d-1\}$, with j relatively prime to d; see Example 7.40(ii) above, where $d = 3$ and $a_2 a_3 = 1$.

For the remainder of this section consider (7.26) with *arbitrary*, i.e., not necessarily positive, coefficients $a_1, a_2, \ldots, a_d$. (Recall that $a_d \neq 0$ throughout.) In this case, the appropriate tools are provided by Theorem 7.21 and Proposition 7.31.

THEOREM 7.41. *For the difference equation* (7.26) *with* $a_1, a_2, \ldots, a_d \in \mathbb{R}$ *and* $a_d \neq 0$ *the following are equivalent:*

(i) *The set* $\{z \in \mathbb{C} : z^d = a_1 z^{d-1} + \ldots + a_{d-1} z + a_d\}$ *is nonresonant;*

(ii) *Every solution* (x_n) *of* (7.26) *is Benford unless* $x_n \equiv 0$.

PROOF. Let A be the matrix associated with (7.26) via (7.27), and recall that A is invertible. If $\sigma(A) = \{z \in \mathbb{C} : z^d = a_1 z^{d-1} + \ldots + a_{d-1} z + a_d\}$ is nonresonant then, for every $h \in \mathcal{L}_d$, the sequence $(h(A^n))$ is Benford or identically zero, by Proposition 7.31. Choosing h as in (7.29), therefore, shows that (x_n) is either Benford or identically zero. Conversely, if $\sigma(A)$ is resonant then, by Theorem 7.21, there exists $h_0 \in \mathcal{L}_d$ for which $(h_0(A^n))$ is neither Benford nor terminating. Recall that $(h_0(A^n))$ solves (7.26), so setting $x_j = h_0(A^j)$ for all $j \in \{1, 2, \ldots, d\}$ yields a solution that is neither Benford nor terminating (let alone identically zero). ∎

EXAMPLE 7.42. The set $\{z \in \mathbb{C} : z^2 = z + 1\} = \{-\varphi^{-1}, \varphi\}$ associated with the Fibonacci recursion (7.7) is nonresonant. This once again shows that, except for $x_n \equiv 0$, every solution of the latter is Benford. For instance, since (F_n) is Benford, so is the sequence (F_n^2), by Theorem 4.4. This can also be seen directly by noticing that $(F_n^2 + \frac{2}{5}(-1)^n)$ solves

$$x_n = 3x_{n-1} - x_{n-2}, \quad n \geq 3,$$

for which the associated set $\{z \in \mathbb{C} : z^2 = 3z - 1\} = \{\varphi^{-2}, \varphi^2\}$ is nonresonant.

✠

EXAMPLE 7.43. Let a_1, a_2 be *integers* with $a_2 > 0$, and consider the second-order equation

$$x_n = a_1 x_{n-1} + a_2 x_{n-2}\,, \quad n \geq 3\,. \tag{7.32}$$

When $a_1 = 0$, clearly the set $\{z \in \mathbb{C} : z^2 = a_1 z + a_2\} = \{\pm\sqrt{a_2}\}$ is resonant. For $a_1 \neq 0$, however, it consists of two real numbers with different absolute value, and in this case it is resonant only if one of these numbers is of the form $\pm 10^k$ for some integer $k \geq 0$. It follows that every solution of (7.32), except for $x_n \equiv 0$, is Benford if and only if

$$a_1 \neq 0 \quad \text{and} \quad |10^{2k} - a_2| \neq |a_1| \cdot 10^k \quad \text{for all } k \in \{0, 1, \ldots, \lfloor \log a_2 \rfloor\}\,. \tag{7.33}$$

For the Fibonacci recursion (7.7), for instance, condition (7.33) reduces to the obviously correct inequalities $1 \neq 0$ and $|1 - 1| \neq 1$. For another simple example consider

$$x_n = -2x_{n-1} + 2x_{n-2}\,, \quad n \geq 3\,,$$

for which (7.33) reads $-2 \neq 0$ and $|1 - 2| \neq 2$, and consequently every nontrivial solution is Benford. On the other hand, for

$$x_n = -2x_{n-1} + 3x_{n-2}\,, \quad n \geq 3\,,$$

(7.33) fails since $|1 - 3| = |-2|$, and $x_n \equiv 1$ is a solution that is neither Benford nor identically zero.

✠

Similarly to the main results in Section 7.3, though it is definitive in theory, Theorem 7.41 may not always be easy to use in practice. Part (iii) of the following example illustrates that practical difficulties may arise in applying Theorem 7.41 even if $d = 2$ and a_1, a_2 are both integers.

EXAMPLE 7.44. Let $0 < a < \sqrt{3}$ be a real number and consider the second-order difference equation

$$x_n = -2a x_{n-1} - 3x_{n-2}\,, \quad n \geq 3\,, \tag{7.34}$$

for which $Z = \{z \in \mathbb{C} : z^2 + 2az + 3 = 0\} = \{-a \pm \imath\sqrt{3 - a^2}\}$. Consider the Benford property for solutions (x_n) of (7.34) for three specific values of the parameter a; Figure 7.3 shows data corresponding to $x_1 = x_2 = 1$.

(i) Let $a = \sqrt{3}\cos\left(\pi\frac{1}{\sqrt{11}}\right) = 1.011$. In this case, $Z = \left\{\sqrt{3}e^{\pm\pi\imath(1 - 1/\sqrt{11})}\right\}$, and $\Delta_Z = \left\{\frac{1}{\sqrt{11}}, 1, 2 - \frac{1}{\sqrt{11}}\right\}$. Since $\log\sqrt{3} \notin \text{span}_\mathbb{Q}\Delta_Z = \text{span}_\mathbb{Q}\left\{1, \frac{1}{\sqrt{11}}\right\}$, the set Z is nonresonant, and every solution of (7.34), except for $x_n \equiv 0$, is Benford.

(ii) Let $a = \sqrt{3}\cos\left(\pi\frac{5}{8}\log 3\right) = 1.025$. Here $\text{span}_\mathbb{Q}\Delta_Z = \text{span}_\mathbb{Q}\{1, \log 3\}$, and since the latter clearly contains $\log 3$, the set Z is resonant. It follows that no solution of (7.34) is Benford in this case.

(iii) Finally, let $a = 1$, in which case (7.34) takes the innocent-looking form (7.9). Since $a_2 = -3 < 0$, the easy-to-check condition (7.33) does not apply.

Moreover, $Z = \{-1 \pm \imath\sqrt{2}\}$, and as already mentioned in Example 7.27, it is unknown whether the set Z is nonresonant. If it were, then every solution of (7.9) except for $x_n \equiv 0$ would be Benford; otherwise, none would. It may not have escaped the reader that the solution (x_n) of (7.34) with $x_1 = x_2 = 1$, i.e., $(x_n) = (1, 1, -5, 7, 1, -23, 43, \ldots)$, is nothing other than the (integer) sequence $(-[A^n]_{11} + [A^n]_{12} + [A^n]_{13})$ considered in Example 7.27. Thus experimental evidence seems to suggest that Z is nonresonant; see Figure 7.3.　✠

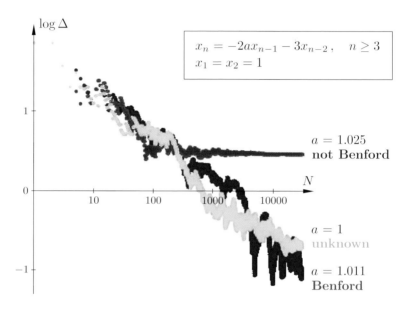

Figure 7.3: For different values of the parameter a, the solutions (x_n) of (7.34) may or may not be Benford; see Example 7.44 and Figure 7.2.

7.6　LINEAR DIFFERENTIAL EQUATIONS

Just like their discrete-time counterparts, linear differential equations are important for both theoretical and practical purposes. In the theory of differential equations, they constitute the simplest and most fundamental class of systems, and for many applications, they provide basic models of real-world processes [77]. As this section demonstrates, they also form a rich source of Benford functions.

EXAMPLE 7.45. With $x_0, v_0 \in \mathbb{R}$, consider initial value problems (IVPs) for three simple linear second-order differential equations. Throughout, $x = x(t)$ denotes a smooth function, and $\dot{x}$, $\ddot{x}$, etc. are, respectively, the first, second, etc. derivative of x with respect to t.

(i) The unique solution $x = x(t)$ of

$$\ddot{x} - x = 0\,, \quad x(0) = x_0\,, \ \dot{x}(0) = v_0\,,$$

is $x(t) = \frac{1}{2}(x_0 + v_0)e^t + \frac{1}{2}(x_0 - v_0)e^{-t}$, which by Example 4.9(i) and Theorem 4.12(ii) is a Benford function unless $x_0 = v_0 = 0$.

(ii) The unique solution of

$$\ddot{x} + x = 0\,, \quad x(0) = x_0\,, \ \dot{x}(0) = v_0\,,$$

is $x(t) = x_0 \cos t + v_0 \sin t$ and hence is not Benford for any x_0, v_0, by Theorem 4.10.

(iii) The unique solution of the IVP

$$\ddot{x} + 2\dot{x} + 3x = 0\,, \quad x(0) = x_0\,, \ \dot{x}(0) = v_0\,,$$

is

$$x(t) = x_0 e^{-t}\cos\!\big(t\sqrt{2}\big) + \frac{x_0 + v_0}{\sqrt{2}} e^{-t}\sin\!\big(t\sqrt{2}\big) = e^{-t}\Re\left(ce^{it\sqrt{2}}\right)\,,$$

with $c = x_0 - \imath(x_0 + v_0)/\sqrt{2}$. If x does not vanish identically, i.e., if $c \neq 0$, fix any $\delta > 0$ and note that $x(n\delta) = a^n f\big(n\delta/(\pi\sqrt{2})\big)$ for all $n \in \mathbb{N}$, with $a = e^{-\delta}$ and $f(t) = \Re(ce^{2\pi it})$. Since the real number $\pi \log e$ is transcendental [158],

$$\frac{\delta}{\pi\sqrt{2}} \notin \mathbb{Q} \quad \text{and} \quad \delta \log e \notin \text{span}_{\mathbb{Q}}\left\{1, \frac{\delta}{\pi\sqrt{2}}\right\}$$

for all but countably many $\delta > 0$, and in that case, Proposition 7.22(i) with $\theta = \delta/\big(\pi\sqrt{2}\big)$, $b = 0$, and $\varepsilon_n \equiv 0$ shows that the sequence $\big(x(n\delta)\big)$ is Benford. By Proposition 4.8, therefore, the function x is Benford as well.   ✠

Using matrix-vector notation, all three IVPs above can be written in the form

$$\frac{\mathrm{d}}{\mathrm{d}t}\begin{bmatrix} \dot{x} \\ x \end{bmatrix} = A\begin{bmatrix} \dot{x} \\ x \end{bmatrix}\,, \quad \begin{bmatrix} \dot{x} \\ x \end{bmatrix}\bigg|_{t=0} = \begin{bmatrix} v_0 \\ x_0 \end{bmatrix}\,, \tag{7.35}$$

with the 2×2-matrix A given, respectively, by

$$\begin{bmatrix} 0 & 1 \\ 1 & 0 \end{bmatrix}\,, \quad \begin{bmatrix} 0 & -1 \\ 1 & 0 \end{bmatrix}\,, \quad \begin{bmatrix} -2 & -3 \\ 1 & 0 \end{bmatrix}\,.$$

The (unique) solution of (7.35) is

$$\begin{bmatrix} \dot{x} \\ x \end{bmatrix} = e^{tA}\begin{bmatrix} v_0 \\ x_0 \end{bmatrix}\,, \tag{7.36}$$

with the matrix exponential $e^{tA} = \sum_{k=0}^{\infty} \frac{t^k}{k!}A^k$. Recall that the $d \times d$-matrix e^{tA} is well-defined for all $t \in \mathbb{R}$ and any $A \in \mathbb{R}^{d \times d}$, and $\frac{\mathrm{d}^k}{\mathrm{d}t^k}e^{tA} = e^{tA}A^k$ for $k = 0, 1, \ldots$. From (7.36) it is clear that

$$x(t) = v_0[e^{tA}]_{21} + x_0[e^{tA}]_{22}\,, \quad t \geq 0\,.$$

In other words, $x(t) = h(e^{tA})$ with $h \in \mathcal{L}_2$ determined by the initial values x_0, v_0, namely, $h = v_0[\,\cdot\,]_{21} + x_0[\,\cdot\,]_{22}$. In complete analogy to the discrete-time context of previous sections, this suggests that the following question be studied: Given $A \in \mathbb{R}^{d \times d}$, is the function $h(e^{tA})$ either Benford or identically zero for every $h \in \mathcal{L}_d$? The main result of this section, Theorem 7.49 below, provides a necessary and sufficient condition for this to be the case. To formulate the result, the following tailor-made terminology is convenient; here, $e^Z = \{e^z : z \in Z\}$ for every $Z \subset \mathbb{C}$.

DEFINITION 7.46. A set $Z \subset \mathbb{C}$ is *exponentially nonresonant* if the set e^{tZ} is nonresonant for some $t \in \mathbb{R}$; otherwise Z is *exponentially resonant*.

EXAMPLE 7.47. The empty set is exponentially nonresonant, and the singleton $\{z\}$ is exponentially nonresonant if and only if $\Re z \neq 0$. Similarly, the set $\{z, \overline{z}\}$ with $z \in \mathbb{C} \setminus \mathbb{R}$ is exponentially nonresonant if and only if $\pi \log e \, \Re z / \Im z \notin \mathbb{Q}$. ✠

In deciding whether a given set $Z \subset \mathbb{C}$ is exponentially nonresonant, the following proposition is often useful; its routine verification is left to the reader. Recall that a set is *countable* if it is either finite (possibly empty) or countably infinite.

PROPOSITION 7.48. *Assume $Z \subset \mathbb{C}$ is symmetric with respect to the real axis, i.e., $Z = \{\overline{z} : z \in Z\}$, and is countable. Then the following are equivalent:*

(i) *Z is exponentially nonresonant;*

(ii) *For every $z \in Z$,*

$$\Re z \notin \mathrm{span}_{\mathbb{Q}} \left\{ \frac{\Im w}{\pi \log e} : w \in Z, \Re w = \Re z \right\}. \tag{7.37}$$

Moreover, if (i) and (ii) hold then the set $\{t \in \mathbb{R} : e^{tZ}$ is resonant$\}$ is countable.

Remark. The symmetry and countability assumptions are essential in Proposition 7.48. If Z is not symmetric with respect to the real axis then the implication (i)⇒(ii) may fail, as is seen for $Z = \{1 + \imath\pi \log e\}$, which is exponentially nonresonant yet does not satisfy (ii). Conversely, if Z is uncountable then (ii)⇒(i) may fail. To see this, take $Z = \mathbb{R} \setminus \{0\}$, which satisfies (ii), and yet e^{tZ} is resonant for every $t \in \mathbb{R}$. Recall, however, that if $Z = \sigma(A)$ for some $A \in \mathbb{R}^{d \times d}$, then clearly Z is both symmetric with respect to the real axis and countable (in fact, Z is finite).

Section 7.3 has demonstrated how the Benford property of sequences $(h(A^n))$ with $h \in \mathcal{L}_d$ depends on whether the spectrum $\sigma(A)$ of the matrix A is nonresonant. The reader may suspect that it is the *exponential* nonresonance of $\sigma(A)$ that guarantees the Benford property for functions $h(e^{tA})$, and indeed this is the case. The following theorem is an analogue of Theorem 7.21 in the continuous-time setting. It extends to arbitrary dimension the simple fact that for $f(t) = e^{ta}$ with $a \in \mathbb{R}$ to be Benford, it is necessary and sufficient that $a \neq 0$ or, equivalently, that the set $\{a\}$ be exponentially nonresonant. As in the case of Theorem 7.21, a proof of Theorem 7.49 below is given here only under an additional assumption on the matrix A, and the reader is referred to [14] for a complete proof not making use of that assumption.

THEOREM 7.49 ([14]). *For every $A \in \mathbb{R}^{d \times d}$ the following are equivalent:*

(i) *The set $\sigma(A)$ is exponentially nonresonant;*

(ii) *For every $h \in \mathcal{L}_d$ the function $h(e^{tA})$ is Benford or identically zero.*

PROOF. Assume first that $\sigma(A)$ is exponentially nonresonant and, given any $h \in \mathcal{L}_d$, define $g : [0, +\infty) \to \mathbb{R}$ as $g(t) = h(e^{tA})$. By Proposition 7.48, the set $\sigma(e^{\delta A}) = e^{\delta \sigma(A)}$ is nonresonant for all but countably many $\delta > 0$. In that case, Theorem 7.21 implies that $\big(g(n\delta)\big) = \big(h(e^{n\delta A})\big)$ is either Benford or identically zero. (Recall that $e^{\delta A}$ is invertible.) With $\inf \varnothing := +\infty$, let

$$\delta_0 = \inf\{\delta > 0 : g(n\delta) = 0 \text{ for all } n \in \mathbb{N}\} \geq 0 \,.$$

If $\delta_0 = 0$ then there exists a strictly decreasing sequence (δ_n) with $\delta_n \searrow 0$ and $g(\delta_n) = 0$ for every n. Since g is real-analytic, it follows that $g(t) \equiv 0$. If, on the other hand, $\delta_0 > 0$ (possibly $\delta_0 = +\infty$) then $(\log |g(n\delta)|)$ is u.d. mod 1 for almost all $0 < \delta < \delta_0$, and Proposition 4.8(i) shows that $\log |g(t)|$ is u.d. mod 1. In other words, the function g is Benford. Thus (i)$\Rightarrow$(ii).

As indicated above, similarly to the discrete-time setting of Section 7.3, the reverse implication (ii)$\Rightarrow$(i) will be proved here only under an additional assumption which, however, is satisfied by *most* $A \in \mathbb{R}^{d \times d}$. Specifically, in analogy to (7.18) assume that

$$\#\big(\sigma(A) \cap (x + \imath \mathbb{R})\big) \leq 2 \quad \text{for all } x \in \mathbb{R}, \tag{7.38}$$

i.e., for every $x \in \mathbb{R}$ the matrix A has at most two eigenvalues λ with $\Re \lambda = x$, and hence either $\sigma(A) \cap (x + \imath \mathbb{R}) = \{x\}$ or else $\sigma(A) \cap (x + \imath \mathbb{R}) = \{x + \imath y, x - \imath y\}$ with some $y > 0$.

Thus, to establish the implication (ii)$\Rightarrow$(i) assume that $\sigma(A)$ is exponentially resonant. By (7.37), there exists $\lambda \in \sigma(A)$ such that

$$\Re \lambda \in \mathrm{span}_{\mathbb{Q}} \left\{ \frac{\Im \mu}{\pi \log e} : \mu \in \sigma(A), \Re \mu = \Re \lambda \right\} . \tag{7.39}$$

Letting $\lambda = x + \imath y$ for convenience, assume first that the set $\sigma(A) \cap (x + \imath \mathbb{R})$ is a singleton. Thus $\lambda = x \in \mathbb{R}$, and in fact $\lambda = 0$, by (7.38). Let $u \in \mathbb{R}^d$ be any unit eigenvector of A corresponding to the eigenvalue $\lambda = 0$. Then $e^{tA} u = u$ for all $t \geq 0$, and, with $h_0 = \sum_{j,k=1}^{d} u_j u_k [\cdot]_{jk}$, the function $g(t) = h_0(e^{tA}) \equiv 1$ is neither Benford nor zero. If, on the other hand, $\sigma(A) \cap (x + \imath \mathbb{R}) = \{x + \imath y, x - \imath y\}$ with $y > 0$, then there exist linearly independent unit vectors $u, v \in \mathbb{R}^d$ with

$$Au = x\,u - y\,v, \quad Au = y\,u + x\,v,$$

and consequently

$$e^{tA}u = e^{xt}\big(\cos(yt)u - \sin(yt)v\big)\,, \quad e^{tA}v = e^{xt}\big(\sin(yt)u + \cos(yt)v\big)\,.$$

Letting $h_0 = \sum_{j,k=1}^d (u_j + v_j)(u_k + v_k)[\,\cdot\,]_{jk}$ yields

$$g(t) = h_0(e^{tA}) = (u+v)^\top e^{tA}(u+v) = 2(1 + u^\top v)e^{xt}\cos(yt)\,, \quad t \geq 0\,.$$

Recall that $1 + u^\top v > 0$ by the Cauchy–Schwarz Inequality, so g is not identically zero. Recall from (7.39) that $qx\pi \log e = py$ with the appropriate $p \in \mathbb{Z}$ and $q \in \mathbb{N}$. Fix $\delta > 0$ such that $y\delta/\pi$ is irrational, and observe that for all $n \in \mathbb{N}$,

$$g(n\delta)^q = e^{qxn\delta}2^q(1 + u^\top v)^q \cos(yn\delta)^q = a^n f(n\theta)\,,$$

with $a = e^{qx\delta}$, $\theta = \frac{1}{2}y\delta/\pi$, and $f(t) = 2^q(1 + u^\top v)^q \cos(2\pi t)^q$. Since by assumption $\log a = qx\delta \log e = 2p\theta \in \mathrm{span}_\mathbb{Q}\{1, \theta\}$ for every $\delta > 0$, Proposition 7.22(ii) with $b = 0$ and $\varepsilon_n \equiv 0$ shows that the sequence $\big(g(n\delta)^q\big)$ is not Benford, and neither is $\big(g(n\delta)\big)$. In fact, as in [16, Lem. 2.6], it is not hard to see that the sequence $\big(\log|g(n\delta)^q|\big)$ is distributed modulo one according to an a.c. probability measure $P \neq \lambda_{0,1}$ on $[0,1)$, which is independent of δ as long as $y\delta/\pi$ is irrational. This means that there exists a non-zero integer k such that

$$0 \neq \widehat{P}(k) = \lim_{N\to\infty} \frac{1}{N}\sum_{n=1}^{N} e^{2\pi\imath k \log|g(n\delta)^q|}$$

for almost all $\delta > 0$. Thus, writing $\log|g(n\delta)^q|$ as $g_n(\delta)$ for convenience,

$$\begin{aligned}
\widehat{P}(k) &= \lim_{N\to\infty} \frac{1}{N}\sum_{n=1}^{N} \int_0^1 e^{2\pi\imath k g_n(\delta)}\,\mathrm{d}\delta \\
&= \lim_{N\to\infty} \frac{1}{N}\sum_{n=1}^{N} \frac{1}{n}\int_0^n e^{2\pi\imath k g_1(t)}\,\mathrm{d}t \qquad\qquad\qquad (7.40) \\
&= \lim_{N\to\infty} \frac{1}{N}\sum_{n=1}^{N} \frac{1}{n}\sum_{i=1}^{n}\int_{i-1}^{i} e^{2\pi\imath k g_1(t)}\,\mathrm{d}t\,,
\end{aligned}$$

where the first equality follows from the Dominated Convergence Theorem. Next, recall from [71, Thm. 92] that if $\lim_{N\to\infty} N^{-1}\sum_{n=1}^{N} n^{-1}\sum_{i=1}^{n} z_i$ exists for a *bounded* sequence (z_n) in $\mathbb{C}$, then $\lim_{N\to\infty} N^{-1}\sum_{n=1}^{N} z_n$ also exists and has the same value. Specifically, since the sequence (z_n) with $z_n = \int_{n-1}^{n} e^{2\pi\imath k g_1(t)}\,\mathrm{d}t$ is clearly bounded, (7.40) implies that

$$0 \neq \widehat{P}(k) = \lim_{N\to\infty} \frac{1}{N}\sum_{n=1}^{N} \int_{n-1}^{n} e^{2\pi\imath k g_1(t)}\,\mathrm{d}t = \lim_{N\to\infty} \frac{1}{N}\int_0^N e^{2\pi\imath k g_1(t)}\,\mathrm{d}t\,,$$

and by the continuous-time version of Weyl's criterion [93, Thm. I.9.2], the function $g_1 = \log|g^q|$ is not u.d. mod 1. Thus the function g^q is not Benford, and neither is g, by Theorem 4.4. In summary, if $\sigma(A)$ is exponentially resonant then (ii) fails. In other words, (ii)$\Rightarrow$(i), and the proof is complete. ∎

Remarks. (i) It follows from (7.37) that $\sigma(A)$ is exponentially resonant whenever $\Re\lambda = 0$ for some $\lambda \in \sigma(A)$, i.e., whenever A has an eigenvalue on the imaginary axis. Moreover, if $\sigma(A) \subset \mathbb{R}$, that is, if all eigenvalues of A are real, then the converse is also true, i.e., $\sigma(A)$ is exponentially resonant *only* if $0 \in \sigma(A)$.

(ii) As in the discrete-time setting of Section 7.3, it is not hard to check that the set

$$\left\{ A \in \mathbb{R}^{d\times d} : \sigma(A) \text{ is exponentially resonant} \right\}$$

is of first category and has Lebesgue measure zero. For *most* $d \times d$-matrices A, therefore, the function $h(e^{tA})$ is either Benford or identically zero for every $h \in \mathcal{L}_d$.

EXAMPLE 7.50. This example reviews Example 7.45 in the light of Theorem 7.49. In each case, the latter is consistent with the observations made in that example.

(i) The matrix $A = \begin{bmatrix} 0 & 1 \\ 1 & 0 \end{bmatrix}$ has eigenvalues $\lambda = \pm 1$; hence $\sigma(A)$ is exponentially nonresonant.

(ii) For $B = \begin{bmatrix} 0 & -1 \\ 1 & 0 \end{bmatrix}$, the spectrum $\sigma(B) = \{\pm\imath\}$ is exponentially resonant.

(iii) The eigenvalues of $C = \begin{bmatrix} -2 & -3 \\ 1 & 0 \end{bmatrix}$ are $\lambda = -1 \pm \imath\sqrt{2}$. Since $\pi \log e$ is transcendental, $-1 \notin \mathrm{span}_{\mathbb{Q}}\left\{ \sqrt{2}/(\pi \log e) \right\}$, and so $\sigma(C)$ is exponentially nonresonant. (By contrast, recall from Example 7.27 that it is apparently unknown whether $\sigma(C)$ is nonresonant.) ✠

EXAMPLE 7.51. The eigenvalues of $A = \begin{bmatrix} 1 & 1 \\ 1 & 0 \end{bmatrix}$ from (7.2) are both real and non-zero. Hence $\sigma(A)$ is exponentially nonresonant, and every function $h(e^{tA})$ is either Benford or zero. This is also evident from the explicit formula

$$e^{tA} = \frac{e^{t\varphi}}{\varphi + 2}(I_2 + \varphi A) + \frac{e^{-t/\varphi}}{\varphi + 2}\left((1 + \varphi)I_2 - \varphi A\right),$$

which shows that $h(e^{tA}) \equiv 0$ if and only if $h(I_2) = h(A) = 0$. ✠

For the nonlinear observables $h(A) = |A|$ and $h(A) = |Au|$, the following analogue of Theorem 7.28 holds.

THEOREM 7.52. *Let $A \in \mathbb{R}^{d\times d}$. If $\sigma(A)$ is exponentially nonresonant then:*

(i) *The function $|e^{tA}|$ is Benford;*

(ii) *For every $u \in \mathbb{R}^d \setminus \{0\}$ the function $|e^{tA}u|$ is Benford.*

PROOF. Since $\sigma(A)$ is exponentially nonresonant, by Proposition 7.48 the set $\sigma(e^{\delta A})$ is nonresonant for all but countably many $\delta > 0$, and $e^{\delta A}$ is invertible whenever $\delta > 0$. By Theorem 7.28(i), the sequence $(|e^{n\delta A}|)$ is Benford in this case, and Proposition 4.8(i) shows that the function $|e^{tA}|$ is Benford as well.

Similarly, for every $u \in \mathbb{R}^d$, the sequence $(|e^{n\delta A}u|)$ is Benford or identically zero, and the latter can happen only if $u = 0$. For $u \neq 0$, therefore, $(|e^{n\delta A}u|)$ is Benford for almost all $\delta > 0$, and, again by Proposition 4.8(i), the function $|e^{tA}u|$ is Benford. ∎

EXAMPLE 7.53. **(i)** For the matrix $A = \begin{bmatrix} 0 & 1 \\ 1 & 0 \end{bmatrix}$ in Example 7.45, the set $\sigma(A)$ is exponentially nonresonant; hence Theorem 7.52 applies, showing that the functions $|e^{tA}|$ and $|e^{tA}u|$ with $u \in \mathbb{R}^2 \setminus \{0\}$ are Benford. This can also be seen directly by means of Example 4.9(i) and Theorem 4.12(ii), since $e^{tA} = I_2 \cosh t + A \sinh t$, so $|e^{tA}| = e^t$, and, for any $u \in \mathbb{R}^2$,

$$|e^{tA}u| = \sqrt{\tfrac{1}{2}(u_1 + u_2)^2 e^{2t} + \tfrac{1}{2}(u_1 - u_2)^2 e^{-2t}}\,.$$

(ii) The spectrum of the matrix $B = \begin{bmatrix} 0 & -1 \\ 1 & 0 \end{bmatrix}$ is exponentially resonant, and $e^{tB} = I_2 \cos t + B \sin t$, so $|e^{tB}| \equiv 1$ is not Benford. Neither is $|e^{tB}u| \equiv |u|$ for any $u \in \mathbb{R}^2$; see Figure 7.4.

(iii) As seen in Example 7.50, the spectrum of $C = \begin{bmatrix} -2 & -3 \\ 1 & 0 \end{bmatrix}$ is exponentially nonresonant. By Theorem 7.52, $|e^{tC}|$ is Benford, and so is $|e^{tC}u|$ for every $u \in \mathbb{R}^2 \setminus \{0\}$. As in (i), this can also easily be confirmed via a direct calculation, since $e^{tC} = I_2 e^{-t} \cos(t\sqrt{2}) + \frac{1}{\sqrt{2}}(C + I_2)e^{-t} \sin(t\sqrt{2})$. ✠

Remarks. (i) As in the discrete-time setting, $|\cdot|$ in Theorem 7.52 can be replaced by any other norm on $\mathbb{R}^{d \times d}$ and $\mathbb{R}^d$, respectively.

(ii) Except for the trivial case $d = 1$, the converse of Theorem 7.52 does not hold in general: Even if $\sigma(A)$ is exponentially resonant, the function $|e^{tA}|$ may be Benford, and so may $|e^{tA}u|$ for every $u \in \mathbb{R}^d \setminus \{0\}$. In fact, continuous-time variants of Example 7.30 show that properties (i) and (ii) in Theorem 7.52 cannot be characterized through properties of $\sigma(A)$ or $\sigma(e^{tA})$ alone.

Recall from Examples 7.27 and 7.44 that it may be difficult in practice to decide whether $\sigma(A)$ is nonresonant for a given $A \in \mathbb{R}^{d \times d}$, even if $d = 2$ and all entries of A are integers. In many cases of practical interest, the situation regarding *exponential* nonresonance is much simpler.

LEMMA 7.54. *Assume that all entries of $A \in \mathbb{R}^{d \times d}$ are rational numbers. Then $\sigma(A)$ is exponentially nonresonant if and only if A has no eigenvalues on the imaginary axis.*

PROOF. The "only if" conclusion is immediate because eigenvalues of A on the imaginary axis always render $\sigma(A)$ exponentially resonant. For the "if" part, assume A has no eigenvalues on the imaginary axis, and suppose $\sigma(A)$ is exponentially resonant. Then, for some $\lambda \in \sigma(A)$,

$$\pi \log e\, \Re\lambda \in \mathrm{span}_{\mathbb{Q}}\{\Im\mu : \mu \in \sigma(A), \Re\mu = \Re\lambda\}\,.$$

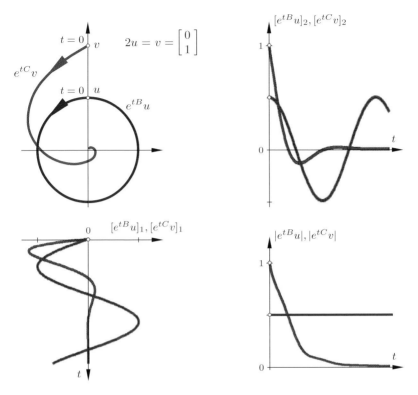

Figure 7.4: With B, C as in Example 7.50, no function $h(e^{tB})$ is Benford, but every nontrivial function $h(e^{tC})$ is; see Example 7.53.

Since $\Re\lambda \neq 0$, this means that $\pi \log e$ is a rational linear combination of the algebraic numbers $\Im\mu/\Re\lambda$. This, however, is impossible because $\pi \log e$ is transcendental. Thus $\sigma(A)$ is exponentially nonresonant. ∎

Remark. As the above proof shows, Lemma 7.54 remains valid if "rational" is replaced by "algebraic" (over $\mathbb{Q}$). Thus, for instance, the spectrum of $\begin{bmatrix} \sqrt{2} & \sqrt{3} \\ \sqrt{5} & \sqrt{7} \end{bmatrix}$ is exponentially nonresonant.

Recall from Section 7.5 that the Benford property for solutions of linear difference equations follows easily from results on sequences $(h(A^n))$. Similarly, Theorem 7.49 can be used to study the Benford property for solutions of scalar linear differential equations of any order. More concretely, consider the linear differential equation

$$x^{(d)} + a_1 x^{(d-1)} + \ldots + a_{d-1}\dot{x} + a_d x = 0, \qquad (7.41)$$

where, as usual, d is a positive integer, referred to as the *order* of (7.41), and $a_1, a_2, \ldots, a_d$ are real numbers with $a_d \neq 0$. (For instance, $d = 2$ for all three differential equations in Example 7.45.) Given any $u \in \mathbb{R}^d$, there exists a unique solution $x = x(t)$ of (7.41) with $x^{(j-1)}(0) = u_j$ for all $j \in \{1, 2, \ldots, d\}$; e.g., see [77]. Using Theorem 7.49, it is straightforward to decide whether all nontrivial solutions of (7.41) are Benford.

THEOREM 7.55. *For the differential equation* (7.41) *with* $a_1, a_2, \ldots, a_d \in \mathbb{R}$ *and* $a_d \neq 0$ *the following are equivalent:*

(i) *The set* $\{z \in \mathbb{C} : z^d + a_1 z^{d-1} + \ldots + a_{d-1} z + a_d = 0\}$ *is exponentially nonresonant;*

(ii) *Every solution* $x = x(t)$ *of* (7.41) *is Benford unless* $x(t) \equiv 0$.

PROOF. In order to make (7.41) amenable to Theorem 7.49, rewrite it in the form

$$
\frac{\mathrm{d}}{\mathrm{d}t}
\begin{bmatrix}
x^{(d-1)} \\
x^{(d-2)} \\
x^{(d-3)} \\
\vdots \\
x
\end{bmatrix}
=
\begin{bmatrix}
-a_1 & -a_2 & \cdots & -a_{d-1} & -a_d \\
1 & 0 & \cdots & 0 & 0 \\
0 & 1 & 0 & \cdots & 0 \\
\vdots & \ddots & \ddots & \ddots & \vdots \\
0 & \cdots & 0 & 1 & 0
\end{bmatrix}
\begin{bmatrix}
x^{(d-1)} \\
x^{(d-2)} \\
x^{(d-3)} \\
\vdots \\
x
\end{bmatrix} .
$$

With the $d \times d$-matrix A given by

$$
A =
\begin{bmatrix}
-a_1 & -a_2 & \cdots & -a_{d-1} & -a_d \\
1 & 0 & \cdots & 0 & 0 \\
0 & 1 & 0 & \cdots & 0 \\
\vdots & \ddots & \ddots & \ddots & \vdots \\
0 & \cdots & 0 & 1 & 0
\end{bmatrix} ,
$$

therefore, any solution $x = x(t)$ of (7.41) can be represented as

$$
x(t) = x^{(d-1)}(0)[e^{tA}]_{d1} + x^{(d-2)}(0)[e^{tA}]_{d2} + \ldots + \dot{x}(0)[e^{tA}]_{dd-1} + x(0)[e^{tA}]_{dd} ,
$$

which is of the form $x(t) = h(e^{tA})$ with the appropriate $h \in \mathcal{L}_d$. Moreover, note that

$$
\det(A - zI_d) = (-1)^d (z^d + a_1 z^{d-1} + \ldots + a_{d-1} z + a_d) ,
$$

and, consequently, $\sigma(A) = \{z \in \mathbb{C} : z^d + a_1 z^{d-1} + \ldots + a_{d-1} z + a_d = 0\}$. Finally, observe that, as in the discrete-time setting, the function $f(t) = h(e^{tA})$ solves (7.41) for *every* $h \in \mathcal{L}_d$ because $f^{(k)}(t) = h(e^{tA} A^k)$ for $k = 0, 1, 2, \ldots$, and so

$$
\begin{aligned}
f^{(d)}(t) &= h(e^{tA} A^d) = h\big(e^{tA}(-a_1 A^{d-1} - a_2 A^{d-2} - \ldots - a_{d-1} A - a_d I_d)\big) \\
&= -a_1 h(e^{tA} A^{d-1}) - a_2 h(e^{tA} A^{d-2}) - \ldots - a_{d-1} h(e^{tA} A) - a_d h(e^{tA}) \\
&= -a_1 f^{(d-1)}(t) - a_2 f^{(d-2)}(t) - \ldots - a_{d-1} \dot{f}(t) - a_d f(t) .
\end{aligned}
$$

With these preparations, the asserted equivalence is now easily established. If $\{z \in \mathbb{C} : z^d + a_1 z^{d-1} + \ldots + a_{d-1} z + a_d = 0\} = \sigma(A)$ is exponentially nonresonant then $h(e^{tA})$ is Benford or trivial for every $h \in \mathcal{L}_d$, by Theorem 7.49. In particular, $x = x(t)$ either is Benford, or else $x(0) = \dot{x}(0) = \ldots = x^{(d-1)}(0) = 0$, in which case $x(t) \equiv 0$. Conversely, if $\{z \in \mathbb{C} : z^d + a_1 z^{d-1} + \ldots + a_{d-1} z + a_d = 0\}$ is exponentially resonant then, again by Theorem 7.49, there exists $h_0 \in \mathcal{L}_d$ such that the function $h_0(e^{tA})$ is neither Benford nor identically zero, and the same is true for the solution $x(t) = h_0(e^{tA})$ of (7.41). ∎

EXAMPLE 7.56. **(i)** Let $a_1, a_2 \in \mathbb{R}$ be rational (or at least algebraic). By Lemma 7.54 and Theorem 7.55, every solution of $\ddot{x} + a_1 \dot{x} + a_2 x = 0$, except $x(t) \equiv 0$, is Benford if and only if $\{z \in \mathbb{C} : z^2 + a_1 z + a_2 = 0\}$ does not intersect the imaginary axis. This is readily confirmed to be equivalent to

$$ a_1 a_2 \neq 0 \quad \text{or} \quad a_2 < 0 \,. \tag{7.42} $$

In Example 7.45(i), $a_2 = -1$, whereas in Example 7.45(iii), $a_1 a_2 = 6$. In both cases, therefore, (7.42) holds and shows again that all nontrivial solutions of the respective IVPs are Benford. On the other hand, for Example 7.45(ii), $a_1 = 0$ and $a_2 = 1$, so (7.42) fails. This is consistent with the fact that no solution of the corresponding IVP is Benford.

(ii) Similarly, if $a_1, a_2, a_3 \in \mathbb{R}$ are rational (or algebraic) and $a_3 \neq 0$ then, except for $x(t) \equiv 0$, every solution of $\dddot{x} + a_1 \ddot{x} + a_2 \dot{x} + a_3 x = 0$ is Benford if and only if

$$ a_1 a_2 \neq a_3 \quad \text{or} \quad a_2 \leq 0 \,. \tag{7.43} $$

Thus every nontrivial solution of $\dddot{x} + x = 0$ or $\dddot{x} + 2\ddot{x} + 3\dot{x} + 4x = 0$, for instance, is Benford. On the other hand, (7.43) fails for $\dddot{x} + \ddot{x} + \dot{x} + x = 0$; for example, $x(t) = \cos t$ is a nontrivial solution that is not Benford, by Theorem 4.10. ✠

EXAMPLE 7.57. For every $a, b \in \mathbb{R}$ with $b \neq 0$, the functions $x(t) = e^{at} \cos(bt)$ and $x(t) = e^{at} \sin(bt)$ both solve $\ddot{x} - 2a\dot{x} + (a^2 + b^2)x = 0$. By (7.42), if a and b are rational (or algebraic) and $a \neq 0$ then x is Benford. In fact, from (7.37) and Theorem 7.55 it is clear that x is Benford if and only if the number $a\pi \log e / b$ is irrational; this is automatically the case whenever a and b are non-zero and rational (or algebraic). ✠

Chapter Eight

Real-valued Random Processes

Benford's law arises naturally in a variety of stochastic settings, including products of independent random variables, mixtures of random samples from different distributions, and iterations of random maps. This chapter provides the reader with concepts and tools to analyze significant digits and significands for these basic random processes. Perhaps not surprisingly, Benford's law also arises in many other important fields of stochastics, such as geometric Brownian motion, random matrices, and Bayesian models, and the present chapter may serve as a preparation for specialized literature on these advanced topics [18, 56, 96, 108, 143, 144]. Throughout the chapter, recall that by Theorem 4.2 a random variable X is Benford if and only if $\log |X|$ is uniformly distributed modulo one.

8.1 CONVERGENCE OF RANDOM VARIABLES TO BENFORD'S LAW

The analysis of sequences of random variables, notably the special case of products and sums of independent, identically distributed (i.i.d.) random variables, constitutes an important topic in classical probability theory. Within this context, the present section studies general scenarios that lead to Benford's law emerging as an "attracting" distribution. The results nicely complement the observations made in previous chapters for deterministic processes.

Recall from Chapter 3 that a (real-valued) random variable X is *Benford* by definition if $\mathbb{P}(S(X) \leq t) = \log t$ for all $t \in [1, 10)$. Also, recall that a sequence $(X_n) = (X_1, X_2, \ldots)$ of random variables converges *in distribution* to a random variable X, symbolically $X_n \xrightarrow{\mathcal{D}} X$, if $\lim_{n \to \infty} \mathbb{P}(X_n \leq x) = \mathbb{P}(X \leq x)$ for every $x \in \mathbb{R}$ for which $\mathbb{P}(X = x) = 0$. Another important concept in limit theorems in probability theory is almost sure convergence. Specifically, the sequence (X_n) converges to X *almost surely* (*a.s.*), in symbols $X_n \xrightarrow{\text{a.s.}} X$, if $\mathbb{P}(\lim_{n \to \infty} X_n = X) = 1$. It is easy to check that $X_n \xrightarrow{\text{a.s.}} X$ implies $X_n \xrightarrow{\mathcal{D}} X$. The reverse implication does not hold in general, as is evident from any i.i.d. sequence $(X_1, X_2, \ldots)$ for which X_1 is not constant: In this case, all X_n have the same distribution, so trivially $X_n \xrightarrow{\mathcal{D}} X_1$ but $\mathbb{P}(\lim_{n \to \infty} X_n \text{ exists}) = 0$. The following definition adapts these general notions to the context of Benford's law.

DEFINITION 8.1. Let $(X_n) = (X_1, X_2, \ldots)$ be a sequence of real-valued random variables.

(i) The sequence (X_n) *converges in distribution to Benford's law* if

$$\lim_{n\to\infty} \mathbb{P}(S(X_n) \leq t) = \log t \quad \text{for all } t \in [1, 10),$$

or, equivalently, if $S(X_n) \xrightarrow{D} S(X)$, where X is a Benford random variable.

(ii) The sequence (X_n) is *Benford with probability one* if

$$\mathbb{P}\big((X_n) \text{ is a Benford sequence}\big) = 1.$$

Note that the distribution function of Benford's law, $F_{\mathbb{B}}(t) = \log t$, is continuous. Consequently, if (X_n) converges in distribution to Benford's law then $|F_{S(X_n)}(t) - \log t| \to 0$ uniformly on $[1, 10)$. The following example shows that the two notions of conformance to Benford's law appearing in Definition 8.1 are generally unrelated.

EXAMPLE 8.2. **(i)** Let X be a Benford random variable and, for every $n \in \mathbb{N}$, define $X_n = \min\{S(X), 10 - n^{-1}\}$. Clearly, (X_n) converges in distribution to Benford's law. To see this formally, notice that $S(X_n) = X_n$ and so, for all $t \in [1, 10)$ and $n \in \mathbb{N}$,

$$|F_{S(X_n)}(t) - \log t| = \begin{cases} 0 & \text{if } 1 \leq t < 10 - n^{-1}, \\ 1 - \log t & \text{if } 10 - n^{-1} \leq t < 10, \end{cases}$$

$$\leq 1 - \log(10 - n^{-1}) < \tfrac{1}{20} n^{-1}.$$

On the other hand, $X_n = S(X)$ whenever $n^{-1} \leq 10 - S(X)$. With probability one, therefore, the sequence (X_n) is eventually constant and hence not Benford. In other words, $\mathbb{P}\big((X_n) \text{ is a Benford sequence}\big) = 0$.

(ii) Let X be a random variable with $\mathbb{P}(X = 2) = 1$ and, for every $n \in \mathbb{N}$, define $X_n = X^n$. Then, by Theorem 4.16, the sequence (X_n) is Benford with probability one but, for every $n \in \mathbb{N}$, the distribution of $S(X_n)$ is concentrated at the single point $S(2^n) = 10^{\langle n \log 2 \rangle}$, i.e., $\mathbb{P}\big(S(X_n) = 10^{\langle n \log 2 \rangle}\big) = 1$, and so $\sup_{t\in[1,10)} |F_{S(X_n)}(t) - \log t| \geq \tfrac{1}{2}$. Thus (X_n) does not converge in distribution to Benford's law. ✠

It is instructive to rephrase several observations made earlier using the terminology of Definition 8.1. When put into this context, Corollary 3.12 simply says that the sequence of powers (X^n) of every Pareto random variable X converges in distribution to Benford's law as $n \to \infty$. Another example of convergence in distribution to Benford's law that follows easily from results in Chapter 3 is the following. Recall that a *median* $\mathrm{med} X$ of a random variable X is any number $m \in \mathbb{R}$ with $\mathbb{P}(X \leq m) \geq \tfrac{1}{2}$ and $\mathbb{P}(X \geq m) \geq \tfrac{1}{2}$.

PROPOSITION 8.3. *Let* (X_n) *be a sequence of Pareto random variables. Then* (X_n) *converges in distribution to Benford's law if and only if* $\text{med} X_n \to +\infty$ *as* $n \to \infty$.

PROOF. For every $n \in \mathbb{N}$, let X_n be Pareto with parameter $\alpha_n > 0$. Denote by $m_n = \text{med} X_n$ the (unique) median of X_n. Then $\mathbb{P}(X_n > m_n) = \frac{1}{2} = m_n^{-\alpha_n}$, so $\alpha_n = \log 2 / \log m_n$. Thus $\alpha_n \to 0$ if and only if $m_n \to +\infty$, and the conclusion follows from Theorem 3.11, with the monotonicity of $F_{S(X_\alpha)}(t) \xrightarrow{\alpha \to 0} \log t$ being used to establish the "only if" part. ∎

For another illustration of the above terminology, recall the characterization of periodic Benford functions provided by Theorem 4.10.

EXAMPLE 8.4. Let the random variable X be uniform on $(0, 1)$ and, for every $n \in \mathbb{N}$, let $X_n = nX$. Clearly, the sequence (X_n) does not converge in distribution or almost surely. Also, since $\max_{1 \le t < 10} |F_{S(X_n)}(t) - \log t| \ge 0.1344$ for all n by Theorem 3.13, (X_n) does not converge in distribution to Benford's law. Similarly, $\mathbb{P}\big((X_n) \text{ is Benford}\big) = 0$, by Example 4.7(ii). With this, consider a (Borel measurable) function $f : [0, +\infty) \to \mathbb{R}$. From

$$\mathbb{P}\big(S\big(f(X_n)\big) \le t\big) = \frac{\lambda\big(\{\tau \in [0, n) : S\big(f(\tau)\big) \le t\}\big)}{n}, \quad n \in \mathbb{N},$$

it is clear that $\big(f(X_n)\big)$ converges in distribution to Benford's law if and only if the function f is Benford. On the other hand, it is easy to see that if f is periodic with period $p > 0$, then $f(X_n) \xrightarrow{D} f(U)$, where U is uniform on $(0, p)$. In this case, as already seen in Theorem 4.10, the random variable $f(U)$ is Benford if and only if f is a Benford function. ✠

8.2 POWERS, PRODUCTS, AND SUMS OF RANDOM VARIABLES

Recall from Theorem 4.16 the perhaps simplest example of a Benford sequence, namely, the sequence $(a^n) = (a, a^2, a^3, \ldots)$, which is Benford if and only if $\log |a|$ is irrational. This example can naturally be interpreted in probabilistic terms: The (deterministic) sequence (a^n) is, with probability one, the same as the (random) sequence $(X_1, X_1 X_2, X_1 X_2 X_3, \ldots)$, where the random variables X_j all have the *same distribution*, namely, $\mathbb{P}(X_j = a) = 1$. Thus a natural random generalization of (a^n) is provided by the class of sequences of random variables

$$\Big(\prod_{j=1}^n X_j\Big) = (X_1, X_1 X_2, \ldots), \quad X_1, X_2, \ldots \text{ identically distributed}. \quad (8.1)$$

Random sequences of the form (8.1) can exhibit a wide variety of different behaviors. This section studies the Benford properties of (8.1) in two important special cases:

(i) The factors X_j are *identical* (and hence dependent *in extremis*), that is, with probability one, $X_j = X_1$ for every j;

(ii) The factors X_j are *independent*, i.e., $X_1, X_2, \ldots$ are i.i.d. random variables.

Keep in mind that both scenarios contain the deterministic sequence (a^n) as a special case.

Powers of a random variable

If all factors $X_1, X_2, \ldots$ are identical then the random sequence (8.1) simply takes the form (X^n), where X is any real-valued random variable. For such sequences, it is straightforward to characterize conformance to Benford's law in the sense of Definition 8.1.

THEOREM 8.5. *Let X be a random variable. Then (X^n) converges in distribution to Benford's law if and only if $\lim_{n\to\infty} \mathbb{E}\left[e^{2\pi i n \log|X|}\right] = 0$.*

PROOF. Define a sequence (Y_n) as $Y_n = \log S(X^n) = \langle n \log|X| \rangle$. Note that (X^n) converges in distribution to Benford's law if and only if $Y_n \xrightarrow{\mathcal{D}} U(0,1)$, and recall from Lemma 4.20(v) that the latter is equivalent to $\lim_{n\to\infty} \widehat{P_{Y_n}}(k) = 0$ for every $k \in \mathbb{Z} \setminus \{0\}$. Thus, if $\lim_{n\to\infty} \mathbb{E}\left[e^{2\pi i n \log|X|}\right] = 0$ then, for every $k > 0$,

$$\widehat{P_{Y_n}}(k) = \mathbb{E}\left[e^{2\pi i k Y_n}\right] = \mathbb{E}\left[e^{2\pi i k n \log|X|}\right] \xrightarrow{n\to\infty} 0 \,.$$

Since Y_n is real-valued, $\widehat{P_{Y_n}}(-k) = \overline{\widehat{P_{Y_n}}(k)}$, hence $Y_n \xrightarrow{\mathcal{D}} U(0,1)$.

Conversely, if $\limsup_{n\to\infty} \left|\mathbb{E}\left[e^{2\pi i n \log|X|}\right]\right| > 0$ then there exist $\delta > 0$ and an increasing sequence (n_j) of positive integers such that $\left|\mathbb{E}\left[e^{2\pi i n_j \log|X|}\right]\right| \geq \delta$ for all j. But then, for every $j \in \mathbb{N}$,

$$\left|\widehat{P_{Y_{n_j}}}(1)\right| = \left|\mathbb{E}\left[e^{2\pi i Y_{n_j}}\right]\right| = \left|\mathbb{E}\left[e^{2\pi i n_j \log|X|}\right]\right| \geq \delta \,,$$

showing that (Y_n) does not converge in distribution to $U(0,1)$, and hence (X^n) does not converge in distribution to Benford's law. ∎

In fact, using Fourier analysis tools as in Lemma 4.20, together with Theorem 4.2 and the observation that

$$\widehat{P_{\langle nY \rangle}}(k) = \widehat{P_{\langle Y \rangle}}(nk) \quad \text{for all } n \in \mathbb{N}, \, k \in \mathbb{Z} \,,$$

it is clear from Theorem 8.5 that (X^n) converges in distribution to Benford's law if and only if $Y = \log|X|$ satisfies $\widehat{P_{\langle Y \rangle}}(k) \to 0$ as $|k| \to \infty$, i.e., precisely if $P_{\langle Y \rangle}$ is a so-called *Rajchman probability*. As Theorem 8.8 below implies, therefore, if a probability on $[0,1)$ is a.c. then it is Rajchman, but the converse is not true. Also, every Rajchman probability is continuous, and again the converse is not true; see [101]. Thus there are continuous random variables whose powers do not converge in distribution to Benford's law; see Example 8.9(ii) below.

THEOREM 8.6. *Let X be a random variable. Then the following are equivalent:*

(i) *The sequence (X^n) is Benford with probability one;*

(ii) $\mathbb{P}(\log|X| \in \mathbb{Q}) = 0$, *i.e.,* $\log|X|$ *is irrational with probability one;*

(iii) $\mathbb{P}(S(X^m) = 1) = 0$ *for every $m \in \mathbb{N}$.*

PROOF. Denote by $(\Omega, \mathcal{A}, \mathbb{P})$ the (abstract) probability space on which X is defined. By Theorem 4.16, for every $\omega \in \Omega$, the sequence

$$\big(X(\omega)^n\big) = (X(\omega), X(\omega)^2, X(\omega)^3, \ldots)$$

is Benford if and only if $\log|X(\omega)|$ is irrational, which shows (i)$\Leftrightarrow$(ii). To see the equivalence of (ii) and (iii), simply note that, for every $m \in \mathbb{N}$ and $t \in [1, 10)$,

$$\big\{\omega \in \Omega : S\big(X(\omega)^m\big) = t\big\} = \left\{\omega \in \Omega : \log|X(\omega)| \in \frac{\log t}{m} + \frac{1}{m}\mathbb{Z}\right\}, \qquad (8.2)$$

and hence $\sup_{m \in \mathbb{N}} \mathbb{P}(S(X^m) = 1) = \mathbb{P}(\log|X| \in \mathbb{Q})$. ∎

Example 8.2(ii) above shows that it is possible for a sequence (X^n) to be Benford with probability one yet not converge in distribution to Benford's law. The reverse implication, however, always holds.

COROLLARY 8.7. *Let X be a random variable. If the sequence (X^n) converges in distribution to Benford's law then it is Benford with probability one.*

PROOF. Suppose that $\mathbb{P}\big((X^n)$ is a Benford sequence$\big) < 1$. By Theorem 8.6 there exist $p \in \mathbb{Z}$ and $q \in \mathbb{N}$ such that $\mathbb{P}(|X| = 10^{p/q}) > 0$. But then

$$F_{S(X^{nq})}(1) = \mathbb{P}(S(X^{nq}) \le 1) \ge \mathbb{P}(|X| = 10^{p/q}) > 0 \quad \text{for every } n \in \mathbb{N},$$

showing that the sequence $\big(F_{S(X^n)}(1)\big)$ does not converge to $0 = \log 1$, and hence (X^n) does not converge in distribution to Benford's law. ∎

Many random variables encountered in practice have a density, and in this case Theorems 8.5 and 8.6 both apply.

THEOREM 8.8. *If the random variable X has a density then the sequence (X^n) converges in distribution to Benford's law and is Benford with probability one.*

PROOF. If X has a density then so has $Y := \langle \log|X| \rangle$. Since Y is integrable, the Riemann–Lebesgue Lemma implies that $\lim_{n \to \infty} \mathbb{E}\big[e^{2\pi i n Y}\big] = 0$, so by Theorem 8.5, (X^n) converges in distribution to Benford's law. The second assertion follows from Corollary 8.7. ∎

EXAMPLE 8.9. **(i)** Let X be uniform on $(0,1)$. Then for every $n \in \mathbb{N}$,

$$F_{S(X^n)}(t) = \sum_{k=1}^{\infty} \mathbb{P}\big(10^{-k/n} \le X \le t^{1/n}10^{-k/n}\big) = \frac{t^{1/n}-1}{10^{1/n}-1}\,,$$

and the proof of Theorem 3.11 shows that $F_{S(X^n)}(t)$ increases to $\log t$ monotonically as $n \to \infty$. In fact, a straightforward calculation yields that

$$n(F_{S(X^n)}(t) - \log t) \stackrel{n\to\infty}{\longrightarrow} -\frac{\log t(1-\log t)}{2\log e} \quad \text{uniformly on } [1,10]\,. \qquad (8.3)$$

Thus $F_{S(X^n)}$ converges to $F_{\mathbb{B}}$ at the (polynomial) rate (n^{-1}); see Figure 8.1.

(ii) Let Y be uniformly distributed on the classical Cantor middle thirds set. In more probabilistic terms, $Y = 2\sum_{j=1}^{\infty}3^{-j}Y_j$, where the Y_j are i.i.d. with $\mathbb{P}(Y_1 = 0) = \mathbb{P}(Y_1 = 1) = \frac{1}{2}$. Let X be the continuous random variable defined by $X = 10^Y$. Then X is not Benford since Y is not $U(0,1)$. But since $\langle Y \rangle = \langle 3^n Y \rangle$ for all $n \in \mathbb{N}$, the distribution of $S(X)$ is the same as the distribution of $S(X^3)$, and of $S\big(X^{3^n}\big)$ for all $n \in \mathbb{N}$. Thus even though X is continuous, (X^n) does not converge in distribution to Benford's law. Recall from Theorem 5.13 that the only continuous positive random variables X which have identically distributed significands $S(X^n)$ for all $n \in \mathbb{N}$ are the Benford random variables. ✠

Note that if X has an atom, i.e., $\mathbb{P}(X = x) = \delta$ for some $x \in \mathbb{R}$ and $\delta > 0$, then (X^n) cannot converge in distribution to Benford's law, since for every $n \in \mathbb{N}$, the probability distribution $P_{S(X^n)}$ has an atom of size at least δ, so either $F_{S(X^n)}$ does not converge as $n \to \infty$, or else it converges to a discontinuous, i.e., non-Benford, limit. Thus if (X^n) converges in distribution to Benford's law then X is continuous, and

$$|F_{S(X^n)}(t) - \log t| \stackrel{n\to\infty}{\longrightarrow} 0 \quad \text{uniformly on } [1,10]\,. \qquad (8.4)$$

The rate of convergence in (8.4) can often be inferred from a Fourier series representation for $F_{S(X^n)}$: With

$$\varphi_k := \widehat{P_{\langle \log |X| \rangle}}(k) = \mathbb{E}\big[e^{2\pi\imath k \log |X|}\big]\,, \quad k \in \mathbb{Z}\,,$$

it follows directly from [139, eq. (3)] that, for every $n \in \mathbb{N}$,

$$F_{S(X^n)}(t) - \log t = \frac{2}{\pi} \sum_{k=1}^{\infty} \frac{\sin(\pi k \log t)}{k} \Re\big(\varphi_{nk}e^{-\pi\imath k \log t}\big)\,, \quad t \in [1,10]\,. \quad (8.5)$$

The next example illustrates the usefulness of (8.5) for three concrete a.c. distributions.

EXAMPLE 8.10. **(i)** Let $X = U(0,1)$ as in Example 8.9(i) above. Then

$$\varphi_k = \int_0^1 e^{2\pi\imath k \log x}\,\mathrm{d}x = \int_{-\infty}^0 \frac{e^{2\pi\imath ky}10^y}{\log e}\,\mathrm{d}y = \frac{1}{1+\imath 2\pi k \log e}\,, \quad k \in \mathbb{Z}\,,$$

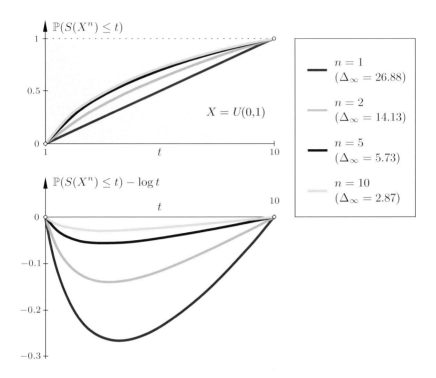

Figure 8.1: If the random variable X is distributed uniformly on $(0,1)$ then the distribution of $S(X^n)$ converges to Benford's law as $n \to \infty$; see Example 8.9(i).

and (8.5) immediately yields the estimate, for every $n \in \mathbb{N}$,

$$n \max_{t \in [1,10)} |F_{S(X^n)}(t) - \log t| \leq \frac{2n}{\pi} \sum_{k=1}^{\infty} \frac{|\varphi_{nk}|}{k}$$

$$= \frac{2n}{\pi} \sum_{k=1}^{\infty} \frac{1}{k\sqrt{1 + 4\pi^2 n^2 k^2 (\log e)^2}} < \frac{1}{\pi^2 \log e} \sum_{k=1}^{\infty} \frac{1}{k^2} = \frac{1}{6 \log e},$$

which is consistent with (8.3) and nearly best possible, as the latter implies that

$$\lim_{n \to \infty} n \max_{t \in [1,10)} |F_{S(X^n)}(t) - \log t| = \frac{1}{8 \log e}.$$

(ii) Let $X > 0$ have the density

$$f_X(x) = \frac{\log e}{\pi x} \cdot \frac{1}{1 + (\log x)^2}, \quad x > 0;$$

in other words, the random variable $\log X$ is standard Cauchy-distributed. In this case,

$$\varphi_k = \int_0^{+\infty} \frac{e^{2\pi \imath k \log x}}{\pi x} \cdot \frac{\log e}{1 + (\log x)^2} \, \mathrm{d}x = \frac{1}{\pi} \int_{-\infty}^{+\infty} \frac{e^{2\pi \imath k y}}{1 + y^2} \, \mathrm{d}s = e^{-2\pi |k|}, \quad k \in \mathbb{Z},$$

and so (8.5) takes the form

$$F_{S(X^n)}(t) - \log t = \frac{1}{\pi} \sum_{k=1}^{\infty} \frac{e^{-2\pi nk}}{k} \sin(2\pi k \log t), \quad t \in [1, 10].$$

From this it is clear that

$$e^{2\pi n}(F_{S(X^n)}(t) - \log t) \overset{n \to \infty}{\longrightarrow} \frac{\sin(2\pi \log t)}{\pi} \quad \text{uniformly on } [1, 10], \qquad (8.6)$$

and hence convergence in (8.4) occurs at the (exponential) rate $(e^{-2\pi n})$.

(iii) Let $X > 0$ be log-normal distributed with parameters $\mu = 0$ and $\sigma^2 = 1$, i.e., $\log X$ is normal with mean 0 and variance $(\log e)^2$. Then

$$\varphi_k = \int_{-\infty}^{+\infty} \frac{e^{2\pi i k y}}{\sqrt{2\pi(\log e)^2}} e^{-\frac{1}{2}(y/\log e)^2} \, \mathrm{d}y = e^{-2\pi^2 k^2 (\log e)^2}, \quad k \in \mathbb{Z},$$

and (8.5) now takes the form

$$F_{S(X^n)}(t) - \log t = \frac{1}{\pi} \sum_{k=1}^{\infty} \frac{e^{-2\pi^2(\log e)^2 n^2 k^2}}{k} \sin(2\pi k \log t),$$

which in turn shows that, uniformly on $[1, 10]$,

$$\lim_{n \to \infty} e^{2\pi^2(\log e)^2 n^2}(F_{S(X^n)}(t) - \log t) = \frac{\sin(2\pi \log t)}{\pi}.$$

For a log-normal random variable X, therefore, convergence in (8.4) occurs at the (super-exponential) rate $(e^{-\gamma n^2})$, with $\gamma = 2\pi^2(\log e)^2 = 3.723$. ✠

From the proofs of Theorems 8.5 and 8.6 as well as Corollary 8.8, it is clear that if X has a density then the sequence (X^{2^n}), for instance, converges in distribution to Benford's law and is Benford with probability one. Note that (X^{2^n}) is not of the form (8.1). However, with $X_n := \frac{1}{10}X^{2^n}$, clearly $X_n = 10X_{n-1}^2$. Thus the map $f(x) = 10x^2$, first studied in Example 6.3 in a purely deterministic context, makes a reappearance in the present stochastic setting. This suggests that several results of Chapter 6 may be extended to the stochastic context. While this topic will not be pursued formally here, the following example gives an informal illustration.

EXAMPLE 8.11. Let (X_n) be a sequence of random variables where X_0 has a density, and

$$X_n = X_{n-1}^2 + 1, \quad n \in \mathbb{N}.$$

In other words, $X_n = f(X_{n-1})$ with the map $f(x) = x^2 + 1$ from Example 6.25. As in the proof of Theorem 6.23, it follows that

$$\log X_n - 2^n Y = \log f^n(X_0) - 2^n Y \overset{\text{a.s.}}{\longrightarrow} 0 \quad \text{as } n \to \infty,$$

with a uniquely defined random variable $Y = \widehat{s}(\log |X_0|)$. Close inspection of the function $\widehat{s}$ shows that Y has a density as well. Hence the sequence (X_n) converges in distribution to Benford's law and is Benford with probability one.

✠

Products of independent random variables

When turning to (8.1) with *independent* factors $X_1, X_2, \ldots$, it is worth noting that in the case of independence, Benford's law already plays a very special role for *finite* products (as opposed to *sequences of products*). Informally put, if one factor in a product of independent factors is Benford, then the whole product is *exactly* Benford, regardless of the the the other factors.

THEOREM 8.12. *Let X and Y be independent random variables, and assume that $\mathbb{P}(XY = 0) = 0$. Then:*

(i) *If X is Benford then so is XY;*

(ii) *If $S(X)$ and $S(XY)$ have the same distribution then either $\log S(Y)$ is rational with probability one, or X is Benford.*

PROOF. Conclusion (i) follows immediately from Theorems 4.2 and 4.21. To see (ii), note first that $\log S(XY) = \langle \log S(X) + \log S(Y) \rangle$, and, since the random variables $X_0 := \log S(X)$ and $Y_0 := \log S(Y)$ are independent, Lemma 4.20(iv) implies that

$$\widehat{P_{\log S(XY)}} = \widehat{P_{\langle X_0 + Y_0 \rangle}} = \widehat{P_{X_0}} \cdot \widehat{P_{Y_0}} \,. \tag{8.7}$$

If $S(X)$ and $S(XY)$ have the same distribution, (8.7) yields

$$\widehat{P_{X_0}}(k)\big(1 - \widehat{P_{Y_0}}(k)\big) = 0 \quad \text{for all } k \in \mathbb{Z} \,.$$

If $\widehat{P_{Y_0}}(k) \neq 1$ for all non-zero k, then $\widehat{P_{X_0}} = \widehat{\lambda_{0,1}}$ by Lemma 4.20(vi), i.e., X is Benford. Otherwise, $\widehat{P_{Y_0}}(k_0) = 1$ for some integer $k_0 \neq 0$, i.e., $\mathbb{E}\big[e^{2\pi i k_0 Y_0}\big] = 1$. It follows that $P_{Y_0}(\frac{1}{k_0}\mathbb{Z}) = 1$; hence $k_0 Y_0 = k_0 \log S(Y)$ is an integer with probability one, so $\log S(Y)$ is rational with probability one. ∎

EXAMPLE 8.13. Let V, W be i.i.d. $U(0, 1)$ random variables. Then $X := 10^V$ and $Y := W$ are independent and, by Theorem 8.12(i), XY is Benford since X is Benford by Example 3.6(i), even though Y is not. If, on the other hand, $X := 10^V$ and $Y := 10^{1-V}$ then X and Y are both Benford, yet XY is not. Hence the independence of X and Y is crucial in Theorem 8.12(i). It is essential in assertion (ii) as well, as can be seen by letting X equal either $10^{\sqrt{2}-1}$ or $10^{2-\sqrt{2}}$ with probability $\frac{1}{2}$ each, and choosing $Y = X^{-2}$. Then $S(X)$ has the same distribution as $S(XY) = S(X^{-1})$, but neither X is Benford nor $\log S(Y)$ is rational with probability one. ✠

COROLLARY 8.14. *Let X be a random variable with $\mathbb{P}(X = 0) = 0$, and let $a \in \mathbb{R}$ be such that $\log |a|$ is irrational. If the significant digits of X and aX are identically distributed then X is Benford.*

Remark. The conclusion of Corollary 8.14 fails under the weaker assumption that the *first* significant digits of X and aX are identically distributed; see Example 5.9.

With the role of Benford's law in finite products clarified, next consider *sequences* of products of independent, identically distributed random variables, that is, consider (8.1) with any i.i.d. sequence (X_n). As with sequences of powers of a random variable above, by employing the tools established in Chapter 4 using Fourier coefficients, it is straightforward to decide whether or not the sequence of products converges in distribution to Benford's law, or is Benford with probability one.

THEOREM 8.15. *Let* $X_1, X_2, \ldots$ *be i.i.d. random variables with* $\mathbb{P}(X_1 = 0) = 0$. *Then the following are equivalent:*

(i) *The sequence* $\left(\prod_{j=1}^n X_j \right)$ *converges in distribution to Benford's law;*

(ii) $\mathbb{P}\left(\log|X_1| \in x + \frac{1}{m}\mathbb{Z} \right) < 1$ *for every* $x \in \mathbb{R}$ *and* $m \in \mathbb{N}$;

(iii) $\mathbb{P}(S(X_1^m) = t) < 1$ *for every* $t \in [1, 10)$ *and* $m \in \mathbb{N}$.

PROOF. Letting $Y_n = \log|X_n|$ for each $n \in \mathbb{N}$, the equivalence of (i) and (ii) follows immediately from Theorems 4.2 and 4.22. The equivalence of (ii) and (iii) follows from (8.2). ∎

The almost sure analogue of Theorem 8.15 for sequences of products of i.i.d. random variables is as follows.

THEOREM 8.16. *Let* $X_1, X_2, \ldots$ *be i.i.d. random variables with* $\mathbb{P}(X_1 = 0) = 0$. *Then the following are equivalent:*

(i) *The sequence* $\left(\prod_{j=1}^n X_j \right)$ *is Benford with probability one;*

(ii) $\mathbb{P}\left(\log|X_1| \in \frac{1}{m}\mathbb{Z} \right) < 1$ *for every* $m \in \mathbb{N}$;

(iii) $\mathbb{P}(S(X_1^m) = 1) < 1$ *for every* $m \in \mathbb{N}$.

PROOF. Letting $Y_n = \log|X_n|$ for $n \in \mathbb{N}$, the equivalence of (i) and (ii) follows immediately from Theorems 4.2 and 4.24. The equivalence of (ii) and (iii) follows as in (8.2), with $t = 1$. ∎

EXAMPLE 8.17. As in Example 8.2(ii), let $\mathbb{P}(X_n = 2) = 1$ for all n. Then $X_1, X_2, \ldots$ are i.i.d. random variables and satisfy Theorem 8.16(ii), so $\left(\prod_{j=1}^n X_j \right)$ is Benford with probability one. On the other hand, the conclusion of Theorem 8.15(ii) fails and, as already seen in Example 8.2, the sequence $\left(\prod_{j=1}^n X_j \right) = (2^n)$ does not converge in distribution to Benford's law. ✠

As the previous example shows, a sequence $\left(\prod_{j=1}^n X_j \right)$ with i.i.d. factors may be Benford with probability one yet may fail to converge in distribution to Benford's law. As in the case of powers (Corollary 8.7), however, the reverse implication always holds.

COROLLARY 8.18. *Let* $X_1, X_2, \ldots$ *be i.i.d. random variables. If the sequence* $\left(\prod_{j=1}^n X_j \right)$ *converges in distribution to Benford's law then it is Benford with probability one.*

PROOF. Simply note that Theorem 8.15(iii) implies Theorem 8.16(iii). ∎

Observe that Theorem 8.15(ii) and Theorem 8.16(ii) hold unless the random variable X_1 is *discrete*, i.e., unless there exists a countable (possibly finite) set $C \subset \mathbb{R}$ with $\mathbb{P}(X_1 \in C) = 1$. Hence Theorems 8.15 and 8.16 together imply the following useful result (cf. Theorem 8.8 and [132]).

COROLLARY 8.19. *Let $X_1, X_2, \ldots$ be i.i.d. random variables. If X_1 is not discrete then the sequence $\left(\prod_{j=1}^{n} X_j\right)$ converges in distribution to Benford's law and is Benford with probability one.*

Note, however, that even if X_1 is discrete, the conclusion of Theorem 8.15(ii) may hold.

EXAMPLE 8.20. Let $X_1 > 0$ be a random variable with $\mathbb{P}(X_1 = 2^j) = 2^{-j}$ for every $j \in \mathbb{N}$. Clearly, X_1 is discrete. However, $\log 2$ is irrational; hence, given any $x \in \mathbb{R}$ and $m \in \mathbb{N}$, the two sets $\{j \log 2 : j \in \mathbb{N}\}$ and $x + \frac{1}{m}\mathbb{Z}$ have at most one element in common, and $\mathbb{P}\left(\log X_1 \in x + \frac{1}{m}\mathbb{Z}\right) \leq \frac{1}{2}$. Thus Theorem 8.15(ii) applies, and the sequence $\left(\prod_{j=1}^{n} X_j\right)$ converges in distribution to Benford's law. By Corollary 8.18, it is also Benford with probability one. ✠

Another immediate consequence of Theorems 8.6 and 8.16 is that the almost sure Benford properties of the sequences (X_1^n) and $\left(\prod_{j=1}^{n} X_j\right)$ are related.

COROLLARY 8.21. *Let $X_1, X_2, \ldots$ be i.i.d. random variables. If the sequence (X_1^n) is Benford with probability one then so is the sequence $\left(\prod_{j=1}^{n} X_j\right)$.*

PROOF. Simply note that $\mathbb{P}(\log |X_1| \in \mathbb{Q}) = 0$ implies Theorem 8.16(ii). ∎

The converse of Corollary 8.21 does not hold in general, as the following example shows.

EXAMPLE 8.22. Let $0 < p < 1$ and let $X_1, X_2, \ldots$ be i.i.d. positive random variables with the common distribution function

$$F_{X_1}(x) = \begin{cases} \dfrac{px}{\sqrt[3]{10}} & \text{if } 0 \leq x < \sqrt[3]{10}, \\ 1 & \text{if } x \geq \sqrt[3]{10}, \end{cases}$$

that is, X_1 is uniformly distributed on $\left(0, \sqrt[3]{10}\right)$ with probability p, and otherwise (i.e., with probability $1 - p$) is equal to $\sqrt[3]{10}$. Since

$$\mathbb{E}\left[e^{6\pi i n \log X_1}\right] = \frac{1 + i6\pi(1-p)n \log e}{1 + i6\pi n \log e} \xrightarrow{n \to \infty} 1 - p \neq 0,$$

Theorem 8.5 shows that the sequence (X_1^n) does not converge in distribution to Benford's law. Since $\mathbb{P}(\log X_1 = \frac{1}{3}) = 1 - p > 0$, it is not Benford with probability one either. On the other hand, X_1 is not discrete, and hence the sequence $\left(\prod_{j=1}^{n} X_j\right)$ converges in distribution to Benford's law and is Benford with probability one. Note that the latter fact is not related to the actual behavior of the sequence $\left(\prod_{j=1}^{n} X_j\right)$: From

$$\mathbb{E}[\log X_1] = \tfrac{1}{3} - p \log e = (p_0 - p) \log e , \quad \text{with } p_0 = \tfrac{1}{3}(\log e)^{-1} = 0.7675 ,$$

it follows that $\prod_{j=1}^{n} X_j \xrightarrow{\text{a.s.}} +\infty$ for $p < p_0$, and $\prod_{j=1}^{n} X_j \xrightarrow{\text{a.s.}} 0$ whenever $p > p_0$. If $p = p_0$ then, with probability one, the sequence $\left(\prod_{j=1}^{n} X_j\right)$ does not converge but attains both arbitrarily large and arbitrarily small positive values. ✠

Finally, note that a particularly important special case of Corollary 8.19 occurs whenever X_1 has a density (cf. Theorem 8.8). In this case, $\mathbb{P}(X_1 \in C) = 0$ for every countable set $C \subset \mathbb{R}$, so X_1 is clearly not discrete.

COROLLARY 8.23. *If the i.i.d. random variables $X_1, X_2, \ldots$ have a density then the sequence $\left(\prod_{j=1}^{n} X_j\right)$ converges in distribution to Benford's law and is Benford with probability one.*

Figure 8.2 illustrates the dependencies between the various Benford properties of random sequences (8.1) for the two scenarios considered so far.

EXAMPLE 8.24. Let $X_1, X_2, \ldots$ be i.i.d. $U(0,1)$ random variables. By Corollary 8.23, the sequence $\left(\prod_{j=1}^{n} X_j\right)$ converges in distribution to Benford's law (a fact apparently first recorded in [2]) and is Benford with probability one. As already seen in Example 3.10(i), $F_{S(X_1)}(t) = \tfrac{1}{9}(t-1)$. For $n \geq 2$, the distribution function of $S\left(\prod_{j=1}^{n} X_j\right)$ can also be computed explicitly, for instance by observing that $-\log X_1$ is gamma-distributed and using the addition property of the gamma distribution. Specifically, it can be shown that

$$F_{S\left(\prod_{j=1}^{n} X_j\right)}(t) = c_{n,1}(t-1) + t \sum_{k=2}^{n} c_{n,k}(-\ln t)^{k-1} , \quad t \in [1, 10) , \quad (8.8)$$

with the appropriate positive real numbers $c_{n,1}, c_{n,2}, \ldots, c_{n,n}$; for example,

$c_{1,1} = \tfrac{1}{9}$;

$c_{2,1} = \tfrac{1}{9} + \tfrac{10}{9^2}(\log e)^{-1} , \quad c_{2,2} = \tfrac{1}{9}$;

$c_{3,1} = \tfrac{1}{9} + \tfrac{10}{9^2}(\log e)^{-1} + \tfrac{55}{9^3}(\log e)^{-2} , \quad c_{3,2} = \tfrac{1}{9} + \tfrac{10}{9^2}(\log e)^{-1} , \quad c_{3,3} = \tfrac{1}{18}$.

As can be seen in Figure 8.3, the distribution of $S(X_1 X_2 X_3)$ is already quite close to Benford's law. Observe, however, that the precise rate of convergence in $|F_{S\left(\prod_{j=1}^{n} X_j\right)}(t) - \log t| \to 0$ is not easily recognizable from (8.8). In the next example, it will be identified by means of Fourier coefficients. ✠

Let $X_1, X_2, \ldots$ be i.i.d. random variables with the common density f_X. Recall that, since $F_{\mathbb{B}}(t) = \log t$ is continuous,

$$|F_{S\left(\prod_{j=1}^{n} X_j\right)}(t) - \log t| \xrightarrow{n \to \infty} 0 \quad \text{uniformly on } [1, 10) . \quad (8.9)$$

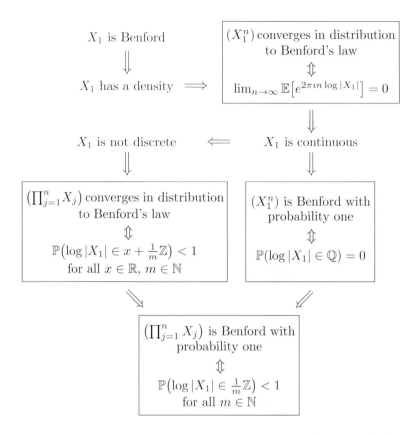

Figure 8.2: The Benford properties of random sequences (X_1^n) and $\left(\prod_{j=1}^n X_j\right)$, where $X_1, X_2, \ldots$ are i.i.d. random variables. Grey boxes indicate main results of this section. Note that none of the conceivable implications not explicitly shown holds in general.

It can be shown that the convergence in (8.9) is actually *exponential* in n; see [139]. The following example illustrates this for the three distributions already considered in Example 8.10, and in each case identifies the exact rate of convergence in (8.9). This is done using a Fourier series representation for the deviation of $F_{S(\prod_{j=1}^n X_j)}$ from Benford's law which, analogously to (8.5), reads, for every $n \in \mathbb{N}$,

$$F_{S(\prod_{j=1}^n X_j)}(t) - \log t = \frac{2}{\pi} \sum_{k=1}^{\infty} \frac{\sin(\pi k \log t)}{k} \Re\left(\varphi_k^n e^{-\pi \imath k \log t}\right) , \quad t \in [1, 10) ;$$

$$(8.10)$$

here, as before, for every $k \in \mathbb{Z}$, the number φ_k is the kth Fourier coefficient of $P_{\langle \log |X_1| \rangle}$, i.e., $\varphi_k = \mathbb{E}\left[e^{2\pi \imath k \log |X_1|}\right]$. This example once again illustrates the great usefulness of Fourier coefficients in the study of Benford's law.

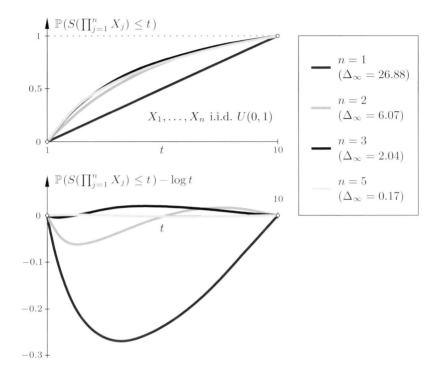

Figure 8.3: If $X_1, X_2, \ldots$ are i.i.d. $U(0,1)$ random variables then the distribution of $S\left(\prod_{j=1}^{n} X_j\right)$ converges to Benford's law as $n \to \infty$; see Example 8.24. Notice that the convergence is much faster here than for $S(X_1^n)$; cf. Figure 8.1.

EXAMPLE 8.25. Consider the three distributions from Example 8.10, i.e., let $X_1, X_2, \ldots$ be i.i.d. random variables where X_1 is uniform on $(0,1)$, log-normal, and log-Cauchy, respectively. Since X_1 has a density, $\lim_{|k| \to \infty} \varphi_k = 0$ by the Riemann–Lebesgue Lemma, and $\varphi_\infty := \max_{k \neq 0} |\varphi_k| < 1$.

(i) If $X_1 = U(0,1)$ then, as seen in Example 8.10(i), $\varphi_k = (1 + \imath 2\pi k \log e)^{-1}$ for every $k \in \mathbb{Z}$. It follows that $\varphi_\infty = |\varphi_{\pm 1}| = (1 + 4\pi^2 (\log e)^2)^{-1/2}$. Using the Dominated Convergence Theorem and $\alpha := \arctan(2\pi \log e) = 1.219$, it is easily deduced from (8.10) that

$$\max_{t \in [1,10)} \left| \frac{F_{S(\prod_{j=1}^{n} X_j)}(t) - \log t}{\varphi_\infty^n} - \frac{2}{\pi} \sin(\pi \log t) \cos(\pi \log t + n\alpha) \right| \stackrel{n \to \infty}{\longrightarrow} 0 \, .$$

Thus, convergence in (8.9) occurs exactly at the rate $(\varphi_\infty^n) = (e^{-\gamma n})$, with

$$\gamma = -\ln(1 + 4\pi^2 (\log e)^2)^{-1/2} = \frac{\log(1 + 4\pi^2 (\log e)^2)}{2 \log e} = 1.066 \, .$$

This convergence is much *faster* than the convergence in $|F_{S(X_1^n)}(t) - \log t| \to 0$, which, as seen in (8.3), is at the rate (n^{-1}).

(ii) If the random variable X_1 is log-normal with $\mu = 0$ and $\sigma^2 = 1$, then $\varphi_k = e^{-2\pi^2 k^2 (\log e)^2}$ for every $k \in \mathbb{Z}$, and hence $\varphi_\infty = |\varphi_{\pm 1}| = e^{-2\pi^2 (\log e)^2}$. The same argument as in (i) applies, showing that again convergence in (8.9) occurs exactly at the rate $(\varphi_\infty^n) = (e^{-\gamma n})$, but now $\gamma = 2\pi^2 (\log e)^2 = 3.723$. This is much *slower* than the convergence in $|F_{S(X_1^n)}(t) - \log t| \to 0$, which is at the rate $(e^{-\gamma n^2})$.

(iii) Finally, if X_1 is a positive random variable where $\log X_1$ is standard Cauchy-distributed, then $\varphi_k = e^{-2\pi|k|}$ for every $k \in \mathbb{Z}$, and (8.10) yields, uniformly in $t \in [1, 10)$,

$$\lim_{n \to \infty} e^{2\pi n} \big(F_{S(\prod_{j=1}^n X_j)}(t) - \log t \big) = \frac{\sin(2\pi \log t)}{\pi},$$

i.e., *exactly* the same limit and rate of convergence as in (8.6). This is not a coincidence: Since $\log X_1$ is Cauchy, the random variables $n \log X_1$ and $\sum_{j=1}^n \log X_j$ have the same distribution for every $n \in \mathbb{N}$, and so have $S(X_1^n)$ and $S\big(\prod_{j=1}^n X_j\big)$. ✠

The reader may want to keep in mind that the scenarios considered in this section represent only two special cases of (8.1), though arguably two important and extreme ones. The tools employed for analyzing these two scenarios may nevertheless be useful in other situations as well.

EXAMPLE 8.26. Let $X_1 = U(0, 1)$ and define a sequence of random variables recursively as

$$X_n = \big\langle X_{n-1} + \tfrac{1}{2} \big\rangle, \quad n \geq 2.$$

Clearly, every X_n is $U(0, 1)$-distributed, so the sequence $\big(\prod_{j=1}^n X_j\big)$ is of the form (8.1). However, the X_n are not identical, since $\mathbb{P}(X_n = X_{n-1}) = 0$. Nor, of course, are they independent, so this example is not covered by either of the two scenarios discussed above. Still, by observing that

$$\prod_{j=1}^n X_j = \begin{cases} X_1^{n/2} \big\langle X_1 + \tfrac{1}{2} \big\rangle^{n/2} & \text{if } n \text{ is even}, \\ X_1^{(n+1)/2} \big\langle X_1 + \tfrac{1}{2} \big\rangle^{(n-1)/2} & \text{if } n \text{ is odd}, \end{cases}$$

and by using arguments similar to those employed in the proofs of Theorems 8.5 and 8.6, it is not hard to see that the sequence $\big(\prod_{j=1}^n X_j\big)$ converges in distribution to Benford's law and is Benford with probability one. ✠

Recalling the fact (mentioned at the very beginning of this chapter) that $X_n \xrightarrow{\text{a.s.}} X$ implies $X_n \xrightarrow{\mathcal{D}} X$, but not vice versa, the reader will notice that, as far as the Benford properties of (8.1) are concerned, this implication appears somewhat reversed, at least in the two scenarios studied here: By Corollaries 8.7 and 8.18, if the sequence $\big(\prod_{j=1}^n X_j\big)$ converges in distribution to Benford's law then it is Benford with probability one, but the converse is false. To put these observations into perspective, it may be helpful to notice that in the even simpler case of an i.i.d. sequence $(X_1, X_2, \ldots)$ the two properties are actually equivalent.

THEOREM 8.27. *If $X_1, X_2, \ldots$ are i.i.d. random variables then the following are equivalent:*

(i) *The sequence (X_n) is Benford with probability one;*

(ii) *The sequence (X_n) converges in distribution to Benford's law;*

(iii) *X_1 is Benford.*

PROOF. Assume that $(X_1, X_2, \ldots)$ is Benford with probability one. This means that for every bounded continuous function $f : [1, 10) \to \mathbb{R}$,

$$\frac{1}{N} \sum\nolimits_{n=1}^{N} f\big(S(X_n)\big) \xrightarrow{\text{a.s.}} \int_1^{10} f(t) t^{-1} \log e \, \mathrm{d}t \quad \text{as } N \to \infty \, .$$

The Dominated Convergence Theorem implies that

$$\lim\nolimits_{N \to \infty} \frac{1}{N} \sum\nolimits_{n=1}^{N} \mathbb{E}\big[f\big(S(X_n)\big) \big] = \int_1^{10} f(t) t^{-1} \log e \, \mathrm{d}t \, ,$$

but $\mathbb{E}\big[f\big(S(X_n)\big) \big] \equiv \mathbb{E}\big[f\big(S(X_1)\big) \big]$, and so $\mathbb{E}\big[f\big(S(X_1)\big) \big] = \int_0^1 f(t) t^{-1} \log e \, \mathrm{d}t$. Since f was arbitrary, $\mathbb{P}(S(X_1) \le t) = \log t$ for all $t \in [1, 10)$, i.e., X_1 is Benford. Thus (i)$\Rightarrow$(iii).

Conversely, assume X_1 is Benford and fix any $t \in [1, 10)$. The Strong Law of Large Numbers applied to the i.i.d. sequence $\big(\mathbb{1}_{[1,t]}(S(X_n)) \big)$ yields

$$\frac{\#\{1 \le n \le N : S(X_n) \le t\}}{N} = \frac{1}{N} \sum\nolimits_{n=1}^{N} \mathbb{1}_{[1,t]}\big(S(X_n) \big)$$

$$\xrightarrow{\text{a.s.}} \mathbb{E}\big[\mathbb{1}_{[1,t]}\big(S(X_1) \big) \big] = \log t \quad \text{as } N \to \infty \, ,$$

showing that (X_n) is Benford with probability one. Thus (i)$\Leftrightarrow$(iii). The equivalence of (ii) and (iii) is clear from Definition 8.1. ∎

Remark. If the random variables $X_1, X_2, \ldots$ are identically distributed but *dependent*, then properties (ii) and (iii) in Theorem 8.27 clearly remain equivalent, and the implication (i)$\Rightarrow$(ii) remains valid also, with the same proof as above. The reverse implication (ii)$\Rightarrow$(i) may fail, however, as the example of a constant sequence $(X_1, X_1, \ldots)$ with X_1 being Benford shows. On the other hand, all equivalences in Theorem 8.27 are valid whenever the stochastic process $\big(S(X_n)\big)_{n \in \mathbb{N}}$ is *stationary* and *ergodic*, for instance, if $X_n = f(X_{n-1})$ with an ergodic map f that preserves the distribution of X_1.

Sums of random variables

The remainder of this section is devoted to the study of Benford's law for *sums* of random variables. The statistical behavior of the significands for sums is much more complex than that for products. The basic reason for this can be seen in the fact that the significand of the sum of two or more numbers depends not only on the significand of each number (as in the case of products), but also on their *exponents*. For example, observe that

$$S\big(3 \cdot 10^3 + 2 \cdot 10^2\big) = 3.2\,, \quad S\big(3 \cdot 10^2 + 2 \cdot 10^2\big) = 5\,, \quad S\big(3 \cdot 10^2 + 2 \cdot 10^3\big) = 2.3\,,$$

are all different, whereas clearly

$$S\big(3 \cdot 10^3 \times 2 \cdot 10^2\big) = S\big(3 \cdot 10^2 \times 2 \cdot 10^2\big) = S\big(3 \cdot 10^2 \times 2 \cdot 10^3\big) = 6\,.$$

From a practical standpoint, this difficulty is reflected in the fact that for any positive real numbers x and y, the value of $\log(x+y)$, relevant for conformance to Benford's law via Theorem 4.2, is not easily expressed in terms of $\log x$ and $\log y$, whereas $\log(xy) = \log x + \log y$.

Recall from Theorem 8.12 that Benford's law is extremely attracting under multiplication of independent random variables, in that the product of positive variables is Benford if any single one of the individual variables is. In contrast, the sum of two independent random variables may not be Benford even if both variables are.

EXAMPLE 8.28. Let X and Y be i.i.d. Benford random variables with values in $[1, 10)$, i.e., $\mathbb{P}(X \leq t) = \mathbb{P}(Y \leq t) = \log t$ for all $1 \leq t < 10$. Then $X + Y$ is not even close to being Benford. Concretely, the density of $S(X+Y)$ equals

$$f_{S(X+Y)}(t) = \frac{2\log e}{t} \cdot \begin{cases} \log(10t - 1) & \text{if } 1 \leq t < 1.1\,, \\[2mm] |\log(t-1)| & \text{if } 1.1 \leq t < 10\,, \end{cases}$$

and hence

$$\max_{d=1}^9 \big|\mathbb{P}(D_1(X+Y) = d) - \log(1 + d^{-1})\big| \geq \log e \max_{d=2}^9 \left| \int_{d-1}^d \frac{2\log t - 1}{t+1}\, dt \right|$$

$$= \log e \int_1^2 \frac{1 - 2\log t}{t+1}\, dt = 0.1205\,;$$

see Figure 8.4.   ✠

More generally, the authors conjecture that if X, Y are independent random variables then X and $X + Y$ can both be Benford only if $Y = 0$ with probability one. Note, however, that if X is Benford then clearly $X + Y$ can be arbitrarily close to a Benford random variable.

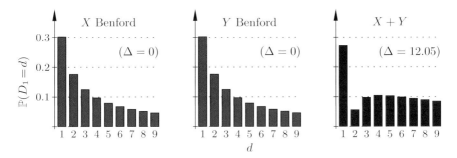

Figure 8.4: The sum of two independent Benford random variables in $[1, 10)$ is not close to being Benford; see Example 8.28.

As an analogue of (8.1), consider the random sequence

$$\left(\sum\nolimits_{j=1}^{n} X_j\right) = (X_1, X_1 + X_2, \ldots), \quad X_1, X_2, \ldots \text{ identically distributed}, \quad (8.11)$$

where, as for products, only the scenarios of *identical* and *independent* terms X_j, respectively, will be discussed in detail.

Assume first that all terms $X_1, X_2, \ldots$ in (8.11) are *identical*. Then the random sequence in (8.11) simply takes the form (nX), where X is any real-valued random variable. In the simplest special case of $\mathbb{P}(X = x) = 1$ for some $x \in \mathbb{R}$, recall from Example 4.7(ii) that (nx) is not a Benford sequence for any value of x. In fact, for the empirical distribution functions F_N associated with the significands of (nx), i.e., for

$$F_N(t) = \frac{\#\{1 \leq n \leq N : S(nx) \leq t\}}{N}, \quad N \in \mathbb{N}, t \in [1, 10),$$

it is easy to show, using Theorem 3.13, that

$$\sup\nolimits_{t\in[1,10)} |F_N(t) - \log t| \geq \tfrac{1}{41} \quad \text{for every } N \in \mathbb{N}.$$

Thus not even any subsequence of (F_N) converges to $F_\mathbb{B}(t) = \log t$. This behavior persists for arbitrary X, as the following analogue to Theorems 8.5 and 8.6 for sums shows.

THEOREM 8.29. *Let X be a random variable. Then:*

(i) *The sequence (nX) converges in distribution to Benford's law if and only if X is Benford;*

(ii) *If X is not Benford then no subsequence of (nX) converges in distribution to Benford's law;*

(iii) *With probability one, (nX) is not a Benford sequence.*

PROOF. Assume first that $\mathbb{P}(X = 0) = 0$. To see (i), let $Y_n = \langle \log|nX|\rangle$ and note that, for every $n \in \mathbb{N}$ and $k \in \mathbb{Z}$,

$$\left|\widehat{P_{Y_n}}(k)\right| = \left|\mathbb{E}\left[e^{2\pi\imath k \log|nX|}\right]\right| = \left|e^{2\pi\imath k \log n}\mathbb{E}\left[e^{2\pi\imath k \log|X|}\right]\right| = \left|\widehat{P_{\langle\log|X|\rangle}}(k)\right|.$$

By Theorem 4.4, if X is Benford then so is nX for every n. For (ii), note that if X is not Benford then $\widehat{P_{\langle\log|X|\rangle}}(k) \neq 0$ for some integer $k \neq 0$, and consequently $|\widehat{P_{Y_n}}(k)| = |\widehat{P_{\langle\log|X|\rangle}}(k)| > 0$ for all n. It follows that not even any subsequence of (Y_n) converges in distribution to $U(0,1)$, that is, no subsequence of (nX) converges in distribution to Benford's law. If $\mathbb{P}(X = 0) = \delta > 0$ then $\mathbb{P}(nX = 0) = \delta$ for every $n \in \mathbb{N}$, from which it is clear that (i) and (ii) hold in this case also. Property (iii) follows immediately from Example 4.7(ii) because $\mathbb{P}\left(\{\omega : \big(nX(\omega)\big) \text{ is a Benford sequence }\}\right) = \mathbb{P}(\varnothing) = 0$. ∎

Finally, consider (8.11) with *independent* terms $X_1, X_2, \ldots$. Here the random variables $X_1, X_1 + X_2, \ldots$ all have different distributions unless $X_1 = 0$ with probability one. Given the discussion above, therefore, the assertions of the following analogue to Theorems 8.15 and 8.16 for sums may not come as a surprise.

THEOREM 8.30. *Let $X_1, X_2, \ldots$ be i.i.d. random variables with finite variance, i.e., $\mathbb{E}\left[X_1^2\right] < +\infty$. Then:*

(i) *No subsequence of $\left(\sum_{j=1}^n X_j\right)$ converges in distribution to Benford's law;*

(ii) *With probability one, $\left(\sum_{j=1}^n X_j\right)$ is not a Benford sequence.*

PROOF. Assume first that $\mathbb{E}[X_1] \neq 0$. By the Strong Law of Large Numbers, $\left(\frac{1}{n}\left|\sum_{j=1}^n X_j\right|\right)$ converges a.s., and hence also in distribution, to the constant $|\mathbb{E}[X_1]|$. Since $\mathbb{P}\left(\sum_{j=1}^n X_j = 0\right) \to 0$ as $n \to \infty$ and

$$\log S\left(\sum\nolimits_{j=1}^n X_j\right) = \left\langle\log\left|\sum\nolimits_{j=1}^n X_j\right|\right\rangle = \left\langle\log\left(\frac{1}{n}\left|\sum\nolimits_{j=1}^n X_j\right|\right) + \log n\right\rangle$$

whenever $\sum_{j=1}^n X_j \neq 0$, any subsequence of $\left(S\big(\frac{1}{n}\sum_{j=1}^n X_j\big)\right)$ either does not converge in distribution at all or converges to a constant; in neither case, therefore, is the limit a Benford random variable. Since $\left|\sum_{j=1}^n X_j\right| \to +\infty$ with probability one, it follows from

$$\log\left|\sum\nolimits_{j=1}^n X_j\right| - \log n = \log\frac{1}{n}\left|\sum\nolimits_{j=1}^n X_j\right| \xrightarrow{\text{a.s.}} |\mathbb{E}[X_1]|,$$

together with Propositions 4.3(i) and 4.6(iii), that with probability one, the sequence $\left(\sum_{j=1}^n X_j\right)$ is not Benford.

It remains to consider the case $\mathbb{E}[X_1] = 0$. Without loss of generality, it can be assumed that $\mathbb{E}\left[X_1^2\right] = 1$. By the Central Limit Theorem, $\left(\frac{1}{\sqrt{n}}\sum_{j=1}^n X_j\right)$

converges in distribution to the standard normal distribution. Thus for every sufficiently large n, and up to a rotation (i.e., an addition mod 1) of $[0, 1)$, the distribution of $\langle \log | \sum_{j=1}^{n} X_j | \rangle$ is arbitrarily close to the distribution of the random variable $Y := \langle \log |Z| \rangle$, where Z is standard normal. Intuitively, it is clear that $P_Y \neq \lambda_{0,1}$, i.e., Y is *not* uniform on $[0, 1)$. To see this more formally, note that

$$F_Y(s) = 2 \sum_{k \in \mathbb{Z}} \left(\Phi\left(10^{s+k}\right) - \Phi\left(10^k\right) \right), \quad 0 \leq s < 1, \qquad (8.12)$$

with $\Phi \, (= F_Z)$ denoting the standard normal distribution function; see Example 3.10(v). Thus

$$|F_Y(s) - s| \geq F_Y(s) - s > 2\left(\Phi\left(10^s\right) - \Phi\left(1\right) \right) - s =: R(s), \quad 0 \leq s < 1,$$

and since R is concave on $[0, 1)$ with $R(0) = 0$ but $R'(0) > 0$,

$$\max_{0 \leq s < 1} |F_Y(s) - s| > \sup_{0 \leq s < 1} R(s) > 0,$$

showing that indeed $P_Y \neq \lambda_{0,1}$; hence the sequence $\left(\sum_{j=1}^{n} X_j \right)$ does not converge in distribution to Benford's law, and neither does any subsequence.

The verification of (ii) in the case $\mathbb{E}[X_1] = 0$ uses an almost sure version of the Central Limit Theorem; see [94]. With the random variables X_n defined on some (abstract) probability space $(\Omega, \mathcal{A}, \mathbb{P})$, let

$$\Omega_1 = \left\{ \omega \in \Omega : \left(\sum_{j=1}^{n} X_j(\omega) \right) \text{ is a Benford sequence} \right\}.$$

By Theorem 4.2 and Proposition 4.6(iii), the sequence $\left(x_n(\omega) \right)$ with

$$x_n(\omega) = \log \frac{1}{\sqrt{n}} \left| \sum_{j=1}^{n} X_j(\omega) \right|, \quad n \in \mathbb{N},$$

is u.d. mod 1 for all $\omega \in \Omega_1$. For every interval $[a, b) \subset [0, 1)$, therefore,

$$\frac{1}{\ln N} \sum_{n=1}^{N} \frac{\mathbb{1}_{[a,b)}\left(\langle x_n(\omega) \rangle \right)}{n} \to b - a \quad \text{as } N \to \infty.$$

(Recall the remarks on logarithmic averaging in Section 3.1.) However, as a consequence of [94, Thm. 2], for every $[a, b) \subset [0, 1)$,

$$\frac{1}{\ln N} \sum_{n=1}^{N} \frac{\mathbb{1}_{[a,b)}\left(\langle x_n \rangle \right)}{n} \xrightarrow{\text{a.s.}} F_Y(b) - F_Y(a),$$

with F_Y given by (8.12). As shown above, $F_Y(s) \not\equiv s$, and therefore $\mathbb{P}(\Omega_1) = 0$. In other words, $\mathbb{P}\left(\left(\sum_{j=1}^{n} X_j \right) \text{ is a Benford sequence} \right) = 0$. ∎

EXAMPLE 8.31. **(i)** Let $X_1, X_2, \ldots$ be i.i.d. with $\mathbb{P}(X_1 = 0) = \mathbb{P}(X_1 = 2) = \frac{1}{2}$. Then $\mathbb{E}[X_1] = 1$, $\mathbb{E}\left[X_1^2\right] = 2$, and by Theorem 8.30(i) neither $\left(\sum_{j=1}^{n} X_j \right)$ nor any of its subsequences converges in distribution to Benford's law. Note that $\frac{1}{2}\sum_{j=1}^{n} X_j$ is binomial with parameters n and $\frac{1}{2}$, i.e., for all $n \in \mathbb{N}$,

$$\mathbb{P}\left(\sum_{j=1}^{n} X_j = 2k\right) = 2^{-n}\binom{n}{k}, \quad k = 0, 1, \ldots, n\,.$$

The Law of the Iterated Logarithm [36, Sec. 10.2] asserts that

$$\sum_{j=1}^{n} X_j = n + Y_n\sqrt{n\ln\ln n} \quad \text{for all } n \geq 3\,, \tag{8.13}$$

where the random sequence (Y_n) is bounded; in fact, $\limsup_{n\to\infty} |Y_n| = \sqrt{2}$ a.s. Thus by Example 4.7(ii), Theorem 4.12(i), and (8.13) it is clear that, with probability one, the sequence $\left(\sum_{j=1}^{n} X_j\right)$ is not Benford.

   (ii) If the i.i.d. random variables $X_1, X_2, \ldots$ are standard normal then so is $\frac{1}{\sqrt{n}}\sum_{j=1}^{n} X_j$ for every $n \in \mathbb{N}$. Hence the distribution function of $S\left(\sum_{j=1}^{n} X_j\right)$ simply equals

$$F_{S\left(\sum_{j=1}^{n} X_j\right)}(t) = 2\sum_{k\in\mathbb{Z}}\left(\Phi\left(\frac{t10^k}{\sqrt{n}}\right) - \Phi\left(\frac{10^k}{\sqrt{n}}\right)\right), \quad t \in [1, 10]\,.$$

By Theorem 8.30, the sequence $\left(\sum_{j=1}^{n} X_j\right)$ does not converge in distribution to Benford's law (not even a subsequence does) and is, with probability one, not Benford. Note that even though $\sum_{j=1}^{n} X_j$ in a sense may be closer to Benford's law than, say, the uniform random variable $U(0, n)$, as seen in Theorem 3.13, the value of $\max_{t\in[1,10)} |F_{S\left(\sum_{j=1}^{n} X_j\right)}(t) - \log t|$ does not change much with growing n and does not go to zero as $n \to \infty$; see Figure 8.5. ✠

   As the final example in this section illustrates, the conclusions of Theorem 8.30 may hold even in cases where X_1 does not have a finite first moment, let alone a finite second moment. The authors conjecture that the hypothesis of finite variance in Theorem 8.30 can be weakened considerably; perhaps even no random walk on the real line at all has Benford paths (in distribution or with probability one).

   EXAMPLE 8.32. Let $X_1, X_2, \ldots$ be i.i.d. random variables with density

$$f_{X_1}(x) = \begin{cases} \frac{1}{\sqrt{2\pi}}e^{-1/(2x)}x^{-3/2} & \text{if } x > 0\,, \\ 0 & \text{if } x \leq 0\,, \end{cases}$$

i.e., the distribution of X_1 is a one-sided stable law with characteristic exponent $\frac{1}{2}$. Note that $\mathbb{E}\left[X_1^p\right]$ is finite if $p < \frac{1}{2}$, and $\mathbb{E}\left[X_1^p\right] = +\infty$ if $p \geq \frac{1}{2}$. Thus $\mathbb{E}[X_1]$ is infinite, and Theorem 8.30 does not apply. However, it is well known (and not hard to check) that for every $n \in \mathbb{N}$, the random variable $n^{-2}\sum_{j=1}^{n} X_j$ has the

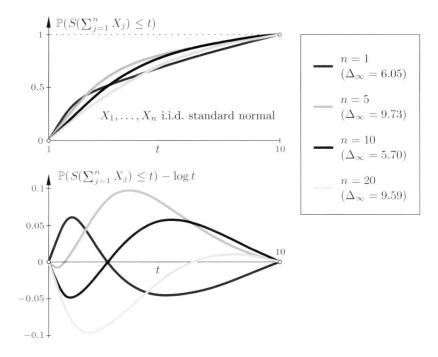

Figure 8.5: The sum of any number of independent standard normal random variables is never close to being Benford; see Example 8.31(ii).

same distribution as X_1. The latter is not Benford because, for all $1 \leq t < 10$,

$$\frac{f_{S(X_1)}(t)}{f_{\mathbb{B}}(t)} = \frac{t}{\log e} \sum_{k \in \mathbb{Z}} f_{X_1}(t 10^k) 10^k = \frac{1}{\log e \sqrt{2\pi t}} \sum_{k \in \mathbb{Z}} e^{-10^{-k}/(2t)} 10^{-k/2}$$

$$\geq \frac{1}{\log e \sqrt{2\pi t}} \left(e^{-5/t} \sqrt{10} + \sum_{n=0}^{\infty} e^{-10^{-n}/(2t)} 10^{-n/2} \right)$$

$$\geq \frac{1}{\log e} \sqrt{\frac{5}{\pi t}} \left(e^{-5/t} + \frac{1}{\sqrt{10}-1} - \frac{5/t}{10\sqrt{10}-1} \right) =: R(t) \, ;$$

in particular, $R(5) = 1.036$. Thus $\delta := \frac{1}{2} \sup_{1 \leq t < 10} |F_{S(X_1)}(t) - \log t| > 0$, and consequently $\sup_{1 \leq t < 10} |F_{S(\sum_{j=1}^{n} X_j)}(t) - \log t| \geq \delta$ for all $n \in \mathbb{N}$. Not even a subsequence of $\left(\sum_{j=1}^{n} X_j \right)$, therefore, converges in distribution to Benford's law. In addition, note that the (random) sequence $\left(\sum_{j=1}^{n} X_j \right)$ is strictly increasing, and so is $\left(\log \sum_{j=1}^{n} X_j \right)$. Recall that $\mathbb{E}\left[X_1^p \right] < +\infty$ for every $p < \frac{1}{2}$; hence $\lim_{n \to \infty} n^{-1/p} \sum_{j=1}^{n} X_j = 0$ almost surely, by [36, Thm. 5.2.2]. Thus

$$\limsup_{n \to \infty} \frac{\log \sum_{j=1}^{n} X_j}{\log n} \leq 2 \quad \text{with probability one} \, ,$$

which, together with Proposition 4.6(iv), implies $\mathbb{P}\big((\sum_{j=1}^{n} X_j)$ is Benford$\big) = 0$. In conclusion, both (i) and (ii) of Theorem 8.30 hold in this example despite the fact that $\mathbb{E}[X_1] = +\infty$.   ✠

Remark. Using alternative notions of conformance to Benford's law (some of which have been discussed briefly in Section 3.1), some positive results can be obtained for random sums; see [140].

8.3   MIXTURES OF DISTRIBUTIONS

The characterizations of the Benford distribution via scale-, base-, and sum-invariance given in Chapter 5, although perhaps mathematically satisfying, hardly help explain the appearance of Benford's law empirically in real-life data. Application of those theorems requires explaining why the underlying data are scale-, base-, or sum-invariant in the first place. Benford's law nevertheless does appear in many real-life datasets. Thus the question arises: What do the population data of three thousand U.S. counties according to the 2010 census have in common with the usage of logarithm tables during the 1880s, numerical data from newspaper articles of the 1930s collected by Benford, universal physical constants examined by Knuth in the 1960s, or numbers on the World Wide Web today (see Figure 1.3)? Why should these data exhibit a logarithmically distributed significand or, equivalently, why should they be scale- or base-invariant?

In fact, most datasets do not follow Benford's law closely. Benford had already observed that while some of his tables conformed to the logarithmic law reasonably well, many others did not. But, as Raimi points out, "what came closest of all, however, was the union of all his tables" ([127, p. 522]; see last row in Table 1.2). Combine the molecular weight tables with baseball statistics and drainage areas of rivers, and *then* there is a very good fit. Many of the previous explanations of Benford's law have first hypothesized some universal table of constants, such as Raimi's [127] "stock of tabular data in the world's libraries," or Knuth's [90] "imagined set of real numbers," and then tried to prove why certain specific sets of real observations were representative of either this mysterious universal table or the set of all real numbers. What seems more natural though is to think of data as coming from many different distributions. This was clearly the case in Benford's original study. After all, he had made an effort "to collect data from as many fields as possible and to include a wide variety of types", noting that "the range of subjects studied and tabulated was as wide as time and energy permitted."

The main goal of this section is to provide a statistical derivation of Benford's law in the form of a central-limit-like theorem that says that if random samples are taken from different distributions, and the results combined, then — provided the sampling is "unbiased" with regard to scale or base — the resulting combined samples will converge to the Benford distribution.

Denote by $\mathcal{M}$ the set of all probability measures on $(\mathbb{R}^+, \mathcal{B}^+)$. Recall that a (positive Borel) *random probability measure*, abbreviated henceforth as *r.p.m.*, is a function $P : \Omega \to \mathcal{M}$, defined on some underlying probability space $(\Omega, \mathcal{A}, \mathbb{P})$, such that for every $B \in \mathcal{B}^+$ the function $\omega \mapsto P(\omega)(B)$ is a random variable, and for every $\omega \in \Omega$, $P(\omega)$ is a probability measure on $(\mathbb{R}^+, \mathcal{B}^+)$, so for any real numbers a, b and any Borel set $B \subset \mathbb{R}^+$,

$$\{\omega : a \le P(\omega)(B) \le b\} \in \mathcal{A} \,;$$

see [86] for an authoritative account of random probability measures. In more abstract conceptual terms, an r.p.m. can be interpreted as follows: When endowed with the topology of convergence in distribution, the set $\mathcal{M}$ becomes a complete and separable metrizable space. Denote by $\mathcal{B}_{\mathcal{M}}$ its Borel σ-algebra, defined as the smallest σ-algebra containing all open subsets of $\mathcal{M}$. Then $\mathbb{P} \circ P^{-1}$ is simply a probability measure on $(\mathcal{M}, \mathcal{B}_{\mathcal{M}})$.

EXAMPLE 8.33. **(i)** Let P be an r.p.m. that is $U(0, 1)$ with probability $\frac{1}{2}$, and otherwise is $\exp(1)$, i.e., exponential with mean 1; see Example 3.10(i, ii). Thus, for every $\omega \in \Omega$, the probability measure P is either $U(0, 1)$ or $\exp(1)$, and $\mathbb{P}(P = U(0, 1)) = \mathbb{P}(P = \exp(1)) = \frac{1}{2}$. For a practical realization of P simply flip a fair coin — if it comes up heads, P is a $U(0, 1)$ distribution, and if it comes up tails, then P is an $\exp(1)$ distribution.

(ii) Let X have an $\exp(1)$ distribution, and let P be an r.p.m. where, for each $\omega \in \Omega$, $P(\omega)$ is the uniform distribution on $(0, X(\omega))$. Here P is continuous, i.e., $\mathbb{P}(P = Q) = 0$ for each probability measure $Q \in \mathcal{M}$. ✠

The following example of an r.p.m. is a variant of a classical construction due to L. Dubins and D. Freedman which, as will be seen below, is an r.p.m. leading to Benford's law.

EXAMPLE 8.34. Let P be an r.p.m. supported on $[1, 10)$, i.e., $P([1, 10)) = 1$ with probability one, defined by its (random) distribution function

$$F_P(t) = F_{P(\omega)}(t) = P(\omega)([1, t]), \quad 1 \le t < 10,$$

as follows: Set $F_P(1) = 0$ and $F_P(10) = 1$. Next pick $F_P(10^{1/2})$ according to the uniform distribution on $[0, 1)$. Then pick $F_P(10^{1/4})$ and $F_P(10^{3/4})$ independently, uniformly on $[0, F_P(10^{1/2}))$ and $[F_P(10^{1/2}), 1)$, respectively, and continue in this manner. This construction is known to generate an r.p.m. almost surely [51, Lem. 9.28], and as can easily be seen, is dense in the set of all probability measures on $([1, 10), \mathcal{B}[1, 10))$, i.e., it generates probability measures that are arbitrarily close to any Borel probability measure on $[1, 10)$. ✠

The next definition formalizes the notion of combining data from different distributions. Essentially, it mimics what Benford did in combining baseball statistics with square-root tables and numbers taken from newspapers. The underlying idea rests upon using an r.p.m. to generate a random sequence of probability distributions, and then successively selecting random samples from each of those distributions.

DEFINITION 8.35. Let m be a positive integer and P an r.p.m. A *sequence of P-random m-samples* is a sequence $(X_1, X_2, \ldots)$ of random variables on $(\Omega, \mathcal{A}, \mathbb{P})$ such that, for all $j \in \mathbb{N}$ and some i.i.d. sequence $(P_1, P_2, \ldots)$ of r.p.m.s with $P_1 = P$, the following two properties hold:

(i) Given $P_j = Q$, the random variables $X_{(j-1)m+1}, X_{(j-1)m+2}, \ldots, X_{jm}$ are i.i.d. with distribution Q;

(ii) The random variables $X_{(j-1)m+1}, X_{(j-1)m+2}, \ldots, X_{jm}$ are independent of P_i, $X_{(i-1)m+1}, X_{(i-1)m+2}, \ldots, X_{im}$ for every $i \neq j$.

In other words, given any sequence $(X_1, X_2, \ldots)$ of P-random m-samples, for each $\omega \in \Omega$ in the underlying probability space, the first m random variables $X_1, X_2, \ldots, X_m$ are a random sample (i.e., i.i.d.) from $P_1(\omega)$, a random probability distribution chosen according to the r.p.m. P; the second m-tuple of random variables $X_{m+1}, \ldots, X_{2m}$ is a random sample from $P_2(\omega)$; and so on. Note the two levels of randomness here: First a probability is selected at random, and then a random sample is drawn from this distribution, and this two-tiered process is continued.

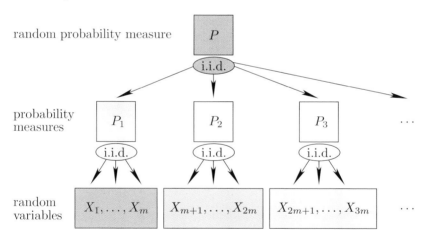

Figure 8.6: Generation of a sequence $(X_1, X_2, \ldots)$ of P-random m-samples (schematic).

EXAMPLE 8.36. Let P be the r.p.m. in Example 8.33(i), and let $m = 3$. Then a sequence of P-random 3-samples is a sequence $(X_1, X_2, \ldots)$ such that with probability $\frac{1}{2}$ the random variables X_1, X_2, X_3 are i.i.d. and uniformly distributed on $(0, 1)$, and otherwise they are i.i.d. with distribution $\exp(1)$; the random variables X_4, X_5, X_6 are again equally likely to be i.i.d. $U(0, 1)$ or $\exp(1)$, and they are independent of X_1, X_2, X_3, etc. Clearly the X_n are all identically distributed since they are all generated by exactly the same process. Note, however, that X_1 and X_2 are dependent: Given that $X_1 > 1$, for example, the random variable X_2 is $\texttt{exp}(1)$-distributed with probability one, whereas the unconditional probability that X_2 is $\texttt{exp}(1)$-distributed is only $\frac{1}{2}$.   ✠

If $(X_1, X_2, \ldots)$ is a sequence of P-random m-samples for some m and some r.p.m. P, then the X_n are a.s. identically distributed with distribution that is the average (expected) distribution of P (see Proposition 8.38 below), but they are not in general independent (see Example 8.36). On the other hand, given $(P_1, P_2, \ldots)$, the X_n are a.s. independent, but clearly are not in general identically distributed.

Although sequences of P-random m-samples have a fairly simple structure, they do not fit into any of the familiar categories of sequences of random variables. For example, they are generally not independent, exchangeable, Markov, martingale, or stationary sequences.

EXAMPLE 8.37. Assume that the r.p.m. P is, with equal probability, the Dirac measure δ_1 concentrated at 1 and the probability measure $\frac{1}{2}(\delta_1 + \delta_2)$, respectively, i.e., $\mathbb{P}(P = \delta_1) = \mathbb{P}\big(P = \frac{1}{2}(\delta_1 + \delta_2)\big) = \frac{1}{2}$. Let $(X_1, X_2, \ldots)$ be a sequence of P-random 3-samples. Then the random variables $X_1, X_2, \ldots$ are:

not independent since

$$\mathbb{P}(X_2 = 2) = \tfrac{1}{4} \neq \tfrac{1}{2} = \mathbb{P}(X_2 = 2 | X_1 = 2)\,;$$

not exchangeable since

$$\mathbb{P}\big((X_1, X_2, X_3, X_4) = (1, 1, 1, 2)\big) = \tfrac{9}{64} \neq \tfrac{3}{64} = \mathbb{P}\big((X_1, X_2, X_3, X_4) = (2, 1, 1, 1)\big)\,;$$

not Markov since

$$\mathbb{P}(X_3 = 1 | X_1 = X_2 = 1) = \tfrac{9}{10} \neq \tfrac{5}{6} = \mathbb{P}(X_3 = 1 | X_2 = 1)\,;$$

not martingale since

$$\mathbb{E}[X_2 | X_1 = 2] = \tfrac{3}{2} \neq \tfrac{5}{4} = \mathbb{E}[X_2]\,;$$

not stationary since

$$\mathbb{P}\big((X_1, X_2, X_3) = (1, 1, 1)\big) = \tfrac{9}{16} \neq \tfrac{15}{32} = \mathbb{P}\big((X_2, X_3, X_4) = (1, 1, 1)\big)\,.   ✠$$

Recall that, given an r.p.m. P and any Borel set B, the quantity $P(B)$ is a random variable with values between 0 and 1. The following property of the expectation of $P(B)$, as a function of B, is easy to check.

PROPOSITION 8.38. *Let P be an r.p.m. Then $\mathbb{E}P$, defined by*

$$(\mathbb{E}P)(B) = \mathbb{E}[P(B)] = \int_\Omega P(\omega)(B)\,d\mathbb{P}(\omega) \quad \text{for all } B \in \mathcal{B}^+\,,$$

is a probability measure on $(\mathbb{R}^+, \mathcal{B}^+)$.

EXAMPLE 8.39. **(i)** Let P be the r.p.m. of Example 8.33(i). Then $\mathbb{E}P$ is the Borel probability measure with density

$$f_{\mathbb{E}P}(x) = \left\{ \begin{array}{ll} \frac{1}{2} + \frac{1}{2}e^{-x} & \text{if } 0 < x < 1 \,, \\ \frac{1}{2}e^{-x} & \text{if } x \geq 1 \,, \end{array} \right\} = \frac{1}{2}\mathbb{1}_{[0,1)}(x) + \frac{1}{2}e^{-x} \,, \quad x > 0 \,.$$

(ii) Let P be the r.p.m. of Example 8.33(ii), that is, $P(\omega)$ is uniform on $\big(0, X(\omega)\big)$, where X is $\texttt{exp}(1)$-distributed. In this case, $\mathbb{E}P$ is also a.c., with density $f_{\mathbb{E}P}(x) = \int_x^{+\infty} t^{-1}e^{-t}\,\mathrm{d}t$ for $x > 0$.

(iii) Even if P is a.c. only with probability zero, it is possible for $\mathbb{E}P$ to be a.c. For a simple example, let $P = \frac{1}{4}\delta_X + \frac{3}{4}\delta_{3X}$, where again X is $\texttt{exp}(1)$-distributed. Then $\mathbb{P}(P \text{ is purely atomic}) = 1$, yet $\mathbb{E}P$ is a.c. with density $f_{\mathbb{E}P}(x) = \frac{1}{2}e^{-2x/3}\cosh\big(\frac{1}{3}x\big)$ for $x > 0$. ✠

Passing from an r.p.m. P to its expectation $\mathbb{E}P$ can be regarded as an averaging step, and since

$$|F_{\mathbb{E}P \circ S^{-1}}(t) - \log t| = \left| \int_\Omega \big(F_{P(\omega) \circ S^{-1}}(t) - \log t\big)\,\mathrm{d}\mathbb{P}(\omega)\right|$$
$$\leq \ \sup_{\omega \in \Omega} \big|F_{P(\omega) \circ S^{-1}}(t) - \log t\big| \,, \quad t \in [1, 10) \,,$$

it is clear that $\mathbb{E}P$ is at least as close to being Benford as *some* $P(\omega)$. In fact, as the next example demonstrates, $\mathbb{E}P$ may be *much closer* to being Benford than *any* $P(\omega)$. Part (ii) of the example shows that it is even possible for $\mathbb{E}P$ to be *exactly* Benford in cases where every individual probability measure $P(\omega)$ is distinctly non-Benford.

EXAMPLE 8.40. **(i)** Consider the random variable $X_T = U(0, T)$, where $T \in [1, 10)$ is itself a random variable with density

$$f_T(t) = \tfrac{2}{81}(10 - t)\,, \quad 1 \leq t < 10 \,.$$

Recall from Theorem 3.13 that $\max_{t \in [1,10)} |F_{S(X_T)}(t) - \log t| \geq 0.1344$ for every value of T. On the other hand, for every $x > 0$,

$$F_{\mathbb{E}X_T}(x) = \mathbb{P}(\mathbb{E}X_T \leq x) = \int_1^{10} \mathbb{P}(X_t \leq x)f_T(t)\,\mathrm{d}t = \int_1^{10} \min\left\{\tfrac{x}{t}, 1\right\}f_T(t)\,\mathrm{d}t$$
$$= \left\{ \begin{array}{ll} \big(\frac{20}{81}(\log e)^{-1} - \frac{2}{9}\big)x & \text{if } 0 < x < 1 \,, \\ \frac{1}{81}x^2 - \frac{20}{81}x\ln x + \frac{20}{81}(\log e)^{-1}x - \frac{19}{81} & \text{if } 1 \leq x < 10 \,, \\ 1 & \text{if } x \geq 10 \,, \end{array} \right.$$

which in turn yields, for $1 \leq t < 10$,

$$F_{S(\mathbb{E}X_T)}(t) = \sum_{k=0}^\infty \big(F_{\mathbb{E}X_T}(t10^{-k}) - F_{\mathbb{E}X_T}(10^{-k})\big)$$
$$= \tfrac{1}{81}\big(t^2 - 20t\ln t + \big(\tfrac{200}{9}(\log e)^{-1} - 2\big)t + 1 - \tfrac{200}{9}(\log e)^{-1}\big) \,.$$

Numerically, $\max_{t \in [1,10)} |F_{S(\mathbb{E}X_T)}(t) - \log t| = 0.03847$, which is significantly smaller than the corresponding value for each individual X_T; see Figure 8.7.

(ii) For an even more extreme example, consider the map $g : [1, 10] \to [1, 10]$, given by

$$g(x) = \tfrac{1}{2}(27 \log x - x + 3) \,.$$

Since $g(1) = 1$, $g(10) = 10$, and $g'(x) > 0$ for all x, the map g is in fact a homeomorphism of $[1, 10]$, and so is $h := g^{-1}$. Moreover, $g(x) > x$, and hence $h(x) < x$ for all $1 < x < 10$. With this, let T be a uniform random variable on $(1, 10)$, i.e., $T = U(1, 10)$, and let X_T be the discrete random variable with the two possible outcomes T and $h(T) < T$, attained with probabilities $\frac{1}{3}$ and $\frac{2}{3}$, respectively, that is,

$$\mathbb{P}(X_T = T) = \tfrac{1}{3} \,, \quad \mathbb{P}\big(X_T = h(T)\big) = \tfrac{2}{3} \,.$$

Clearly, no individual random variable X_T is Benford since, for every T, it has an atom of size $\frac{2}{3}$, and so $\sup_{t \in [1,10)} |F_{S(X_T)}(t) - \log t| \geq \frac{1}{3}$. On the other hand, for every $t \in [1, 10)$,

$$\mathbb{P}(\mathbb{E}X_T \leq t) = \tfrac{1}{3}\mathbb{P}(T \leq t) + \tfrac{2}{3}\mathbb{P}(h(T) \leq t) = \tfrac{1}{27}(t - 1) + \tfrac{2}{27}(g(t) - 1) = \log t \,,$$

which shows that $\mathbb{E}X_T$ is *exactly* Benford; see Figure 8.8. ✠

The next lemma shows that the limiting proportion of times that a sequence of P-random m-samples falls in a (Borel) set B is, with probability one, the average $\mathbb{P}$-value of the set B, i.e., the limiting proportion equals $\mathbb{E}P(B)$. Note that this is not simply a direct corollary of the classical Strong Law of Large Numbers since the random variables in the sequence are not necessarily independent; see Examples 8.36 and 8.37.

LEMMA 8.41. *Let P be an r.p.m., and let $(X_1, X_2, \ldots)$ be a sequence of P-random m-samples for some $m \in \mathbb{N}$. Then, for every $B \in \mathcal{B}^+$,*

$$\frac{\#\{1 \leq n \leq N : X_n \in B\}}{N} \xrightarrow{a.s.} \mathbb{E}P(B) \quad \text{as } N \to \infty \,.$$

PROOF. Fix $B \in \mathcal{B}^+$ and $j \in \mathbb{N}$, and let $Y_j = \#\{1 \leq i \leq m : X_{(j-1)m+i} \in B\}$. It is clear that

$$\lim_{N \to \infty} \frac{\#\{1 \leq n \leq N : X_n \in B\}}{N} = \frac{1}{m} \lim_{n \to \infty} \frac{1}{n} \sum_{j=1}^{n} Y_j \,, \qquad (8.14)$$

whenever the limit on the right exists. By Definition 8.35(i), given P_j, the random variable Y_j is binomially distributed with parameters m and $\mathbb{E}[P_j(B)]$; so, with probability one,

$$\mathbb{E}[Y_j] = \mathbb{E}\big[\mathbb{E}[Y_j | P_j]\big] = \mathbb{E}[mP_j(B)] = m\,\mathbb{E}P(B) \,, \qquad (8.15)$$

since P_j has the same distribution as P. By Definition 8.35(ii), the Y_j are independent. They are also uniformly bounded, since $0 \leq Y_j \leq m$ for all j, and

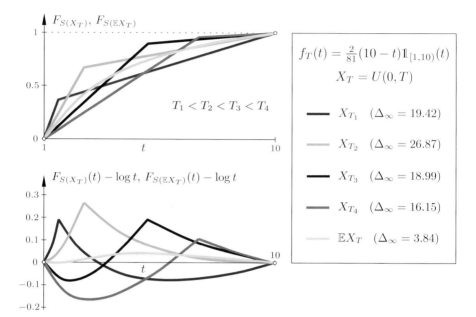

Figure 8.7: Although every value X_T of a random probability measure may be distinctively non-Benford, the distribution of $\mathbb{E}X_T$ may be much closer to Benford's law; see Example 8.40(i) and also Figure 3.8. Recall that Theorem 3.13 implies $\Delta_\infty \geq 13.44$ for every X_T.

hence $\sum_{j=1}^\infty \mathbb{E}\big[Y_j^2\big]/j^2 < +\infty$. Moreover, by (8.15) all Y_j have the same mean value $m\mathbb{E}P(B)$. Thus, by [36, Cor. 5.1],

$$\frac{1}{n}\sum_{j=1}^n Y_j \xrightarrow{a.s.} m\,\mathbb{E}P(B) \quad \text{as } n \to \infty, \tag{8.16}$$

and the conclusion follows from (8.14) and (8.16).   ∎

The assumption that each P_j is sampled exactly m times is not essential: The above argument can easily be modified to show that the same conclusion holds if the jth r.p.m. is sampled M_j times, where $(M_1, M_2, \ldots)$ is a sequence of independent, uniformly bounded $\mathbb{N}$-valued random variables that are also independent of the rest of the process.

The stage is now set to give a statistical limit law (Theorem 8.44 below) that is the central-limit-like theorem for significant digits mentioned above. Roughly speaking, this law says that if probability distributions are selected at random, and random samples are then taken from each of these distributions in such a way that the overall process is scale- or base-neutral, then the significant digit frequencies of the combined sample will converge to the logarithmic distribution. This theorem may help explain and predict the appearance of Benford's law in

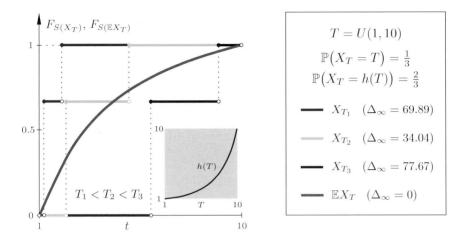

Figure 8.8: Every random variable X_T has only two possible outcomes, namely, T and $h(T) < T$, which occur with probabilities $\frac{1}{3}$ and $\frac{2}{3}$, respectively. Thus $\Delta_\infty \geq 33.33$ for every X_T. Yet the random variable $\mathbb{E}X_T$ is Benford; see Example 8.40(ii).

significant digits in mixtures of tabulated data such as the combined data from Benford's tables (as well as his individual dataset of numbers gleaned from newspapers), and the appearance of Benford's law in experiments designed to estimate the distribution of significant digits of all numbers on the World Wide Web; see Figure 1.6 and [97].

In order to draw any conclusions concerning Benford's law for the process of sampling from different distributions, clearly there must be *some* restriction on the underlying r.p.m. that generates the sampling procedure. Otherwise, if the r.p.m. is, say, $U(0,1)$ with probability one, then any resulting sequence of P-random m-samples will be i.i.d. $U(0,1)$, and hence a.s. not Benford, by Example 3.10(i). Similarly, it can easily be checked that sequences of P-random m-samples from the r.p.m.s in Example 8.33(i) and (ii) will not generate Benford sequences. A natural assumption to make concerning an r.p.m. in this context is that *on average* the r.p.m. is unbiased (i.e., invariant) with respect to changes in scale or base. (Analogous assumptions and conclusions concerning *sum-invariance* are left to the interested reader.) Recall that $\mathcal{S}$ denotes the significand σ-algebra (Definition 2.9).

DEFINITION 8.42. An r.p.m. P has *scale-unbiased significant (decimal) digits* if, for every significand event A, i.e., for every $A \in \mathcal{S}$, the expected value of $P(A)$ is the same as the expected value of $P(aA)$ for every $a > 0$, that is, if

$$\mathbb{E}P(aA) = \mathbb{E}P(A) \quad \text{for all } a > 0 \text{ and } A \in \mathcal{S},$$

or, equivalently, if the Borel probability measure $\mathbb{E}P$ has scale-invariant significant digits.

Similarly, P has *base-unbiased significant (decimal) digits* if, for every $A \in \mathcal{S}$ the expected value of $P(A)$ is the same as the expected value of $P(A^{1/n})$ for every $n \in \mathbb{N}$, that is, if

$$\mathbb{E}P(A^{1/n}) = \mathbb{E}P(A) \quad \text{for all } A \in \mathcal{S} \text{ and } n \in \mathbb{N},$$

i.e., if $\mathbb{E}P$ has base-invariant significant digits.

An immediate consequence of Theorems 5.3 and 5.13 is as follows.

PROPOSITION 8.43. *For every r.p.m. P, the following are equivalent:*

(i) *P has scale-unbiased significant digits;*

(ii) *$P \circ S^{-1}(\{1\}) = 0$ with probability one, or, equivalently, $\mathbb{E}P(\{10^k : k \in \mathbb{Z}\}) = 0$, and P has base-unbiased significant digits;*

(iii) *$\mathbb{E}P(A) = \mathbb{B}(A)$ for all $A \in \mathcal{S}$, i.e., $\mathbb{E}P$ is Benford.*

Random probability measures with scale- or base-unbiased significant digits are easy to construct mathematically; see Example 8.46 below. In real-life examples, however, scale- or base-unbiased significant digits should not be taken for granted. For instance, picking beverage-producing companies in Europe at random and looking at the metric volumes of samples of m products from each company is not likely to produce data with scale-unbiased significant digits, since the volumes in this case are probably closely related to liters. Conversion of the data to another unit such as gallons would likely yield a radically different set of first-digit frequencies. On the other hand, if species of mammals in Europe are selected at random and their metric volumes sampled, it seems more likely that this process is unrelated to the choice of human units.

The question of base-unbiasedness of significant digits is most interesting when the units in question are universally agreed upon, such as the numbers of things, as opposed to sizes. For example, picking cities at random and looking at the number of leaves of m-samples of trees from those cities is certainly less base-dependent than looking at the number of fingers of m-samples of people from those cities.

As will be seen in the next theorem, scale- or base-unbiasedness of an r.p.m. implies that sequences of P-random samples are Benford a.s. A crucial point in the definition of an r.p.m. P with scale- or base-unbiased significant digits is that individual realizations of P are not required to have scale- or base-invariant significant digits. In fact, it is often the case (see Benford's original data in [9] and Example 8.46 below) that a.s. *none* of the random probabilities has either scale- or base-invariant significant digits, and it is only on average that the sampling process does not favor one scale or base over another. Recall that $P \circ S^{-1}(\{1\}) = 0$ is the event $\{\omega \in \Omega : P(\omega)(S = 1) = 0\}$.

THEOREM 8.44 ([74]). *Let P be an r.p.m. and assume that P either has scale-unbiased significant digits or has base-unbiased significant digits, and that*

$P \circ S^{-1}(\{1\}) = 0$ *with probability one. Then, for all* $m \in \mathbb{N}$, *every sequence* $(X_1, X_2, \ldots)$ *of* P-*random* m-*samples is Benford with probability one, that is, for all* $t \in [1, 10)$,

$$\frac{\#\{1 \le n \le N : S(X_n) \le t\}}{N} \xrightarrow{a.s.} \log t \quad \text{as } N \to \infty \,.$$

PROOF. Assume first that P has scale-unbiased significant digits, i.e., the probability measure $\mathbb{E}P$ has scale-invariant significant digits. By Theorem 5.3, $\mathbb{E}P$ is Benford. Consequently, Lemma 8.41 implies that for every sequence $(X_1, X_2, \ldots)$ of P-random m-samples and every $t \in [1, 10)$,

$$\frac{\#\{1 \le n \le N : S(X_n) \le t\}}{N} = \frac{\#\left\{1 \le n \le N : X_n \in \bigcup_{k \in \mathbb{Z}} 10^k[1, t]\right\}}{N}$$
$$\xrightarrow{a.s.} \mathbb{E}P\left(\bigcup_{k \in \mathbb{Z}} 10^k[1, t]\right) = \log t \quad \text{as } N \to \infty \,.$$

Assume in turn that $P \circ S^{-1}(\{1\}) = 0$ with probability one, and that P has base-unbiased significant digits. Then

$$\mathbb{E}P\big(S^{-1}(\{1\})\big) = \int_{\Omega} P \circ S^{-1}(\omega)(\{1\}) \, \mathrm{d}\mathbb{P}(\omega) = 0 \,.$$

Hence $q = 0$ holds in (5.6) with P replaced by $\mathbb{E}P$, proving that $\mathbb{E}P$ is Benford, and the remaining argument is the same as before. ∎

COROLLARY 8.45. *If an r.p.m.* P *has scale-unbiased significant digits, then for every* $m \in \mathbb{N}$, *every sequence* $(X_1, X_2, \ldots)$ *of* P-*random* m-*samples, and every* $d \in \{1, 2, \ldots, 9\}$,

$$\frac{\#\{1 \le n \le N : D_1(X_n) = d\}}{N} \xrightarrow{a.s.} \log(1 + d^{-1}) \quad \text{as } N \to \infty \,.$$

A main point of Theorem 8.44 is that there are many natural sampling procedures that lead to the same logarithmic distribution. This may help explain how the different empirical evidence of Newcomb, Benford, Knuth, and Nigrini all led to the same law. It may also help explain why samples of numbers from newspaper front pages [9] and the World Wide Web [97] (see Figure 1.3) or extensive accounting data [112] often tend toward Benford's law, since in each of these cases, various distributions are being sampled in a presumably unbiased way. In a newspaper, perhaps the first article contains statistics about population growth, the second article about stock prices, the third about forest acreage. None of these individual distributions itself may be unbiased, but the mixture may well be.

Justification of the hypothesis of scale- or base-unbiasedness of significant digits in practice is akin to justification of the hypothesis of independence (and identical distribution) when applying the Strong Law of Large Numbers or the Central Limit Theorem to real-life processes: Neither hypothesis can be formally proved, yet in many real-life sampling procedures, they appear to be reasonable assumptions.

Many standard constructions of an r.p.m. automatically have scale- and base-unbiased significant digits, and thus satisfy Benford's law in the sense of Theorem 8.44.

EXAMPLE 8.46. Recall the classical Dubins–Freedman construction of an r.p.m. P described in Example 8.34. It follows from Theorem 4.2 and [51, Lem. 9.28] that $\mathbb{E}P$ is Benford, so P has scale- and base-unbiased significant digits. Note, however, that with probability one P will *not* have scale- or base-invariant significant digits. It is only *on average* that these properties hold, but, as demonstrated by Theorem 8.44, this is enough to guarantee that random sampling from P will generate Benford sequences a.s.

In the Dubins–Freedman construction, the fact that $F_P(10^{1/2})$, $F_P(10^{1/4})$, $F_P(10^{3/4})$, etc. are chosen *uniformly* from the appropriate intervals is not crucial: If Q is any probability measure on $(0,1)$, and the values of $F_P(10^{1/2})$, etc. are chosen independently according to an appropriately scaled version on Q, then, for the r.p.m. thus generated, $\mathbb{E}P$ will still be Benford, provided that Q is *symmetric* with respect to $\frac{1}{2}$; see [51, Thm. 9.29]. As a matter of fact, real-world processes often exhibit this symmetry in a natural way: Many data may be equally apt to be recorded using certain units or their reciprocals, e.g., in miles per gallon vs. gallons per mile, or Benford's "candles per watt" vs. "watts per candle." This suggests that the distribution of $\log S$ should be symmetric with respect to $\frac{1}{2}$.   ✠

Data having scale- or base-unbiased significant digits may be produced in many ways other than through random samples. If such data are combined with unbiased random m-samples then the result will again conform to Benford's law in the sense of Theorem 8.44. (Presumably, this is what Benford did when combining mathematical tables with numerical data from newspapers.) For example, consider the sequence (2^n), which may be thought of as the result of a periodic sampling from a (deterministic) geometric process. Since (2^n) is Benford, any mixture of this sequence with a sequence of unbiased random m-samples will again be Benford. Finally, it is important to note that many r.p.m.s and sampling processes do *not* conform to Benford's law, and hence necessarily are scale- and base-biased.

EXAMPLE 8.47. **(i)** Let P be the constant r.p.m. $P \equiv \delta_1$. Since $\mathbb{E}P = \delta_1$ has base-invariant significant digits, P has base-unbiased significant digits. Nevertheless, for every sequence $(X_1, X_2, \ldots)$ of P-random m-samples, the sequence of first significant digits is constant, namely, $D_1(X_n) \equiv 1$.

Similarly, if $P = \lambda_{0,1}$ with probability one, then $\mathbb{E}P = \lambda_{0,1}$ does not have scale- or base-invariant significant digits. Consequently, every sequence of P-random m-samples is an i.i.d. $U(0,1)$-sequence and hence not Benford, by Example 3.10(i).

(ii) The r.p.m.s considered in Example 8.33 do not have scale- or base-unbiased significant digits, simply because $\mathbb{E}P$ is not Benford.

(iii) As another variant of the classical construction in [51], consider the following way of generating an r.p.m. on $[1, 10)$: First let $X_{1/2}$ be uniform on $[1, 10)$ and set $F_P(X_{1/2}) = \frac{1}{2}$. Next let $X_{1/4}$ and $X_{3/4}$ be independent and uniform on $[1, X_{1/2})$ and $[X_{1/2}, 10)$, respectively, and set $F_P(X_{1/4}) = \frac{1}{4}$ and $F_P(X_{3/4}) = \frac{3}{4}$, etc. It follows from [51, Thm. 9.21] that

$$F_{\mathbb{E}P}(t) = \frac{2}{\pi} \arcsin \log t, \quad 1 \leq t < 10,$$

and hence $\mathbb{E}P$ is not Benford. By Proposition 8.43, therefore, the r.p.m. P has scale- and base-biased significant digits. ✠

Remark. The authors do not know of any results pertaining to the *rate* of convergence in Theorem 8.44 based, say, on m and properties of P; such results may be of practical use also (see Sections 10.1–3).

8.4 RANDOM MAPS

The purpose of this brief section is to illustrate and prove one simple basic theorem that combines deterministic aspects of Benford's law studied in Chapter 6 with stochastic considerations of the present chapter. Specifically, it is shown how applying randomly selected deterministic maps successively may generate Benford sequences with probability one. Random maps constitute a wide and intensely studied field, and for stronger results than the ones discussed here the interested reader is referred to [10].

Recall that for any (deterministic or random) sequence $(f_1, f_2, \ldots)$ of maps mapping $\mathbb{R}$ or a subset of $\mathbb{R}$ into itself and for every $x_0 \in \mathbb{R}$, $(f^n(x_0))$ denotes the sequence $(f_1(x_0), f_2(f_1(x_0)), \ldots)$. The following example illustrates several of the key Benford ideas regarding random maps.

EXAMPLE 8.48. Let $f : \mathbb{R}^+ \to \mathbb{R}^+$ be the map $f(x) = \sqrt{x}$. The sequence $(f^n(x_0))$ is not Benford for any $x_0 > 0$ because $f^n(x_0) = x_0^{2^{-n}} \to 1$ as $n \to \infty$. More generally, consider the randomized map

$$f(x) = \begin{cases} \sqrt{x} & \text{with probability } p, \\ x^3 & \text{with probability } 1 - p, \end{cases}$$

and assume that, at each step, the iteration of f is independent of the entire past process. If $p = 1$, this is simply the constant map $f(x) = \sqrt{x}$, and hence for every $x_0 \in \mathbb{R}^+$, the sequence $(f^n(x_0))$ is not Benford. On the other hand, if $p = 0$ then Theorem 6.23 and Corollary 6.28 imply that, for almost every $x_0 \in \mathbb{R}^+$, the sequence $(f^n(x_0))$ is Benford. It is plausible to expect that the latter situation persists for small $p > 0$. As the following theorem shows, this is indeed the case even in situations where the map $\sqrt{x}$ occurs more than half the time: If

$$p < \frac{\log 3}{\log 2 + \log 3} = 0.6131, \tag{8.17}$$

then, for a.e. $x_0 \in \mathbb{R}^+$, the (random) sequence $(f^n(x_0))$ is Benford with probability one.   ✠

THEOREM 8.49 ([10]). *Let $W_1, W_2, \ldots$ be i.i.d. positive random variables, and assume that $\log W_1$ has finite variance, i.e., $\mathbb{E}\big[(\log W_1)^2\big] < +\infty$. For the sequence (f_n) of random maps given by $f_n(x) = x^{W_n}$ and a.e. $x_0 \in \mathbb{R}^+$, the sequence $(f^n(x_0))$ is Benford with probability one or zero, depending on whether $\mathbb{E}[\log W_1] > 0$ or $\mathbb{E}[\log W_1] \leq 0$.*

PROOF. For every $x \in \mathbb{R}^+$ and $n \in \mathbb{N}$,

$$\log\big(f_n \circ \ldots \circ f_1(x)\big) = \Big(\prod\nolimits_{j=1}^n W_j\Big)\log x = 10^{B_n}\log x \,,$$

where $B_n = \sum_{j=1}^n \log W_j$. Assume first that $\mathbb{E}[\log W_1] > 0$. By the Strong Law of Large Numbers, $B_n/n \xrightarrow{a.s.} \log W_1$ as $n \to \infty$, and it can be deduced from Proposition 4.14 that, with probability one, the sequence $(10^{B_n}y)$ is u.d. mod 1 for a.e. $y \in \mathbb{R}^+$. Since $\log$ maps the family of (Lebesgue) nullsets into itself, with probability one, $(f^n(x_0))$ is Benford for a.e. $x_0 \in \mathbb{R}^+$. More formally, with $(\Omega, \mathcal{A}, \mathbb{P})$ denoting the underlying (abstract) probability space, there exists $\Omega_1 \in \mathcal{A}$ with $\mathbb{P}(\Omega_1) = 1$ such that for every $\omega \in \Omega_1$, the sequence $(f^n(x_0))$ is Benford for all $x_0 \in \mathbb{R}^+ \backslash B_\omega$, where $B_\omega \in \mathcal{B}$ with $\lambda(B_\omega) = 0$. Denote by $N \subset \mathbb{R}^+ \times \Omega$ the set of all (x_0, ω) for which $(f^n(x_0))$ is *not* Benford, and let

$$N_x = \{\omega \in \Omega : (x, \omega) \in N\}\,, \quad x \in \mathbb{R}^+\,,$$
$$N^\omega = \{x \in \mathbb{R}^+ : (x, \omega) \in N\}\,, \quad \omega \in \Omega\,.$$

Then $N_x \in \mathcal{A}$ and $N^\omega \in \mathcal{B}$ for all $x \in \mathbb{R}^+$ and $\omega \in \Omega$, respectively, and $\lambda(N^\omega) = 0$ for all $\omega \in \Omega_1$. By Fubini's Theorem,

$$0 = \int_\Omega \lambda(N^\omega)\,\mathrm{d}\mathbb{P}(\omega) = \int_{\mathbb{R}^+ \times \Omega} \mathbb{1}_N\,\mathrm{d}(\lambda \times \mathbb{P}) = \int_{\mathbb{R}^+} \mathbb{P}(N_x)\,\mathrm{d}\lambda(x)\,,$$

showing that $\mathbb{P}(N_x) = 0$ for a.e. $x \in \mathbb{R}^+$. Thus $\mathbb{P}\big((f^n(x_0)) \text{ is Benford}\big) = 1$ for a.e. $x_0 \in \mathbb{R}^+$.

Next assume that $\mathbb{E}[\log W_1] < 0$. In this case, $f_n \circ \ldots \circ f_1(x) \xrightarrow{a.s.} 1$ as $n \to \infty$ for every $x > 0$, and hence $(f^n(x))$ is not Benford. (Note, however, that $(f_n \circ \ldots \circ f_1(x) - 1)$ may be Benford.)

Finally, it remains to consider the case $\mathbb{E}[\log W_1] = 0$. It follows from the Law of the Iterated Logarithm that, for every $t \in \mathbb{R}^+$,

$$\limsup\nolimits_{N \to \infty} \frac{\#\{1 \leq n \leq N : B_n \leq t\}}{N} \geq \frac{1}{2} \quad \text{with probability one}\,.$$

Clearly, this implies $\mathbb{P}\big((f^n(x_0)) \text{ is Benford}\big) = 0$ for every $x_0 \in \mathbb{R}^+$.   ∎

EXAMPLE 8.50. **(i)** For the random map in Example 8.48,

$$\mathbb{P}\left(W_1 = \tfrac{1}{2}\right) = p = 1 - \mathbb{P}(W_1 = 3),$$

and the condition $\mathbb{E}[\log W_1] = -p \log 2 + (1 - p) \log 3 > 0$ is equivalent to (8.17). Note that $\mathbb{E}[\log W_1] > 0$ is not generally equivalent to the equally plausible condition $\mathbb{E}[W_1] > 1$; in the present example, the latter simply reads $p < \tfrac{4}{5}$.

(ii) Let $(f_1, f_2, \ldots)$ be the sequence of random maps $f_n(x) = x^{10^{2n + Z_n}}$, where $Z_1, Z_2, \ldots$ are i.i.d. standard Cauchy random variables. Note that Theorem 8.49 does not apply since $\mathbb{E}\left[(\log W_1)^2\right] = \mathbb{E}\left[(2n + Z_1)^2\right] = +\infty$. Nevertheless, it follows from $B_n = n(n + 1) + \sum_{j=1}^{n} Z_j$ and [36, Thm. 5.2.2] that $B_n / n^2 \xrightarrow{a.s.} 1$ as $n \to \infty$. The latter is enough to deduce from Proposition 4.14 that, with probability one, $(10^{B_n} y)$ is u.d. mod 1 for a.e. $y \in \mathbb{R}$. The same argument as that in the above proof shows that $\mathbb{P}\left(\left(f^n(x_0)\right) \text{ is Benford}\right) = 1$ for a.e. $x_0 \in \mathbb{R}^+$. Thus the conclusions of Theorem 8.49 may hold under weaker assumptions. Statements in the spirit of Theorem 8.49 are true also for random maps more general than monomials [10]. ✠

Chapter Nine

Finitely Additive Probability and Benford's Law

Benford's law, as was seen in earlier chapters, arises naturally in a wide variety of standard mathematical contexts. It is the purpose of this short chapter to illustrate that Benford's law is robust even with regard to basic underlying mathematical hypotheses. Recall that throughout the earlier chapters of this book, Benford's law has been studied exclusively in the standard countably additive measure-theoretic setting of probability, where a probability measure $\mathbb{P}$ is a real-valued function defined on a σ-algebra $\mathcal{A}$ of subsets of a set Ω satisfying the following three basic axioms:

$$\mathbb{P}(A) \geq 0 \quad \text{for all } A \in \mathcal{A}\,; \tag{9.1}$$

$$\mathbb{P}(\Omega) = 1\,; \tag{9.2}$$

$$\mathbb{P}\left(\bigcup\nolimits_{n\in\mathbb{N}} A_n\right) = \sum\nolimits_{n\in\mathbb{N}} \mathbb{P}(A_n) \quad \text{for all disjoint } A_1, A_2, \ldots \in \mathcal{A}\,. \tag{9.3}$$

Working within this countably additive framework has several important implications for the theory of Benford's law. First, the σ-algebra $\mathcal{A}$ in (9.1) usually does not include all subsets A of Ω, so there are sets $C \subset \mathbb{R}$ whose Benford probability $\mathbb{B}(C)$ does not exist; see Definition 3.7. Second, there are no countably additive Benford probability distributions on the positive integers $\mathbb{N}$: If each $n \in \mathbb{N}$ had probability zero then the total probability of $\mathbb{N}$ would be zero, whereas if some integer n had strictly positive probability, that would contradict the basic equation (1.2), which implies that the probability of a number having exactly the same first m significant digits as n goes to zero as $m \to \infty$; more formally,

$$\mathbb{P}(\{n\}) \leq \mathbb{P}\big(\{k \in \mathbb{N} : D_j(k) = D_j(n) \text{ for } j = 1, \ldots, m\}\big) \overset{m\to\infty}{\longrightarrow} 0\,,$$

contradicting $\mathbb{P}(\{n\}) > 0$. There are even no countably additive Benford probability measures on the significand σ-algebra $\mathcal{S}_{\mathbb{N}}$ on $\mathbb{N}$ defined in Example 2.14, as was shown prior to Example 3.9. Similarly, there are no such Benford probability measures on the power σ-algebra (i.e., the set of all subsets) of $\mathbb{T}^+$, the set of positive terminating decimals (10-adic numbers) defined by

$$\mathbb{T}^+ = \big\{t \in \mathbb{R}^+ : 10^n t \in \mathbb{N} \text{ for some } n \in \mathbb{N}\big\}\,.$$

Raimi wrote that $\mathbb{T}^+$ seemed "particularly appropriate as the model for tabular data" [127, p. 530], since "actual tables of constants are generally written (or approximated) as decimal fractions" [126, p. 347]. Thus, if one views the tables of constants in Benford's original data or the "stock of tabular data in the world's libraries" [128, p. 217] as a sample from an underlying universal distribution on all of $\mathbb{T}^+$, then there is no countably additive probability explanation for this model.

Generalizing probabilities to the finitely additive framework, on the other hand, guarantees the existence of Benford probability distributions that are defined on *all* subsets of positive numbers. Similarly, there are finitely additive Benford probability distributions defined on the class of all subsets of $\mathbb{N}$, and on all subsets of $\mathbb{T}^+$. It is perhaps reassuring to know that such Benford probability distributions do exist in some natural sense, even though they may not exist in the theory of countably additive probability. This chapter records, without proof, several of the main theorems pertinent to a theory of Benford's law in the context of finitely additive probability, and points the interested reader to further references on that theory in the literature. The chapter may be read independently of the rest of the book, or, of course, be skipped entirely.

9.1   FINITELY ADDITIVE PROBABILITIES

A *finitely additive probability measure* $\mathbb{P}^*$ differs from the traditional probability measure $\mathbb{P}$ defined in (9.1)–(9.3) in two important respects. First, $\mathbb{P}^*$ is defined on *all* subsets of Ω, and second (the finitely additive part), the fundamental conditions (9.1) and (9.2) are again required, but (9.3) is replaced by its much simpler special case

$$\mathbb{P}^*(A \cup B) = \mathbb{P}^*(A) + \mathbb{P}^*(B) \quad \text{if } A \cap B = \varnothing, \, A, B \subset \Omega \, . \qquad (9.3')$$

From a philosophical standpoint, finitely additive probability measures have sometimes been viewed as preferable to countably additive ones since (9.3′) is simpler and more natural than (9.3). L. Dubins and L. Savage reported that "De Finetti always insisted that countable additivity is not an integral part of the probability concept but is rather in the nature of a regularity hypothesis," and they themselves viewed "countably additive measures much as one views analytic functions — as a particularly important case" [52, p. 10]. Moreover, since it is impossible to verify (9.3) in a finite number of operations, it is impossible to distinguish between a finitely additive and a countably additive probability measure in practice [125].

There are, of course, both mathematical advantages and disadvantages to replacing (9.3) with (9.3′). One practical advantage of using finitely additive probabilities is that they are defined on all subsets of Ω, thus eliminating tedious, cumbersome, and often ultimately unnecessary measurability technicalities typically encountered in the countably additive theory. On the other hand, without (9.3), many of the most basic analytic tools in traditional probability — tools repeatedly used in the proofs in the rest of this book — are not available. For example, the fundamental continuity theorem of probability,

$$\mathbb{P}\left(\bigcup_{n\in\mathbb{N}} A_n\right) = \lim_{N\to\infty} \mathbb{P}\left(\bigcup_{n=1}^{N} A_n\right),$$

may fail for finitely additive probability measures, so many basic limiting arguments may fail as well.

Recall that in standard (i.e., countably additive) measure theory, a measure on a σ-algebra $\mathcal{A}$ on a set Ω is often constructed as follows: First a countably additive nonnegative set function (also referred to as a *pre-measure*) is defined only on an *algebra* $\mathcal{F}$ on Ω. (A non-empty family of subsets of Ω is an *algebra* if it is closed under complements and *finite* unions; note that this notion is reminiscent of, but weaker than that of, a σ-algebra, because the latter, by definition, is closed even under countable unions.) This pre-measure is then extended to a bona fide measure on $\mathcal{A} = \sigma(\mathcal{F})$, the σ-algebra on Ω generated by $\mathcal{F}$. Under mild conditions, this procedure yields a unique measure on $\sigma(\mathcal{F})$, but this σ-algebra does not generally include all subsets of Ω; e.g., see [27]. A classic example is the extension of the notion of length from finite intervals to $\mathcal{B}$, the family of all Borel subsets of $\mathbb{R}$, but not to *all* subsets of $\mathbb{R}$. Finitely additive measures defined on an algebra of sets $\mathcal{F}$, on the other hand, extend to finitely additive measures defined on all subsets of Ω, and this allows strong conclusions regarding Benford's law. Note that in the following key result due to A. Horn and A. Tarski, recorded here for ease of reference, the extension need not be unique, in contrast to the countably additive context.

PROPOSITION 9.1. [81, Thm. 1.22] *Let $\Omega \neq \varnothing$ be a set and $\mathcal{F}$ an algebra on Ω. Every finitely additive probability measure defined on $\mathcal{F}$ may be extended to a finitely additive probability measure on all subsets of Ω.*

The next three examples illustrate the power and generality of Proposition 9.1, and contrast it with the countably additive extension theory. These ideas help illustrate some of the key differences between finitely and countably additive probability measures, and will be applied directly to Benford's law in the next section. In analogy to their countably additive counterparts, finitely additive probability measures on $\Omega \subset \mathbb{R}$ are denoted by P^*, Q^*, etc. throughout.

EXAMPLE 9.2. There exists a finitely additive probability measure P^* defined on all subsets of $[0,1]$ that agrees with Lebesgue measure λ on all Borel sets, that is, $P^*([a,b]) = \lambda([a,b]) = b - a$ for all intervals $[a,b] \subset [0,1]$ and, more generally, $P^*(B) = \lambda(B)$ for all Borel subsets B of $[0,1]$. To see this, simply note that Lebesgue measure λ on $[0,1]$ is a finitely (in fact, countably) additive probability measure defined on the algebra (in fact, σ-algebra) $\mathcal{B}[0,1]$ on $[0,1]$. Hence by Proposition 9.1, there is a finitely additive probability measure P^*, defined on all subsets of $[0,1]$, that agrees with λ on all Borel subsets of $[0,1]$. (In fact, it can be shown that P^* may even be chosen to be *countably* additive — provided that large (measurable) cardinals exist [83].)   ✠

EXAMPLE 9.3. (Attributed in [52] to B. De Finetti.) There exists a finitely additive probability measure Q^* defined on all subsets of $[0,1] \cap \mathbb{Q}$, the set of rational numbers in $[0,1]$, such that

$$Q^*([a,b] \cap \mathbb{Q}) = b - a \quad \text{for all } 0 \leq a \leq b \leq 1. \tag{9.4}$$

To see this, apply Proposition 9.1 to extend the finitely additive probability measure Q^* satisfying (9.4), originally defined only on the algebra of finite (disjoint) unions of intervals $[a,b] \cap \mathbb{Q}$ with $[a,b] \subset [0,1]$. In contrast to Example 9.2, however, it clearly is impossible to make Q^* countably additive, because $[0,1] \cap \mathbb{Q}$ is countable and by (9.4) each singleton must have measure zero. ✠

Recall the notion of (*natural*) *density* ρ of sets $C \subset \mathbb{N}$ appearing in Sections 3.1 and 5.1, defined as $\rho(C) = \lim_{N \to \infty} \#\{n \leq N : n \in C\}/N$ whenever this limit exists. The next example, not an immediate consequence of Proposition 9.1, relates this concept of density to finitely additive probability measures.

EXAMPLE 9.4 ([84]). There is a finitely additive probability measure P^* defined on all subsets of $\mathbb{N}$ that agrees with density on all subsets of $\mathbb{N}$ that have density, that is $P^*(A) = \rho(A)$ for all $A \subset \mathbb{N}$ for which $\rho(A)$ exists. In particular, $P^*(A) = 0$ for every finite set A of positive integers, and every arithmetic progression $\{l + m, l + 2m, \ldots\} = \{l + nm : n \in \mathbb{N}\}$ with $l \in \mathbb{N}_0$ and $m \in \mathbb{N}$ has P^*-probability m^{-1}. ✠

The theory of finitely additive measures is a well-established subfield of functional analysis, since there is a natural one-to-one correspondence between finitely additive measures and certain linear functionals; for a detailed treatise of this material, the reader is referred to Chapter III of the classic text [53].

9.2   FINITELY ADDITIVE BENFORD PROBABILITIES

From a historical standpoint, it is in the framework of finitely additive probability measures that the first rigorous mathematical results on Benford's law appeared. In their work on Benford's law, R. Bumby and E. Ellentuck wrote that among the conditions (9.1)–(9.3), "countable additivity [(9.3)] seems least necessary," and they replaced it in their theorems with the finite additivity hypothesis (9.3′) [30, p. 34]. Similarly, Raimi wrote "There is no reason a priori why all things subjectively regarded as 'probabilities' should be countably additive, though there are of course many mathematical conveniences," and he concluded that finite additivity is a "closer model to reality" [126, pp. 344–5].

Moreover, finitely additive probabilities also help overcome some of the "shortcomings" of countably additive probabilities with respect to Benford's law. For example, the countably additive scale-invariance characterization (Theorem 5.3) holds only for a proper subset of the Borel σ-algebra (namely, the significand σ-algebra of Definition 2.9), and that subset does not even include intervals of real numbers. As Knuth pointed out, there is no scale-invariant countably additive probability measure on the Borel subsets of the positive real numbers, i.e., on $\mathcal{B}^+$, since existence of such a measure would imply the existence of a unique smallest median, and rescaling would change that median [90]. Thus hypothesizing the existence of a scale-invariant countably additive probability measure on $\mathcal{B}^+$, or even on $\mathcal{B}$ (as in [124]), leads to a vacuous conclusion, in particular with regard to a Benford distribution.

In this section, certain aspects of finitely additive probability theory summarized in the previous section will be applied to Benford's law. Even without the standard assumption of countable additivity, direct analogues of some of the most important and basic conclusions hold in the finitely additive setting as well. As described below, analogues of the fundamental uniform distribution characterization (Theorem 4.2) also hold in the finitely additive framework, as does the analogue of the countably additive conclusion that scale-invariance implies Benford's law (Theorem 5.3).

Recall the definitions of the significand function S and the significand σ-algebra $\mathcal{S}$, Definitions 2.3 and 2.9, respectively. A natural finitely additive generalization of the countably additive notion of a Benford probability measure (Definition 3.4) is as follows.

DEFINITION 9.5. A finitely additive probability measure P^* on a non-empty set $E \subset \mathbb{R}^+$ is *Benford* if it is defined on all subsets of E, and satisfies

$$P^*\big(\{x \in E : S(x) \le t\}\big) = \log t \quad \text{for all } t \in [1, 10) \,.$$

Note that if P^* is Benford then in particular

$$P^*\big(\{x \in E : D_1(x) = d\}\big) = \log(d+1) - \log d \quad \text{for all } d \in \{1, 2, \ldots, 9\} \,.$$

Recall from Section 3.3 that there are no countably additive Benford probability measures defined on all subsets of the positive integers $\mathbb{N}$ or on the set of positive terminating decimals $\mathbb{T}^+$. Nor are there countably additive probability measures on $\mathcal{B}^+$ that are scale-invariant. In contrast to this fact, however, there exist *finitely* additive Benford probability measures defined on all subsets of $\mathbb{N}$ and $\mathbb{T}^+$, respectively, as well as finitely additive Benford probability measures defined on *all* subsets of the significand range $[1, 10)$.

EXAMPLE 9.6. There is a finitely additive Benford probability measure defined on all subsets of the positive integers $\mathbb{N}$. To see this, let P^* be the finitely additive probability measure on the algebra of finite (disjoint) unions of subsets of $\mathbb{N}$ of the form $\{n \in \mathbb{N} : S(n) \in [a, b)\}$, defined by

$$P^*\big(\{n \in \mathbb{N} : S(n) \in [a, b)\}\big) = \log b - \log a \quad \text{for all } 1 \le a \le b < 10 \,,$$

and then, via Proposition 9.1, extend P^* to a finitely additive Benford probability measure defined on all subsets of $\mathbb{N}$.   ✠

There also exist finitely additive Benford probability measures defined on all subsets of the positive terminating decimals $\mathbb{T}^+$. One such example, of course, is to extend P^* of Example 9.6 to all subsets of $\mathbb{T}^+$ by setting $P^*(\{t\}) = 0$ for all $t \in \mathbb{T}^+ \setminus \mathbb{N}$. Another, which is quite different from this one, is the following.

EXAMPLE 9.7. Consider the finitely additive probability measure Q^* on the algebra of finite unions of intervals of $\mathbb{T}^+$ of the form $\bigcup_{k \in \mathbb{Z}} 10^k[a, b) \cap \mathbb{T}^+$ defined by

$$Q^* \left(\bigcup_{k \in \mathbb{Z}} 10^k[a, b) \cap \mathbb{T}^+ \right) = \log b - \log a \quad \text{for all } 1 \leq a \leq b < 10,$$

and extend Q^* to the power set of $\mathbb{T}^+$ by means of Proposition 9.1. ✠

Recall from Example 3.6(i) the canonical countably additive Benford probability measure P_0 with density function $t^{-1} \log e$ on $[1, 10)$, that is,

$$P_0(B) = \int_B t^{-1} \log e \, dt \quad \text{for all } B \in \mathcal{B}[1, 10). \tag{9.5}$$

Countably additive measure theory does not extend absolutely continuous probability distributions (i.e., distributions with densities) to all subsets of $\mathbb{R}$, and so in particular does not extend (9.5) to all subsets of $[1, 10)$. Using finitely additive probabilities, on the other hand, allows extension of the canonical Benford probability measure P_0 in (9.5) to all subsets of $[1, 10)$, as the next example shows.

EXAMPLE 9.8. There is a finitely additive Benford probability measure P^* on $[1, 10)$ that extends (9.5) to all subsets of $[1, 10)$, that is, $P^*(B) = P_0(B)$ for all Borel subsets B of $[1, 10)$. To see this, simply apply Proposition 9.1 to extend the countably additive P_0 in (9.5) to a finitely additive P^* defined on all subsets of $[1, 10)$. ✠

The next result is a finitely additive analogue due to T. Jech of both the uniform distribution characterization (Theorem 4.2) of Benford's law and the corresponding characterization by means of Fourier coefficients, obtained via Theorem 4.2 and Lemma 4.20(vi). The hypothesis that the probability is defined on all subsets of E can easily be weakened via Proposition 9.1. Note that, as explained in [30, 52], if a finitely additive probability measure P^* is defined on all subsets of E, then the linear and order-preserving nature of the integral uniquely determine the value of $\int f \, dP^*$ for all bounded functions $f : E \to \mathbb{R}$; also recall that $\langle x \rangle = x - \lfloor x \rfloor$.

THEOREM 9.9 ([83]). *Let P^* be a finitely additive probability measure defined on all subsets of a set $E \subset \mathbb{R}^+$. Then the following are equivalent:*

(i) *P^* is Benford;*

(ii) *$\int_E f(\langle \log x \rangle) \, dP^*(x) = \int_0^1 f(s) \, ds$ for every Riemann-integrable function $f : [0, 1] \to \mathbb{R}$;*

(iii) *$\int_E e^{2\pi i k \log x} \, dP^*(x) = 0$ for every integer $k \neq 0$.*

The next theorem contains a finitely additive analogue of Theorem 5.3, the characterization of Benford's law as the unique scale-invariant countably additive probability measure defined on the significand σ-algebra. Although there are no scale- or translation-invariant countably additive probability measures on $\Omega = \mathbb{N}$, $\mathbb{T}^+$, or $\mathbb{R}^+$, there are finitely additive probability measures on all three sets that are both scale- and translation-invariant, and all of those are necessarily Benford. Though defined on all subsets, the finitely additive version, on the other hand, is not unique.

THEOREM 9.10 ([126, 127]). *There exist finitely additive probability measures* P^* *defined on all subsets of* $\mathbb{R}^+$ *such that, for all* $A \subset \mathbb{R}^+$ *and all* $a > 0$, *both*

$$P^*(aA) = P^*(A) \quad (\text{scale-invariance}) \tag{9.6}$$

and

$$P^*(a + A) = P^*(A) \quad (\text{translation-invariance}). \tag{9.7}$$

Moreover, every finitely additive probability P^* *satisfying* (9.6) *is Benford.*

In addition to (9.6) and (9.7) it may also be required in Theorem 9.10 that $P^*((0, t)) = 0$ for all $t \in \mathbb{R}^+$, thus "avoiding the philosophically awkward [midpoint] of all physical constants since the resulting measure is concentrated in the neighborhood of infinity [or zero]" [127, p. 530].

The analogous conclusion that scale-invariance of a finitely additive probability measure on $\mathbb{N}$ implies that the measure is Benford is given in [30]. The scale- and translation-invariance conditions (9.6) and (9.7) in Theorem 9.10 can be replaced (see [38] or [83]) by the single condition

$$P^*(2A \cup (1 + 2A)) = P^*(A) \quad \text{for all } A \subset \mathbb{N}.$$

As observed by Raimi, in view of Theorem 9.10, the additional translation-invariance requirement (9.7) together with scale-invariance (9.6), "means that an affine change of scale, as from Fahrenheit to Celsius, will preserve Benford's law" [127, p. 530]. R. Scozzafava [145] provides an additional finitely additive explanation of Benford's law based on a conditional probability concept ("non-conglomerability") of De Finetti.

Although the finitely additive Benford theorems in this chapter have conclusions identical in spirit to those in the countably additive theory (e.g., that scale-invariance implies Benford's law), their proofs are quite different. The proof of Theorem 9.10, for example, is a "fairly technical exercise in the theory of amenable semigroups" [126, p. 346], and the existence of scale-invariant finitely additive probability measures is "guaranteed by the Markov–Kakutani fixed point theorem" [30, p. 40]. As noted by Raimi, this aspect of finitely additive probability theory also relies on results in the theories of divergent sequences, invariant measures, and the Stone-Čech compactification [126, p. 349].

Chapter Ten

Applications of Benford's Law

The purpose of this chapter is to present a brief overview of practical applications of Benford's law, thereby complementing the theory developed in previous chapters. While the interplay between theory and applications of Benford's law has generally proved very fruitful in both directions, several applications in particular have played a prominent role in the development of the theory. For example, in the use of Benford's law to help detect tax fraud, the discovery by Nigrini that many tables of data are approximately sum-invariant [112] inspired the sum-invariance characterization of the law (Theorem 5.18). The question "Do dynamical systems follow Benford's law?", raised in [154] and subsequently addressed in [151], led directly to the discovery of Theorems 6.13 and 6.23, and indirectly to most of the results in Chapters 6 and 7. Observations about the prevalence and implications of Benford's law in scientific computing [90] inspired Theorem 6.35 as well as Propositions 7.34 and 7.35. Empirical evidence of Benford's law in the stock market [98] helped motivate the study of products of random variables in the context of Benford's law (Theorems 8.15 and 8.16). In his original 1938 paper [9], Benford combined data from radically different datasets, and this led to the discovery of the mixing-of-distributions theorem (Theorem 8.44). Conversely, it is hoped that the new theoretical results presented in earlier chapters, notably those relevant to (deterministic or random) dynamical processes, will serve as catalysts for further applications.

While the main goal of this book has been to develop a solid and comprehensive theoretical basis for the appearance of Benford's law in a broad range of mathematical settings, it is the empirical ubiquity of this logarithmic distribution of significant digits that has captured the interest and imagination of a surprisingly wide audience. From natural science to medicine, from social science to economics, from computer science to theology, Benford's law, even in its most basic form, provides a simple analytical tool that invites anyone with numerical tables to look for this easily recognizable pattern in their own data. A glance at the online database [24] quickly reveals the magnitude, breadth, and recent growth of interest in both applications and theory of the significant-digit phenomenon.

In contrast to the rest of the book, this chapter is necessarily expository and informal. It has been organized into a handful of ad hoc categories, which the authors hope will help illuminate the main ideas. None of the conclusions of the experiments or data presented here have been scrutinized or verified by the authors of this book, since the intent here is not to promote or critique any specific application. Rather the goal is to offer a representative cross-section of the related scientific literature, in the hopes that this might continue to facilitate research in both the theory and practical applications of Benford's law. For useful guidelines on conducting and applying significant-digit analyses in general, the reader may also wish to consult [66].

10.1 FRAUD DETECTION

The most well-known and widespread application of Benford's law is in the field of forensic auditing, in particular, in the statistical detection of accounting fraud, where "fraud" means both data fabrication (inventing data) and data falsification (manipulating and/or altering data values). The underlying idea is this: If certain classes of valid financial datasets have been observed to closely follow a Benford distribution, then fabricated or falsified datasets can sometimes be identified simply by comparing their leading digits against the logarithmic distribution.

The main impetus for this line of research was the discovery by Nigrini [113, 114] that certain verified IRS tax data follow Benford's law very closely, but fraudulent data often do not. It is widely accepted that authentic data are difficult to fabricate [75]. Thus, standard goodness-of-fit tests, such as a chi-squared test for first and/or second significant digits, or the Ones-scaling test (Example 5.9), provide simple "red flag" tests for fraud. Whether given data are close to Benford's law or are not close proves nothing, but if true data approximately follow Benford's law, then a poor fit raises the level of suspicion, at which time independent (non-Benford) tests or monitoring may be applied. For cases where specific deviations from true data might be expected, such as in tax returns when data are falsified in order to lower tax liabilities, specialized one-sided goodness-of-fit tests have also been developed [114].

The first success of this method was reported by the Brooklyn District Attorney's office in New York, where the chief financial investigator used an elementary goodness-of-fit test to the first-digit law (1.1) to identify and successfully prosecute seven companies for theft [26]. Similar tests are now routinely employed by many American state revenue services, as well as by the federal IRS and many foreign government tax agencies [114].

The detection of fraud in macroeconomic data can also benefit from similar testing. R. Abrantes-Metz et al. studied data from the Libor (London Interbank Offered Rate), and applied goodness-of-fit to Benford's law as a statistical test to identify specific industries and banks with competition issues such as price-fixing rate manipulation or collusion [1]. Similarly, T. Michalski and G. Stoltz examined financial data from the International Monetary Fund, and used Benford's law to "find evidence supporting the hypothesis that countries at times misreport their economic data strategically" [107, p. 591].

Due to the simplicity and effectiveness of goodness-of-fit tests based on Benford's law, investigators from a wide variety of domains are now routinely including them in their arsenal of fraud detection techniques. In fact, workshops on Benford's law are appearing in annual fraud conferences such as that of the Association of Certified Fraud Examiners [120]. In the field of medicine, applications range from assessing clinical trial data for new drugs [32] to identifying fraudulent scientific publications in the field of anesthesiology [72]. Political scientists use modified goodness-of-fit tests to study the validity of voting results. W. Mebane found that while low-level vote counts rarely have first digits that satisfy Benford's law, they often have second digits that are a close fit, and using this observation, he applied chi-squared statistical tests to challenge the validity of the 2009 Presidential election results in Iran [105]. Similar analyses have also been performed for voting data from Argentina [33], Germany [28], and Venezuela [153], but the reader should note that there is some debate about the validity of applying tests based on Benford's law to election results [43, 106, 149].

The growing field of image and computer graphics forensics uses Benford's law to detect fraud in visual information processing. E. Del Acebo and M. Sbert report that light intensities in certain classes of natural images and in synthetic images generated using "physically realistic" methods both follow Benford's law closely, whereas other types of images fail to do so [44]. D. Fu et al. developed a "novel statistical model based on Benford's law" for analysis of JPEG images, and observed that this tool is effective for detection of compression and double compression of the images [61]. J.-M. Jolion examined the frequencies of digits in both the gradients and the Laplace transforms of "non particular" images (i.e., images that are not pure noise, fine texture, or synthetic), and compared them to those of artificial images; the former obey Benford's law, whereas the latter do not [85]. F. Pérez-González et al. applied goodess-of-fit tests to the discrete cosine transforms of images to help determine whether certain natural digital images contain hidden messages [122], and B. Xu et al. used a similar technique to help separate computer generated or "tampered" images from photographic images. They concluded that tests based on Benford's law are comparable to other state-of-the-art methods, but have considerably lower dimension and much lower computational complexity [160].

10.2   DETECTION OF NATURAL PHENOMENA

Just as goodness-of-fit to Benford's law is used to help detect fraud in tax and other human-generated data, it is also being used to detect changes in natural phenomena. The general approach here is essentially the same as that found in the case of fraud — if the significant digits of a certain process in nature exhibit a Benford distribution when the process is in one particular state, but do not when the process is in a different state, then simple goodness-of-fit tests can help identify when changes in the state of the process occur.

Recent studies in earthquake science illustrate this idea. M. Sambridge et al. found that the set of values of the ground displacements before an earthquake, when small shifts are due to background noise, do not closely follow Benford's law, but goodness-of-fit to Benford's law increases sharply at the beginning of an earthquake. Observing that the "onset of an earthquake can clearly be manifested in the digit distribution alone" [135, p. 4], they reported detection of a local earthquake near Canberra using first significant digit information alone, without the necessity of seeing the complete details of the seismogram. They also concluded that goodness-of-fit to Benford's law may prove useful for comparing the characteristics of earthquakes between regions, or over time [136]. G. Sottili et al. subsequently studied data from approximately 17,000 seismic events in Italy from 2006–2012, and also reported close conformity to Benford's law for recurrence times of consecutive seismic events [152].

Reporting a similar phenomenon at the quantum level, A. Sen and U. Sen found that the relative frequencies of the significant digits are different before and after quantum phase transitions, and concluded that goodness-of-fit to Benford's law can be used effectively to detect phase transitions in quantum many-body problems, adding that use of Benford's law tests thus also offers new ways to tackle problems at the interface of quantum information science [147]. In the field of medical science, B. Horn et al. applied digital analysis to recorded electroencephalogram signals and found that different states of anesthesia can be detected by goodness-of-fit to Benford's law [80], and more recently, M. Kreuzer et al. similarly noted that "Benford's law can be successfully used to detect . . . signal modulations in electrophysiological recordings" [92, p. 183]. Digit analysis applied to certain datasets in interventional radiology by S. Cournane et al. was found to be potentially useful for monitoring and identifying system output changes [40].

Goodness-of-fit to Benford's law has also been used to help detect data that originate from several different natural sources as opposed to data that originate from a single source. R. Seaman analyzed field errors in geopotential analysis and forecasting data in Australia, and associated closeness of fit to Benford's law with the extent of mixing of data from different sources in the underlying distribution of those errors. He reported that goodness-of-fit tests provided useful confirmation with other methods designed to determine when background field errors are composed of a mixture of different distributions, rather than a single distribution [146].

10.3　DIAGNOSTICS AND DESIGN

In situations where Benford's law is expected or is known to arise, goodness-of-fit tests can also be used to evaluate the predicted outcomes of a mathematical model or the quality of a detection instrument's output. For example, since the 1990, 2000, and 2010 census statistics of populations of all the counties in the United States follow Benford's law fairly closely (see Figure 1.3 and also

[102, 113]), a simple diagnostic test to evaluate a given mathematical model's prediction of future populations is the following: Enter current values as input, and then check to see how closely the output of that model agrees with Benford's law. H. Varian, for instance, proposed Benford's law as a criterion to help evaluate the "reasonableness" of a forecasting model for land usage in the San Francisco Bay area, and tested the idea using simulation. As with the goodness-of-fit tests described above for detecting fraud, this "Benford-in, Benford-out" principle provides a red flag test; as Varian put it, "if the input data did obey Benford's law while the output didn't ... well, that would make one a trifle uneasy about actually using that output" [156, p. 65]. More recently, C. Tolle et al. found empirical evidence of Benford's law in certain gas phase and condensed phase experimental molecular dynamics, and concluded that the law can provide a useful diagnostic for selecting dynamical models, since "if the data follow Benford's law, then the model dynamical system should do so as well" [154, p. 331].

Goodness-of-fit tests to Benford's law can contribute to the explicit design of algorithms or mathematical models. J. Dickinson used this approach to assess the appropriateness of various algorithms employed in business simulation games, noting that their design should generate output close to real-world business data, which he found were a good fit to Benford's law [48]. In another design application, M. Chou et al. examined the problem of assigning payoffs in fixed-odds numbers games, where the prizes and odds of winning are known at the time of placing the wager. Their objective was to determine how a game operator should set the sales limit, and empirical evidence showed that players have a tendency to bet on small numbers that are closely related to events around them. Viewing these numbers as being drawn from many different data sets (birth dates, addresses, etc.), they then used the connection between mixing datasets and Benford's law to propose guidelines for setting appropriate sales limits in those games [35].

Benford's law is also making an appearance as a diagnostic tool for quality control of datasets produced by measurement instruments and other devices. P. Manoochehrnia et al., appealing to scale-invariance and mixing-of-distributions characterizations of Benford's law (Theorems 5.13 and 8.44, respectively), applied significant-digit analysis to evaluation of the detection efficiency of lightning location networks. Studying cloud-to-ground flashes detected in Switzerland from 1997 to 2007, they found the data "in very good agreement with Benford's law," and that the data for detection of low energy lightning strikes (absolute peak currents less than 5 kA) were in significantly poorer agreement. Past methods of performance evaluation for cloud-to-ground flashes, which included erection of instrument towers and artificially triggered lightning, were assumed to be of higher quality than that for the low energy strikes, and thus closeness to Benford's law was associated with accuracy of detection. Their studies concluded that goodness-of-fit to Benford's law could be used to evaluate detection efficiencies, and suggested subsequent test applications for analyzing lightning data over regions close to the boundaries of lightning location networks, where the efficiency of the network is also expected to decrease significantly [103].

M. Orita et al. demonstrated that certain classes of drug discovery data follow Benford's law, and they "propose a protocol based on Benford's law which will provide researchers with an efficient method for data quality assessment" [118, 119]. Docampo et al. propose goodness-of-fit to Benford tests as a quality control for aerobiological datasets [49], and in the field of public health, A. Idrovo et al. found that a significant-digit analysis of the A(H1N1) influenza pandemic proved a useful tool for evaluating the performance of public health surveillance systems [82].

10.4 COMPUTATIONS AND COMPUTER SCIENCE

The appearance of Benford's law in real-life scientific computations is now widely accepted as an empirical fact (e.g., see [7, 57, 62, 70, 78, 90, 141]), and theoretical support for this is found in its pervasiveness in large classes of deterministic and random processes (see Chapters 6, 7, and 8). Thus, from both theoretical and practical standpoints, Benford's law plays an important role in the analysis of scientific computations using digital computers.

In floating-point arithmetic, for example, Benford's law can be used to help analyze various types of inherent computational errors. Even using the IEEE Standard "Unbiased" Rounding, calculations invariably contain round-off errors, and the magnitudes of possible errors may not be equally likely in practice. As pointed out by Knuth, an assumption of uniformly distributed fraction parts in calculations using floating-point arithmetic tends to underestimate the average relative round-off error in cases where the actual statistical distribution of fraction parts is skewed toward smaller leading significant digits [90]. And Benford's law has exactly this property — the leading digits tend to be small.

In using an algorithm to estimate an unknown value of some desired parameter, a rule for stopping the algorithm must be specified, such as "stop when $n = 10^6$" or "stop when the difference between successive approximations is less than 10^{-6}." Every stopping rule will invariably result in some unknown round-off error, and if its true statistical distribution follows Benford's law, but the error is assumed instead to be uniformly distributed, this can lead to a considerable underestimate of the round-off error. In particular, if X denotes the absolute round-off error, and Y denotes the significand of the approximation at the time of stopping, then the relative error is X/Y. Assuming that X and Y are independent, the average (i.e., expected) relative error is simply $\mathbb{E}[X]\mathbb{E}[1/Y]$. If the significand Y is naïvely assumed to be uniformly distributed on $[1, 10)$, then the average relative error is $\int_1^{10} \frac{1}{9}t^{-1}\,dt = 0.2558$ times the expected absolute error $\mathbb{E}[X]$, whereas if the true distribution of Y is actually Benford, then the true average relative error is $\int_1^{10} t^{-2}\log e\,dt = 0.3908$ times the expected absolute error. Thus, in numerical algorithms based on Newton's method, or in numerical solutions of the many classes of deterministic or random processes where Benford's law is known to appear, ignoring the fact that Y is Benford creates an average *under*estimation of the relative error by more than one third.

A second type of error in scientific calculations using digital computers is overflow (or underflow), which occurs when the running calculations exceed the largest (or smallest, in absolute value) floating-point number allowed by the computer. Feldstein and Turner found that "under the assumption of the logarithmic distribution of numbers [i.e., Benford's Law], floating-point addition and subtraction can result in overflow or underflow with alarming frequency" [57, p. 241], and they suggest that special attention be given to overflow and underflow errors in any computer algorithm where the output is expected to follow a Benford distribution, such as that for estimating roots by means of Newton's method; see Theorem 6.35. To help address this issue, they proposed "a long word format which will reduce the risks [of overflow and underflow errors] to acceptable levels" [57, p. 241].

Because of its prevalence in scientific calculations, Benford's law may also have implications for the design of computer hardware and software. For example, nearly half a century ago, Hamming provided a number of applications including the theoretical design problem of where to place the decimal point in order to minimize the number of normalization shifts in the calculations of products, and he showed that a Benford distribution of significands is balanced in the sense that it yields equal probability of requiring a shift to the right or to the left. He also applied his empirically observed Benford distribution of significands in computations to estimate the maximum relative and average relative representation errors (in order to obtain the mean and variance of the estimated propagation errors) and to optimize a library routine for minimizing running time [70]. P. Schatte, similarly hypothesizing that the significands tend toward the Benford distribution after a long series of computations, addressed the question of finding the best base $b = 2^n$ for digital computers. Under the assumption of Benford distribution of significant digits, he found that from the standpoint of effective storage use ("effektive Bitnutzung"), $b = 2^3 = 8$ is optimal, and, under the same Benford hypothesis, he showed that the speed of floating-point multiplication and division can also be determined [137, 141].

More recently, B. Ravikumar studied the problem of computing the leading significant digits of integral powers of positive integers, and used the theory of Benford's law, especially the uniform distribution characterization (Theorem 4.2), to prove that certain natural formal languages are not unambiguous and context-free [129]. Also in the field of software development, Jolion suggested an application to entropy-based image coding, which he found to be efficient and useful for data transmission [85], and E. Goldoni et al. developed a source coding technique for wireless sensor networks based on Benford's law [65]. Though outside the scope of this book, a survey of the impact of these findings on the evolution of digital computer hardware and software would be of interest to both theoreticians and practitioners.

10.5   PEDAGOGICAL TOOL

The increasing adoption of Benford's law techniques by auditors, scientists, and other researchers has made analysis of significant digits an attractive subject for educators, and it is now being incorporated into curricula at several academic levels, both as a concrete statistical tool and as a general instructional tool. In a workshop for the Australian Association of Mathematics Teachers, P. Turner described lessons on Benford's law for high school students [155], and D. Aldous and T. Phan used Benford's law to teach university undergraduate students the conceptually difficult task of testing proposed explanations of statistical phenomena [3].

A. Gelman and D. Nolan found that concepts related to Benford's law help college students learn about statistical sampling procedures precisely because these concepts are both simple to state and counterintuitive [64]. M. Linville observed that class assignments requiring students to create fictitious data proved very effective after the instructor had detected the students' fabrications using goodness-of-fit to Benford's law. He found that their curiosity was then satisfied by a discussion of digital analysis and Benford's law, and the usefulness of the techniques was "abundantly clear to the students without the necessity of instructor emphasis"[99, p. 56] — students were motivated to learn about Benford's law, digital analysis, and goodness-of-fit after seeing just how effective it was in ferreting out their own fabrications.

In addition to statistics classroom use, Benford's law material is also being employed as an educational tool in other areas of mathematics. K. Ross's article on first digits of squares and cubes included a section on "suggested undergraduate research" on Benford's law [133]. R. Nillsen used the ideas underlying the uniform distribution and dynamical systems aspects of Benford's law in his attempt to "bridge a gap between undergraduate teaching and the research level in mathematics" [115]; and S. Wagon utilized computer graphic demonstrations of Benford's law to illustrate the effectiveness and versatility of *Mathematica* in the classroom as well as in research [157].

List of Symbols

$\mathbb{N}, \mathbb{N}_0, \mathbb{Z}$	set of positive integers, nonnegative integers, integers
$\mathbb{Q}, \mathbb{R}^+, \mathbb{R}, \mathbb{C}$	set of rational, positive real, real, complex numbers
$[a, b)$	(half-open) interval $[a, b) = \{x \in \mathbb{R} : a \le x < b\}$ with $a < b$
D_1, D_2, etc.	first, second, etc. significant decimal digit
$D_m^{(b)}$	m^{th} significant digit base b, $b \ge 2$
$\log x$	logarithm base 10 of $x \in \mathbb{R}^+$
$\log_b x$	logarithm base b of $x \in \mathbb{R}^+$
$\ln x$	natural logarithm (base e) of $x \in \mathbb{R}^+$
Δ	deviation (in percent) from first-digit law; $\Delta = 100 \cdot \max_{d=1}^9 \left\|\mathsf{Prob}(D_1 = d) - \log(1 + d^{-1})\right\|$
$\#A$	cardinality (number of elements) of finite set A
$U(a, b)$	random variable uniformly distributed on (a, b) with $a < b$
S	(decimal) significand function
$\lfloor x \rfloor$	largest integer not larger than $x \in \mathbb{R}$
$\langle x \rangle$	fractional part of $x \in \mathbb{R}$; $\langle x \rangle = x - \lfloor x \rfloor$
$\mathcal{A}$	σ-algebra, on some non-empty set Ω
$\sigma(\mathcal{E})$	σ-algebra generated by collection $\mathcal{E}$ of subsets of Ω
$\mathcal{B}$	Borel σ-algebra on $\mathbb{R}$ or parts thereof
$\sigma(f)$	σ-algebra generated by function $f : \Omega \to \mathbb{R}$
$\mathcal{S}$	significand σ-algebra; $\mathcal{S} = \mathbb{R}^+ \cap \sigma(S)$
$(\Omega, \mathcal{A}, \mathbb{P})$	probability space
A^c	complement of A in set Ω; $A^c = \{\omega \in \Omega : \omega \notin A\}$
$A \backslash B$	set of elements in A but not in B; $A \backslash B = A \cap B^c$

$A \Delta B$	symmetric difference of A and B; $A \Delta B = (A \backslash B) \cup (B \backslash A)$
λ	Lebesgue measure on $(\mathbb{R}, \mathcal{B})$ or parts thereof
$\lambda_{a,b}$	normalized Lebesgue measure (uniform distribution) on $\big([a,b), \mathcal{B}[a,b)\big)$
δ_ω	Dirac probability measure concentrated at $\omega \in \Omega$
$\rho(C)$	(natural) density of $C \subset \mathbb{N}$
$\mathbb{1}_A$	indicator function of set A
(F_n)	sequence of Fibonacci numbers; $(F_n) = (1, 1, 2, 3, 5, \ldots)$
(p_n)	sequence of prime numbers; $(p_n) = (2, 3, 5, 7, 11, \ldots)$
i.i.d.	independent, identically distributed
$X, Y, \ldots$	(real-valued) random variables $\Omega \to \mathbb{R}$
$\mathbb{E}[X]$	expected (or mean) value of random variable X
$\mathrm{var}\, X$	variance of random variable X with $\mathbb{E}[|X|] < +\infty$; $\mathrm{var}\, X = \mathbb{E}\big[(X - \mathbb{E}X)^2\big]$
P	probability measure on $(\mathbb{R}, \mathcal{B})$, possibly random
P_X	distribution of random variable X
F_P, F_X	distribution function of P, X
$\widehat{P}(k)$	k^{th} Fourier coefficient of probability P on $\mathcal{B}[0,1)$; $\widehat{P}(k) = \int_0^1 e^{2\pi i k s} \, \mathrm{d}P(s)$, $k \in \mathbb{Z}$
$\mathbb{B}$	Benford distribution on $(\mathbb{R}^+, \mathcal{S})$
Δ_∞	deviation ($100 \times$ sup-norm) from Benford's law; $\Delta_\infty = 100 \cdot \sup_{1 \leq t < 10} |F(t) - \log t|$
Φ	standard normal distribution function; $\Phi(x) = \frac{1}{\sqrt{2\pi}} \int_{-\infty}^x e^{-t^2/2} \, \mathrm{d}t$
$\mathbb{P}_f, \mathbb{P} \circ f^{-1}$	probability measure on $\mathbb{R}$ induced by $\mathbb{P}$ and measurable function $f : \Omega \to \mathbb{R}$, via $\mathbb{P}_f(\bullet) = \mathbb{P}\big(f^{-1}(\bullet)\big)$
u.d. mod 1	uniformly distributed modulo one
a.a.	(Lebesgue) almost all
a.s.	almost surely, i.e., with probability one
$\nearrow, \searrow$	increasing, decreasing (not necessarily strictly)

$\left(f^n(x_0)\right)$	orbit of x_0 (under map f); $\left(f^n(x_0)\right) = \left(f(x_0), f \circ f(x_0), \dots\right)$
$\mathcal{O}$	$f = \mathcal{O}(g)$ as $x \to +\infty$ if $\limsup_{x \to +\infty} |f(x)/g(x)| < +\infty$
o	$f = o(g)$ as $x \to +\infty$ if $\lim_{x \to +\infty} f(x)/g(x) = 0$
$\Re z, \Im z$	real, imaginary part of $z \in \mathbb{C}$
$\overline{z}, |z|$	conjugate, absolute value (modulus) of $z \in \mathbb{C}$
$\arg z$	argument of $z \in \mathbb{C} \setminus \{0\}$; $\arg z \in (-\pi, \pi]$ and $z = |z| e^{\iota \arg z}$
$\mathbb{S}$	unit circle in $\mathbb{C}$; $\mathbb{S} = \{z \in \mathbb{C} : |z| = 1\}$
$\mathrm{span}_{\mathbb{Q}} Z$	rational span of $Z \subset \mathbb{C}$; $\mathrm{span}_{\mathbb{Q}} Z = \left\{\sum_{k=1}^{n} \rho_k z_k : n \in \mathbb{N}, \rho_k \in \mathbb{Q}, z_k \in Z\right\}$
C^k	set of all k times continuously differentiable functions
C^∞	set of all smooth (i.e., infinitely differentiable) functions
N_g	Newton map associated with differentiable function g
$\mathbb{R}^{d \times d}$	set (linear space) of all real $d \times d$-matrices
I_d	$d \times d$-identity matrix
$\mathcal{L}_d$	set (linear space) of all linear observables on $\mathbb{R}^{d \times d}$
$[A]_{jk}$	entry in j^{th} row, k^{th} column of $A \in \mathbb{R}^{d \times d}$
$\sigma(A)$	spectrum (set of all eigenvalues) of $A \in \mathbb{R}^{d \times d}$
$\rho(A)$	spectral radius of $A \in \mathbb{R}^{d \times d}$; $\rho(A) = \max\{|\lambda| : \lambda \in \sigma(A)\}$
$|x|, |A|$	Euclidean norm of $x \in \mathbb{R}^d$, $A \in \mathbb{R}^{d \times d}$
$\dot{x}, \ddot{x}, x^{(k)}$	first, second, k^{th} derivative of $x = x(t)$ with respect to t
$X_n \xrightarrow{\mathcal{D}} X$	(X_n) converges in distribution to X
$X_n \xrightarrow{a.s.} X$	(X_n) converges to X almost surely
r.p.m.	random probability measure
$\mathbb{E}P$	expectation of r.p.m. P
$\mathbb{T}^+$	set of all positive terminating decimals; $\mathbb{T}^+ = \bigcup_{m \in \mathbb{N}} 10^{-m} \mathbb{N}$
P^*	finitely additive probability measure on $\Omega \subset \mathbb{R}$
■	end of PROOF
✠	end of EXAMPLE

Bibliography

[1]   R. M. Abrantes-Metz, S. B. Villas-Boas, and G. Judge. Tracking the Libor rate. *Applied Economics Letters*, 18(10):893–899, 2011.

[2]   A. K. Adhikari and B. P. Sarkar. Distribution of most significant digit in certain functions whose arguments are random variables. *Sankhyā Ser. B*, 30:47–58, 1968.

[3]   D. Aldous and T. Phan. When can one test an explanation? Compare and contrast Benford's law and the fuzzy CLT. *Amer. Statist.*, 64(3):221–227, 2010.

[4]   P. C. Allaart. An invariant-sum characterization of Benford's law. *J. Appl. Probab.*, 34(1):288–291, 1997.

[5]   R. C. Baker, G. Harman, and J. Pintz. The difference between consecutive primes, II. *Proc. London Math. Soc.*, 83(3):532–562, 2001.

[6]   R. B. Bapat and T.E.S. Raghavan. *Nonnegative matrices and applications*, volume 64 of *Encyclopedia of Mathematics and its Applications*. Cambridge University Press, Cambridge, 1997.

[7]   J. L. Barlow and E. H. Bareiss. On roundoff error distributions in floating point and logarithmic arithmetic. *Computing*, 34(4):325–347, 1985.

[8]   P. W. Becker. Patterns in listings of failure-rate and MTTF values and listings of other data. *IEEE Transactions on Reliability*, 31(2):132–134, 1982.

[9]   F. Benford. The law of anomalous numbers. *Proc. Amer. Philosophical Soc.*, 78(4):551–572, 1938.

[10] A. Berger. Benford's law in power-like dynamical systems. *Stoch. Dyn.*, 5(4):587–607, 2005.

[11] A. Berger. Multi-dimensional dynamical systems and Benford's law. *Discrete Contin. Dyn. Syst.*, 13(1):219–237, 2005.

[12] A. Berger. Large spread does not imply Benford's Law. Preprint, Department of Mathematical and Statistical Sciences, University of Alberta, Edmonton, AB, 2010.

[13] A. Berger. Some dynamical properties of Benford sequences. *J. Difference Equ. Appl.*, 17(2):137–159, 2011.

[14] A. Berger. Most linear flows on $\mathbb{R}^d$ are Benford. In preparation, 2014.

[15] A. Berger, L. A. Bunimovich, and T. P. Hill. One-dimensional dynamical systems and Benford's law. *Trans. Amer. Math. Soc.*, 357(1):197–219, 2005.

[16] A. Berger and G. Eshun. Benford solutions of linear difference equations. In *Theory and Applications of Difference Equations and Discrete Dynamical Systems (ICDEA 2013)*, volume 102 of *Proceedings in Mathematics & Statistics*, pp. 23–60. Springer-Verlag, Berlin, 2014.

[17] A. Berger and G. Eshun. A characterization of Benford's Law in discrete-time linear systems. To appear in *J. Dynam. Differential Equations*.

[18] A. Berger and S. N. Evans. A limit theorem for occupation measures of Lévy processes in compact groups. *Stoch. Dyn.*, 13(1), 16 pp., 2013.

[19] A. Berger and T. P. Hill. Newton's method obeys Benford's law. *Amer. Math. Monthly*, 114(7):588–601, 2007.

[20] A. Berger and T. P. Hill. A basic theory of Benford's law. *Probab. Surv.*, 8:1–126, 2011.

[21] A. Berger and T. P. Hill. Benford's law strikes back: No simple explanation in sight for mathematical gem. *Math. Intelligencer*, 33(1):85–91, 2011.

[22] A. Berger, T. P. Hill, B. Kaynar, and A. Ridder. Finite-state Markov chains obey Benford's law. *SIAM J. Matrix Anal. Appl.*, 32(3):665–684, 2011.

[23] A. Berger, T. P. Hill, and K. E. Morrison. Scale-distortion inequalities for mantissas of finite data sets. *J. Theoret. Probab.*, 21(1):97–117, 2008.

[24] A. Berger, T. P. Hill, and E. Rogers. Benford Online Bibliography. `http://www.benfordonline.net`, 2009.

[25] A. Berman and R. J. Plemmons. *Nonnegative Matrices in the Mathematical Sciences*. Academic Press, New York, 1979.

[26] L. Berton. He's got their number: Scholar uses math to foil financial fraud. *The Wall Street Journal*, July 10, 1995.

[27] P. Billingsley. *Probability and Measure*. Wiley Series in Probability and Mathematical Statistics. John Wiley & Sons, New York, third edition, 1995.

[28] C. Breunig and A. Goerres. Searching for electoral irregularities in an established democracy: Applying Benford's law tests to Bundestag elections in unified Germany. *Electoral Studies*, 30(3):534–545, 2011.

[29] B. Buck, A. C. Merchant, and S. M. Perez. An illustration of Benford's first digit law using alpha decay half lives. *Eur. J. Phys.*, 14:59–63, 1993.

[30] R. Bumby and E. Ellentuck. Finitely additive measures and the first digit problem. *Fund. Math.*, 65:33–42, 1969.

[31] J. Burke and E. Kincanon. Benford's law and physical constants: The distribution of initial digits. *Amer. J. Phys.*, 59(10):952, 1991.

[32] M. Buyse, S. L. George, S. Evans, N. L. Geller, J. Ranstam, B. Scherrer, E. Lesaffre, G. Murray, L. Edler, J. Hutton, T. Colton, P. Lachenbruch, and B. L. Verma. The role of biostatistics in the prevention, detection and treatment of fraud in clinical trials. *Statist. Med.*, 18(24):3435–3451, 1999.

[33] F. Cantú and S. M. Saiegh. Fraudulent democracy? An analysis of Argentina's infamous decade using supervised machine learning. *Political Analysis*, 19(4):409–433, 2011.

[34] N. Chernov. Decay of correlations. *Scholarpedia*, 3(4):4862, 2008.

[35] M. C. Chou, Q. Kong, C.-P. Teo, Z. Wang, and H. Zheng. Benford's law and number selection in fixed-odds numbers game. *Journal of Gambling Studies*, 25(4):503–521, 2009.

[36] Y. S. Chow and H. Teicher. *Probability Theory. Independence, Interchangeability, Martingales*. Springer Texts in Statistics. Springer, New York, third edition, 1997.

[37] K. L. Chung. *A Course in Probability Theory*. Academic Press, San Diego, CA, third edition, 2001.

[38] D.I.A. Cohen. An explanation of the first digit phenomenon. *J. Combinatorial Theory Ser. A*, 20(3):367–370, 1976.

[39] E. Costas, V. López-Rodas, F. Javier Toro, and A. Flores-Moya. The number of cells in colonies of the cyanobacterium *Microcystis aeruginosa* satisfies Benford's law. *Aquatic Botany*, 89(3):341–343, 2008.

[40] S. Cournane, N. Sheehy, and J. Cooke. The novel application of Benford's second order analysis for monitoring radiation output in interventional radiology. *Physica Medica*, 30:413–418, 2014.

[41] P. Cull, M. Flahive, and R. Robson. *Difference Equations. From Rabbits to Chaos*. Undergraduate Texts in Mathematics. Springer, New York, 2005.

[42] K. Dajani and C. Kraaikamp. *Ergodic Theory of Numbers*, volume 29 of *Carus Mathematical Monographs*. Mathematical Association of America, Washington, DC, 2002.

[43] J. Deckert, M. Myagkov, and P. C. Ordeshook. Benford's law and the detection of election fraud. *Political Analysis*, 19(3):245–268, 2011.

[44] E. del Acebo and M. Sbert. Benford's law for natural and synthetic images. In *Proceedings of Computational Aesthetics in Graphics, Visualizations and Imaging*, pp. 169–176, 2005.

[45] R. L. Devaney. *An Introduction to Chaotic Dynamical Systems*. Addison-Wesley Studies in Nonlinearity. Addison-Wesley, Redwood City, CA, second edition, 1989.

[46] P. Diaconis. The distribution of leading digits and uniform distribution mod 1. *Ann. Probability*, 5(1):72–81, 1977.

[47] P. Diaconis and D. Freedman. On Rounding Percentages. *J. Amer. Statist. Assoc.*, 74(366):359–364, 1979.

[48] J. R. Dickinson. A universal mathematical law criterion for algorithmic validity. *Developments in Business Simulation and Experiential Learning*, 29:26–33, 2002.

[49] S. Docampo, M. del Mar Trigo, M. J. Aira, B. Cabezudo, and A. Flores-Moya. Benford's law applied to aerobiological data and its potential as a quality control tool. *Aerobiologia*, 25(4):275–283, 2009.

[50] M. Drmota and R. F. Tichy. *Sequences, Discrepancies and Applications*, volume 1651 of *Lecture Notes in Mathematics*. Springer-Verlag, Berlin, 1997.

[51] L. E. Dubins and D. A. Freedman. Random distribution functions. In *Proc. Fifth Berkeley Sympos. Math. Statist. and Probability (Berkeley, Calif., 1965/66)*, Vol. II: Contributions to Probability Theory, Part 1, pp. 183–214. University of California Press, Berkeley, 1967.

[52] L. E. Dubins and L. J. Savage. *Inequalities for Stochastic Processes: How to Gamble if You Must*. Dover, 1976.

[53] N. Dunford and J. T. Schwartz. *Linear Operators. Vol. 1: General theory*, volume 7 of *Pure and applied mathematics*. Wiley-Interscience, 1967.

[54] M. Einsiedler. What is ... measure rigidity? *Notices Amer. Math. Soc.*, 56(5):600–601, 2009.

[55] S. Elaydi. *An Introduction to Difference Equations*. Undergraduate Texts in Mathematics. Springer, New York, third edition, 2005.

[56] H.-A. Engel and C. Leuenberger. Benford's law for exponential random variables. *Statist. Probab. Lett.*, 63(4):361–365, 2003.

[57] A. Feldstein and P. Turner. Overflow, underflow, and severe loss of significance in floating-point addition and subtraction. *IMA J. Numer. Anal.*, 6(2):241–251, 1986.

[58] W. Feller. *An Introduction to Probability Theory and Its Applications. Vol. II.* John Wiley & Sons, New York, second edition, 1966.

[59] B. J. Flehinger. On the probability that a random integer has initial digit A. *Amer. Math. Monthly*, 73:1056–1061, 1966.

[60] J. L. Friar, T. Goldman, and J. Pérez-Mercader. Genome Sizes and the Benford Distribution. *PLoS ONE*, 7(5):e36624, 2012.

[61] D. Fu, Y. Q. Shi, and W. Su. A generalized Benford's law for JPEG coefficients and its applications in image forensics. In *Proceedings of SPIE, Volume 6505, Security, Steganography and Watermarking of Multimedia Contents IX*, 2007.

[62] F. Gambarara and O. Nagy. Benford distribution in science. Technical report, ETH Zürich, 2004.

[63] M. Gardner. Mathematical games: Problems involving questions of probability and ambiguity. *Scientific American*, 201:174–182, 1959.

[64] A. Gelman and D. Nolan. Some statistical sampling and data collection activities. *The Mathematics Teacher*, 95(9):688–693, 2002.

[65] E. Goldoni, P. Savazzi, and P. Gamba. A novel source coding technique for wireless sensor networks based on Benford's law. In *2012 IEEE Workshop on Environmental Energy and Structural Monitoring Systems (EESMS)*, pp. 32–34, 2012.

[66] W. M. Goodman. Reality checks for a distributional assumption: The case of "Benford's law." In *JSM Proceedings*, pp. 2789–2803. American Statistical Association, 2013.

[67] S. A. Goudsmit and W. H. Furry. Significant figures of numbers in statistical tables. *Nature*, 154(3921):800–801, 1944.

[68] G. Grekos. On various definitions of density (survey). *Tatra Mt. Math. Publ.*, 31:17–27, 2005.

[69] S. J. Gustafson and I. M. Sigal. *Mathematical Concepts of Quantum Mechanics.* Universitext. Springer-Verlag, Berlin, 2003.

[70] R. W. Hamming. On the distribution of numbers. *Bell System Tech. J.*, 49:1609–1625, 1970.

[71] G. H. Hardy. *Divergent Series*. Clarendon Press, Oxford, 1949.

[72] J. Hein, R. Zobrist, C. Konrad, and G. Schüpfer. Scientific fraud in 20 falsified anesthesia papers. *Der Anaesthesist*, 61(6):543–549, 2012.

[73] T. P. Hill. Base-invariance implies Benford's law. *Proc. Amer. Math. Soc.*, 123(3):887–895, 1995.

[74] T. P. Hill. A statistical derivation of the significant-digit law. *Statist. Sci.*, 10(4):354–363, 1995.

[75] T. P. Hill. The difficulty of faking data. *Chance*, 12(3):27–31, 1999.

[76] T. P. Hill and K. Schürger. Regularity of digits and significant digits of random variables. *Stochastic Process. Appl.*, 115(10):1723–1743, 2005.

[77] M. W. Hirsch, S. Smale, and R. L. Devaney. *Differential Equations, Dynamical Systems, and an Introduction to Chaos*. Academic Press, Waltham, MA, 2013.

[78] J. M. Horgan. *Probability with R. An Introduction with Computer Science Applications*, section 9.4, pp. 142–144. John Wiley & Sons, 2009.

[79] R. A. Horn and C. R. Johnson. *Matrix Analysis*. Cambridge University Press, Cambridge, 1985.

[80] B. Horn, M. Kreuzer, E. F. Kochs, and G. Schneider. Different states of anesthesia can be detected by Benford's Law. *J. Neurosurg. Anesthesiol.*, 18(4):328–329, 2006.

[81] A. Horn and A. Tarski. Measures in Boolean algebras. *Trans. Amer. Math. Soc.*, 64:467–497, 1948.

[82] A. J. Idrovo, J. A. Fernández-Niño, I. Bojórquez-Chapela, and J. Moreno-Montoya. Performance of public health surveillance systems during the influenza A(H1N1) pandemic in the Americas: Testing a new method based on Benford's law. *Epidemiol. Infect.*, 139(12):1827–1834, 2011.

[83] T. Jech. The logarithmic distribution of leading digits and finitely additive measures. *Discrete Math.*, 108(1-3):53–57, 1992.

[84] T. Jech and K. Prikry. On projections of finitely additive measures. *Proc. Amer. Math. Soc.*, 74(1):161–165, 1979.

[85] J.-M. Jolion. Images and Benford's law. *J. Math. Imaging Vision*, 14(1):73–81, 2001.

[86] O. Kallenberg. *Random Measures*. Academic Press, London; Akademie-Verlag, Berlin, fourth edition, 1986.

[87] S. Kanemitsu, K. Nagasaka, G. Rauzy, and J.-S. Shiue. On Benford's law: The first digit problem. In *Probability theory and mathematical statistics (Kyoto, 1986)*, volume 1299 of *Lecture Notes in Math.*, pp. 158–169. Springer-Verlag, Berlin, 1988.

[88] A. Katok and B. Hasselblatt. *Introduction to the Modern Theory of Dynamical Systems*, volume 54 of *Encyclopedia of Mathematics and its Applications*. Cambridge University Press, Cambridge, 1995.

[89] P. E. Kloeden and M. Rasmussen. *Nonautonomous Dynamical Systems*, volume 176 of *Mathematical Surveys and Monographs*. American Mathematical Society, Providence, RI, 2011.

[90] D. E. Knuth. *The Art of Computer Programming*. Addison-Wesley, Reading, Mass.-London-Amsterdam, second edition, 1975.

[91] A. V. Kontorovich and S. J. Miller. Benford's law, values of L-functions and the $3x + 1$ problem. *Acta Arith.*, 120(3):269–297, 2005.

[92] M. Kreuzer, D. Jordan, B. Antkowiak, B. Drexler, E. F. Kochs, and G. Schneider. Brain electrical activity obeys Benford's law. *Anesth. Analg.*, 118(1):183–91, 2014.

[93] L. Kuipers and H. Niederreiter. *Uniform distribution of sequences*. John Wiley & Sons, New York, 1974.

[94] M. T. Lacey and W. Philipp. A note on the almost sure central limit theorem. *Statist. Probab. Lett.*, 9(3):201–205, 1990.

[95] J. C. Lagarias and K. Soundararajan. Benford's law for the $3x+1$ function. *J. London Math. Soc. (2)*, 74(2):289–303, 2006.

[96] L. M. Leemis, B. W. Schmeiser, and D. L. Evans. Survival distributions satisfying Benford's law. *Amer. Statist.*, 54(4):236–241, 2000.

[97] G. Leibon. Google numbers. *Chance News*, 13(03), 2004.

[98] E. Ley. On the peculiar distribution of the US stock indexes' digits. *Amer. Statist.*, 50(4):311–313, 1996.

[99] M. Linville. Introducing digit analysis with an interactive class exercise. *Academy of Educational Leadership Journal*, 12(3):55–69, 2008.

[100] M. Loève. *Probability Theory. II*, volume 46 of *Graduate Texts in Mathematics*, Springer, New York, fourth edition, 1978.

[101] R. Lyons. Seventy years of Rajchman measures. In *Proceedings of the Conference in Honor of Jean-Pierre Kahane (Orsay, 1993)*, Special Issue of *J. Fourier Anal. Appl.*, pp. 363–377, 1995.

[102] D. Ma. Benford's Law and US Census Data, Parts I & II. Word-Press blog, http://introductorystats.wordpress.com/2011/11/24/benfords-law-and-us-census-data-part-i/, 2011.

[103] P. Manoochehrnia, F. Rachidi, M. Rubinstein, W. Schulz, and G. Diendorfer. Benford's law and its application to lightning data. *IEEE Transactions on Electromagnetic Compatibility*, 52(4):956–961, 2010.

[104] B. Massé and D. Schneider. A survey on weighted densities and their connection with the first digit phenomenon. *Rocky Mountain J. Math.*, 41(5):1395–1415, 2011.

[105] W. R. Mebane. Fraud in the 2009 presidential election in Iran? *Chance*, 23(1):6–15, 2010.

[106] W. R. Mebane. Comment on "Benford's Law and the Detection of Election Fraud." *Political Analysis*, 19(3):269–272, 2011.

[107] T. Michalski and G. Stoltz. Do countries falsify economic data strategically? Some evidence that they might. *The Review of Economics and Statistics*, 95(2):591–616, 2013.

[108] S. J. Miller and M. J. Nigrini. Order statistics and Benford's law. *Int. J. Math. Math. Sci.*, Art. ID 382948, 19 pp., 2008.

[109] K. E. Morrison. The multiplication game. *Math. Mag.*, 83(2):100–110, 2010.

[110] K. Nagasaka. On Benford's law. *Ann. Inst. Statist. Math.*, 36(2):337–352, 1984.

[111] S. Newcomb. Note on the frequency of use of the different digits in natural numbers. *Amer. J. Math.*, 4(1-4):39–40, 1881.

[112] M. J. Nigrini. *The Detection of Income Tax Evasion through an Analysis of Digital Distributions*. PhD dissertation, Department of Accounting, University of Cincinnati, Cincinnati, OH, 1992.

[113] M. J. Nigrini. A taxpayer compliance application of Benford's law. *The Journal of the American Taxation Association*, 18(1):72–91, 1996.

[114] M. J. Nigrini. *Benford's Law: Applications for Forensic Accounting, Auditing, and Fraud Detection*. John Wiley & Sons, Hoboken, NJ, 2012.

[115] R. Nillsen. *Randomness and Recurrence in Dynamical Systems*, volume 31 of *Carus Mathematical Monographs*. Mathematical Association of America, Washington, DC, 2010.

[116] J. R. Norris. *Markov Chains*. Cambridge University Press, Cambridge, 1997.

[117] The On-line Encyclopedia of Integer Sequences. http://oeis.org, 2011.

[118] M. Orita, Y. Hagiwara, A. Moritomo, K. Tsunoyama, T. Watanabe, and K. Ohno. Agreement of drug discovery data with Benford's law. *Expert Opin. Drug Discov.*, 8(1):1–5, 2013. PMID: 23121309.

[119] M. Orita, A. Moritomo, T. Niimi, and K. Ohno. Use of Benford's law in drug discovery data. *Drug Discovery Today*, 15(9–10):328–331, 2010.

[120] G. Overhoff. The impact and reality of fraud auditing — Benford's law: Why and how to use it. Course for the 22nd Annual ACFE Fraud Conference and Exhibition, 2011.

[121] K. Palmer. *Shadowing in Dynamical Systems*, volume 501 of *Mathematics and its Applications*. Kluwer Academic Publishers, Dordrecht, 2000.

[122] F. Pérez-González, G. L. Heileman, and C. T. Abdallah. Benford's law in image processing. In *IEEE International Conference on Image Processing (ICIP 2007)*, volume 1, pp. I-405–I-408. IEEE, 2007.

[123] S. Y. Pilyugin. *Shadowing in Dynamical Systems*, volume 1706 of *Lecture Notes in Mathematics*. Springer-Verlag, Berlin, 1999.

[124] R. S. Pinkham. On the distribution of first significant digits. *Ann. Math. Statist.*, 32(4):1223–1230, 1961.

[125] R. A. Purves and W. D. Sudderth. Some finitely additive probability. *Ann. Probability*, 4(2):259–276, 1976.

[126] R. A. Raimi. On the distribution of first significant figures. *Amer. Math. Monthly*, 76(4):342–348, 1969.

[127] R. A. Raimi. The first digit problem. *Amer. Math. Monthly*, 83(7):521–538, 1976.

[128] R. A. Raimi. The first digit phenomenon again. *Proc. Amer. Philosophical Soc.*, 129(2):211–219, 1985.

[129] B. Ravikumar. The Benford-Newcomb distribution and unambiguous context-free languages. *Internat. J. Found. Comput. Sci.*, 19(3):717–727, 2008.

[130] H. Rindler. Ein Problem aus der Theorie der Gleichverteilung. II. *Math. Z.*, 135:73–92, 1973/74.

[131] H. Robbins. On the equidistribution of sums of independent random variables. *Proc. Amer. Math. Soc.*, 4:786–799, 1953.

[132] K. A. Ross. Benford's law, a growth industry. *Amer. Math. Monthly*, 118(7):571–583, 2011.

[133] K. A. Ross. First digits of squares and cubes. *Math. Mag.*, 85(1):37–43, 2012.

[134] C. Ryll-Nardzewski. On the ergodic theorems. III. The random ergodic theorem. *Studia Math.*, 14:298–301, 1954/55.

[135] M. Sambridge, H. Tkalčić, and P. Arroucau. Benford's law of first digits: From mathematical curiosity to change detector. *Asia Pac. Math. Newsl.*, 1(4):1–6, 2011.

[136] M. Sambridge, H. Tkalčić, and A. Jackson. Benford's law in the natural sciences. *Geophysical Research Letters*, 37(22):L22301, 2010.

[137] P. Schatte. Zur Verteilung der Mantisse in der Gleitkommadarstellung einer Zufallsgröße. *Z. Angew. Math. Mech.*, 53:553–565, 1973.

[138] P. Schatte. On H_∞-summability and the uniform distribution of sequences. *Math. Nachr.*, 113:237–243, 1983.

[139] P. Schatte. On the asymptotic uniform distribution of sums reduced mod 1. *Math. Nachr.*, 115:275–281, 1984.

[140] P. Schatte. On the asymptotic behaviour of the mantissa distributions of sums. *J. Inform. Process. Cybernet.*, 23(7):353–360, 1987.

[141] P. Schatte. On mantissa distributions in computing and Benford's law. *J. Inform. Process. Cybernet.*, 24(9):443–455, 1988.

[142] P. Schatte. On the uniform distribution of certain sequences and Benford's law. *Math. Nachr.*, 136:271–273, 1988.

[143] K. Schürger. Extensions of Black-Scholes processes and Benford's law. *Stochastic Process. Appl.*, 118(7):1219–1243, 2008.

[144] K. Schürger. Lévy processes and Benford's Law. Preprint, 2011.

[145] R. Scozzafava. Un esempio concreto di probabilita non σ–additiva: La distribuzione della prima cifra significativa dei dati statistici. *Boll. Un. Mat. Ital. A(5)*, 18(3):403–410, 1981.

[146] R. S. Seaman. The relevance of Benford's Law to background field errors in data assimilation. *Aust. Met. Mag.*, 51:25–33, 2002.

[147] A. Sen and U. Sen. Benford's law detects quantum phase transitions similarly as earthquakes. *Europhysics Letters*, 95(5):50008, 2011.

[148] Z. Shengmin and W. Wenchao. Does Chinese stock indices agree with Benford's law? In *2010 International Conference on Management and Service Science (MASS)*, pp. 1–3, 2010.

[149] S. Shikano and V. Mack. When does the second-digit Benford's law-test signal an election fraud? Facts or misleading test results. *Jahrbücher f. Nationalökonomie u. Statistik*, 231(5+6):719–732, 2011.

[150] S. W. Smith. *The Scientist and Engineer's Guide to Digital Signal Processing*, chapter 34 — Explaining Benford's Law. California Technical Publishing, San Diego, CA, 1997. Republished by Newnes, 2002.

[151] M. A. Snyder, J. H. Curry, and A. M. Dougherty. Stochastic aspects of one-dimensional discrete dynamical systems: Benford's law. *Phys. Rev. E*, 64:026222, 2001.

[152] G. Sottili, D. M. Palladino, B. Giaccio, and P. Messina. Benford's Law in Time Series Analysis of Seismic Clusters. *Mathematical Geosciences*, 44(5):619–634, 2012.

[153] J. Taylor. Too many ties? An empirical analysis of the Venezuelan recall referendum counts. Technical report, Stanford University, 2005.

[154] C. R. Tolle, J. L. Budzien, and R. A. LaViolette. Do dynamical systems follow Benford's law? *Chaos*, 10(2):331–336, 2000.

[155] P. Turner. A classroom exploration of Benford's Law and some error finding tricks in accounting. In K. Milton, H. Reeves, and T. Spencer (eds.), *Proceedings of the 21st Biennial Conference of the Australian Association of Mathematics Teachers Inc.*, *Mathematics: Essential for Learning, Essential for Life*, pp. 250–259, 2007.

[156] H. R. Varian. Benford's law. *Amer. Statist.*, 26(3):65–66, 1972.

[157] S. Wagon. *Mathematica in Action: Problem Solving Through Visualization and Computation*, chapter 12 — Benford's Law of First Digits. Springer, New York, third edition, 2010.

[158] M. Waldschmidt. *Diophantine Approximation on Linear Algebraic Groups. Transcendence Properties of the Exponential Function in Several Variables*, volume 326 of *Grundlehren der Mathematischen Wissenschaften*. Springer-Verlag, Berlin, 2000.

[159] W. Weaver. *Lady Luck: The Theory of Probability*. Doubleday Anchor Series, New York, 1963. Republished by Dover, 1982.

[160] B. Xu, J. Wang, G. Liu, and Y. Dai. Photorealistic computer graphics forensics based on leading digit law. *Journal of Electronics (China)*, 28(1):95–100, 2011.

[161] R. Zweimüller. *S*-unimodal Misiurewicz maps with flat critical points. *Fund. Math.*, 181(1):1–25, 2004.

Index